T0203007

Lecture Notes in Computer Science   11504

Commenced Publication in 1973
Founding and Former Series Editors:
Gerhard Goos, Juris Hartmanis, and Jan van Leeuwen

Editorial Board Members

David Hutchison
  Lancaster University, Lancaster, UK
Takeo Kanade
  Carnegie Mellon University, Pittsburgh, PA, USA
Josef Kittler
  University of Surrey, Guildford, UK
Jon M. Kleinberg
  Cornell University, Ithaca, NY, USA
Friedemann Mattern
  ETH Zurich, Zurich, Switzerland
John C. Mitchell
  Stanford University, Stanford, CA, USA
Moni Naor
  Weizmann Institute of Science, Rehovot, Israel
C. Pandu Rangan
  Indian Institute of Technology Madras, Chennai, India
Bernhard Steffen
  TU Dortmund University, Dortmund, Germany
Demetri Terzopoulos
  University of California, Los Angeles, CA, USA
Doug Tygar
  University of California, Berkeley, CA, USA

Yves Coudière · Valéry Ozenne ·
Edward Vigmond · Nejib Zemzemi (Eds.)

Functional Imaging and Modeling of the Heart

10th International Conference, FIMH 2019
Bordeaux, France, June 6–8, 2019
Proceedings

 Springer

Editors
Yves Coudière
University of Bordeaux
Talence, France

Valéry Ozenne
IHU Liryc – Hôpital Xavier Arnozan
Pessac Cedex, France

Edward Vigmond
IHU Liryc – Hôpital Xavier Arnozan
Pessac Cedex, France

Nejib Zemzemi
Inria Bordeaux Sud-Ouest
Talence, France

ISSN 0302-9743   ISSN 1611-3349   (electronic)
Lecture Notes in Computer Science
ISBN 978-3-030-21948-2   ISBN 978-3-030-21949-9   (eBook)
https://doi.org/10.1007/978-3-030-21949-9

LNCS Sublibrary: SL6 – Image Processing, Computer Vision, Pattern Recognition, and Graphics

This Springer imprint is published by the registered company Springer Nature Switzerland AG
The registered company address is: Gewerbestrasse 11, 6330 Cham, Switzerland

Preface

FIMH 2019 was the 10th International Conference on Functional Imaging and Modeling of the Heart. It was held in Bordeaux, France, during June 6–8, 2019. This year edition of FIMH followed the past nine editions held in Helsinki (2001), Lyon (2003), Barcelona (2005), Salt Lake City (2007), Nice (2009), New York (2011), London (2013), Maastricht (2015), and Toronto (2017). FIMH 2019 provided a unique forum for the discussion of the latest developments in the areas of functional cardiac imaging as well as computational modeling of the heart. The topics of the conference included (but were not limited to) advanced cardiac imaging and image processing techniques, construction of computational meshes from images, myocardial tissue characterization and perfusion, computational fluid dynamics, forward and inverse problems in electrophysiology, cardiac growth, computational physiology and biomechanics of the heart, parameterization of mathematical models from data, as well as the pre-clinical and clinical applicability of these methods.

FIMH 2019 drew many submissions from around the world. From the initial registered papers, 46 selected papers were invited to be published by Springer in this *Lecture Notes in Computer Science* proceedings volume. All submitted papers were peer-reviewed by two or three Program Committee members or additional reviewers, who were international experts in the field. The review process was double-blinded. When preparing the final version of their manuscripts, authors addressed specific concerns and issues raised by reviewers, and improved the scientific content and the quality of the manuscripts.

The conference was greatly enhanced by invited keynote lectures given by four world experts in various fields related to the use of imaging to guide clinical treatment, ultrasound for mechanical and electromechanical characterization of the heart, patient-specific hemodynamic simulations for interventional planning, as well as uncertainty quantification to improve robustness when constructing patient-specific models. We are extremely grateful to Dr. Hubert Cochet (Hopital Haut-Lévêque, Pessac, France), Dr. Elisa Konofagou (Columbia University, New York City, USA), Dr. Irène Vignon-Clémentel (Inria, Paris, France), and Dr. Richard Gray (US Food and Drug Administration, Washington DC, USA) for their exceptional lectures.

We would like to thank the service congrès from the University of Bordeaux, the service de communication et médiation from Inria Bordeaux Sud-Ouest, specifically Flavie Attiguie, the administration of the Institut de Mathématique de Bordeaux, and namely Anne-France Contentin and Élodie Gaillacq from Liryc Institute for their kind and helpful support all along the preparation of this conference.

We hope that all these papers, along with the keynotes' contributions and fruitful discussions during the conference, will act to accelerate progress in the important areas of functional imaging and modeling of the heart.

June 2019

Yves Coudière
Valéry Ozenne
Edward Vigmond
Nejib Zemzemi

Organization

We would like to thank all organizers, additional reviewers, contributing authors, and sponsors for their time, effort, and financial support in making FIMH 2019 a successful event.

General Chair

Yves Coudière	University of Bordeaux and Liryc Institute, France
Valery Ozenne	Liryc Institute and University of Bordeaux, France
Edward J. Vigmond	Liryc Institute and University of Bordeaux, France
Nejib Zemzemi	Inria Bordeaux Sud-Ouest and Liryc Institute, France

Program Committee

Leon Axel	New York University, USA
Peter Bovendeerd	Eindhoven University of Technology, The Netherlands
Oscar Camara	University Pompeu Fabra, Barcelona, Spain
Dorin Comaniciu	Siemens, USA
Yves Coudière	Bordeaux University, France
Cesare Corrado	King's College London, UK
Patrick Clarysse	CNRS, Lyon, France
Tammo Delhaas	Maastricht University, The Netherlands
Nicolas Duchateau	University of Lyon, France
Jean-Frederic Gerbeau	Inria, France
Arun Holden	Leeds University, UK
Pablo Lamata	King's College London, UK
Cristian Linte	RIT-Rochester, USA
Herve Lombaert	Inria, France
Cristian Lorenz	Philips Research, Germany
Rob MacLeod	University of Utah, USA
Isabelle Magnin	University of Lyon, France
Tommaso Mansi	Siemens, USA
Dimitris Metaxas	Rutgers University, USA
Martyn Nash	University of Auckland, New Zealand
Valery Ozenne	Liryc Institute, France
Mihaela Pop	Sunnybrook, University of Toronto, Canada
Daniel Rueckert	Imperial College London, UK
Frank Sachse	University of Utah, USA
Maxime Sermesant	Inria, France
Kaleem Siddiqi	McGill University, Canada
Larry Staib	Yale University, USA

Regis Vaillant	GE Healthcare, France
Edward Vigmond	Liryc Institute, France
Jurgen Weese	Phillips Research, Germany
Graham Wright	Sunnybrook, University of Toronto, Canada
Guang Yang	Imperial College, London, UK
Alistair Young	University of Auckland, New Zealand
Nejib Zemzemi	Inria, France
Xiahai Zhuang	Fudan University, China

Additional Reviewers

Abhirup Banerjee	Oxford University, UK
Jason Bayer	Liryc Institute, France
Yves Bourgault	University of Ottawa, Canada
Radomir Chabiniok	Inria Saclay, France
Dominique Chapelle	Inria Saclay, France
Martin Genet	École Polytechnique, France
Vicente Grau	Oxford University, UK
Luigi Perotti	University of Central Florida, USA
Mark Potse	Université de Bordeaux, France
Caroline Roney	King's College of London, UK
Zhinuo Wang	Oxford University, UK

Sponsors

We are extremely grateful for the industrial and institutional funding support from the following sponsors (https://fimh2019.sciencesconf.org/resource/sponsors):

Siemens Healthineers

Siemens Healthineers enables health-care providers to increase value by expanding precision medicine, transforming care delivery, improving patient diagnostic.

inHEART

inHEART, a spin-off from IHU Liryc and Inria, has the vision to bridge the gap between radiology and cardiac electrophysiology by providing cloud-based solutions for image-guided diagnosis, therapy planning, and navigation for patients with arrhythmias.

Liryc Institute

The Electrophysiology and Heart Modeling Institute (French: L'Institut de RYthmologie et Modélisation Cardiaque, LIRYC) is one of six French university hospital institutions created in 2011 as part of the Investments for the Future Program to boost medical research and innovation. LIRYC is a basic research, clinical, and teaching center focusing on the understanding, care, and treatment of cardiac electrical diseases that lead to heart failure and sudden death. It includes national and international doctors and researchers in cardiology, imaging, and signal processing and modeling, who have overlapping interests and skills in cardiac bio-electricity.

L'INSTITUT DE RYTHMOLOGIE
ET MODÉLISATION CARDIAQUE

Université de Bordeaux

The University of Bordeaux is a public scientific, cultural, and professional institution. Run by a president who is elected by its executive board, it is composed of governing bodies, administrative components, and departments. The University of Bordeaux is ranked among the top French universities for the quality of its education and research. A multidisciplinary, research-focused, international institution, it leads an ambitious development program with its partners to further promote Bordeaux as a Campus of Excellence.

université ^{de}BORDEAUX

CNRS, UMR 5251 Institut de Mathématiques de Bordeaux and GDR Mamovi

The French National Center for Scientific Research (French: Centre national de la recherche scientifique, CNRS) is the largest governmental research organization in France and the largest fundamental science agency in Europe. In 2016, it employed 31,637 staff, including 11,137 tenured researchers, 13,415 engineers and technical staff, and 7,085 contractual workers.

Inria

The National Institute for Research in Computer Science and Automation (Inria; French: Institut national de recherche en informatique et en automatique) is a French national research institution focusing on computer science and applied mathematics. It was created under the name Institut de recherche en informatique et en automatique (IRIA) in 1967 at Rocquencourt near Paris, part of Plan Calcul. Its first site was the historical premises of SHAPE (central command of NATO military forces). In 1979 IRIA became Inria. [1] Since 2011, it has been styled Inria. Inria is a Public Scientific and Technical Research Establishment (EPST) under the double supervision of the French Ministry of National Education, Advanced Instruction and Research, and the Ministry of Economy, Finance, and Industry.

inventeurs du monde numérique

France Life Imaging

The France Life Imaging – FLI – network was launched in 2012 to ensure high technological innovation in biomedical imaging and to offer open access for the academic, clinician, and industrial community to state-of-the-art in vivo imaging technologies and integrated services. FLI's mission is to increase French visibility in Europe and worldwide. This infrastructure is coordinated by the CEA (French Alternative Energies and Atomic Energy Commission).

Bordeaux INP

Bordeaux INP groups together eight of the region's engineering graduate schools. Its development strategy is based on enhancing the synergy between its three missions of training, research, and technology transfer via a range of high-level scientific and technical training courses, backed by 11 research laboratories of excellence and in permanent contact with the socio-economic world.

Région Nouvelle Aquitaine

Contents

Electrophysiology: Mapping and Biophysical Modelling

Transcriptomic Approaches to Modelling Long Term Changes in Human
Cardiac Electrophysiology 3
    Furkan Bayraktar, Alan P. Benson, Arun V. Holden,
    and Eleftheria Pervolaraki

Virtual Catheter Ablation of Target Areas Identified from Image-Based
Models of Atrial Fibrillation 11
    Aditi Roy, Marta Varela, Henry Chubb, Robert S. MacLeod,
    Jules Hancox, Tobias Schaeffter, Mark O'Neill, and Oleg Aslanidi

Deep Learning Formulation of ECGI for Data-Driven Integration
of Spatiotemporal Correlations and Imaging Information 20
    Tania Bacoyannis, Julian Krebs, Nicolas Cedilnik, Hubert Cochet,
    and Maxime Sermesant

Spatial Downsampling of Surface Sources in the Forward Problem
of Electrocardiography 29
    Steffen Schuler, Jess D. Tate, Thom F. Oostendorp, Robert S. MacLeod,
    and Olaf Dössel

GRÖMeR: A Pipeline for Geodesic Refinement of Mesh Registration 37
    Jake A. Bergquist, Wilson W. Good, Brian Zenger, Jess D. Tate,
    and Robert S. MacLeod

A Numerical Method for the Optimal Adjustment of Parameters
in Ionic Models Accounting for Restitution Properties 46
    Jacob Pearce-Lance, Mihaela Pop, and Yves Bourgault

EP-Net: Learning Cardiac Electrophysiology Models for Physiology-Based
Constraints in Data-Driven Predictions 55
    Ibrahim Ayed, Nicolas Cedilnik, Patrick Gallinari,
    and Maxime Sermesant

Pipeline to Build and Test Robust 3D T1 Mapping-Based Heart Models
for EP Interventions: Preliminary Results 64
    Mengyuan Li, Maxime Sermesant, Sebastian Ferguson, Fumin Guo,
    Jen Barry, Xiuling Qi, Peter Lin, Matthew Ng, Graham Wright,
    and Mihaela Pop

Maximal Conductances Ionic Parameters Estimation in Cardiac
Electrophysiology Multiscale Modelling.   73
   Yassine Abidi, Julien Bouyssier, Moncef Mahjoub, and Nejib Zemzemi

Standard Quasi-Conformal Flattening of the Right and Left Atria   85
   Marta Nuñez-Garcia, Gabriel Bernardino, Ruben Doste, Jichao Zhao,
   Oscar Camara, and Constantine Butakoff

A Spatial Adaptation of the Time Delay Neural Network for Solving ECGI
Inverse Problem   94
   Amel Karoui, Mostafa Bendahmane, and Nejib Zemzemi

On Sampling Spatially-Correlated Random Fields for Complex Geometries . . .   103
   Simone Pezzuto, Alessio Quaglino, and Mark Potse

Interpolating Low Amplitude ECG Signals Combined with Filtering
According to International Standards Improves Inverse Reconstruction
of Cardiac Electrical Activity   112
   Ali Rababah, Dewar Finlay, Laura Bear, Raymond Bond, Khaled Rjoob,
   and James Mclaughlin

Model Assessment Through Data Assimilation of Realistic Data
in Cardiac Electrophysiology   121
   Antoine Gérard, Annabelle Collin, Gautier Bureau, Philippe Moireau,
   and Yves Coudière

Fibrillation Patterns Creep and Jump in a Detailed Three-Dimensional
Model of the Human Atria.   131
   Mark Potse, Alain Vinet, Ali Gharaviri, and Simone Pezzuto

Tissue Drives Lesion: Computational Evidence of Interspecies Variability
in Cardiac Radiofrequency Ablation   139
   Argyrios Petras, Massimiliano Leoni, Jose M. Guerra, Johan Jansson,
   and Luca Gerardo-Giorda

Correcting Undersampled Cardiac Sources in Equivalent Double Layer
Forward Simulations   147
   Jess D. Tate, Steffen Schuler, Olaf Dössel, Robert S. MacLeod,
   and Thom F. Oostendorp

**Novel Imaging Tools and Analysis Methods for Myocardial Tissue
Characterization and Remodelling**

Left Ventricular Shape and Motion Reconstruction Through a Healthy
Model for Characterizing Remodeling After Infarction.   159
   Mathieu De Craene, Paolo Piro, Nicolas Duchateau, Pascal Allain,
   and Eric Saloux

Towards Automated Quantification of Atrial Fibrosis in Images from
Catheterized Fiber-Optics Confocal Microscopy Using Convolutional
Neural Networks.   168
   Chao Huang, Stephen L. Wasmund, Takanori Yamaguchi,
   Nathan Knighton, Robert W. Hitchcock, Irina A. Polejaeva,
   Kenneth L. White, Nassir F. Marrouche, and Frank B. Sachse

High-Resolution *Ex Vivo* Microstructural MRI After Restoring Ventricular
Geometry via 3D Printing   177
   Tyler E. Cork, Luigi E. Perotti, Ilya A. Verzhbinsky, Michael Loecher,
   and Daniel B. Ennis

Synchrotron X-Ray Phase Contrast Imaging and Deep Neural Networks
for Cardiac Collagen Quantification in Hypertensive Rat Model   187
   Hector Dejea, Christine Tanner, Radhakrishna Achanta,
   Marco Stampanoni, Fernando Perez-Cruz, Ender Konukoglu,
   and Anne Bonnin

3D High Resolution Imaging of Human Heart for Visualization
of the Cardiac Structure.   196
   Kylian Haliot, Julie Magat, Valéry Ozenne, Emma Abell,
   Virginie Dubes, Laura Bear, Stephen H. Gilbert, Mark L. Trew,
   Michel Haissaguerre, Bruno Quesson, and Olivier Bernus

Investigating the 3D Local Myocytes Arrangement in the Human
LV Mid-Wall with the Transverse Angle   208
   Shunli Wang, Iulia Mirea, François Varray, Wan-Yu Liu,
   and Isabelle E. Magnin

Biomechanics: Modelling and Tissue Property Measurements

Development of a Computational Fluid Dynamics (CFD)-Model
of the Arterial Epicardial Vasculature   219
   Johannes Martens, Sabine Panzer, Jeroen P. H. M. van den Wijngaard,
   Maria Siebes, and Laura M. Schreiber

Mesh Based Approximation of the Left Ventricle Using a Controlled
Shrinkwrap Algorithm   230
   Faniry H. Razafindrazaka, Katharina Vellguth, Franziska Degener,
   Simon Suendermann, and Titus Kühne

A Computational Approach on Sensitivity of Left Ventricular Wall Strains
to Geometry.   240
   Luca Barbarotta and Peter Bovendeerd

A Simple Multi-scale Model to Evaluate Left Ventricular Growth Laws.   249
   Emanuele Rondanina and Peter Bovendeerd

Modeling Cardiac Growth: An Alternative Approach.................. 258
Nick van Osta, Loes van der Donk, Emanuele Rondanina,
and Peter Bovendeerd

Minimally-Invasive Estimation of Patient-Specific End-Systolic Elastance
Using a Biomechanical Heart Model............................ 266
Arthur Le Gall, Fabrice Vallée, Dominique Chapelle,
and Radomír Chabiniok

Domain Adaptation via Dimensionality Reduction for the Comparison
of Cardiac Simulation Models................................. 276
Nicolas Duchateau, Kenny Rumindo, and Patrick Clarysse

Large Scale Cardiovascular Model Personalisation for Mechanistic
Analysis of Heart and Brain Interactions 285
Jaume Banus, Marco Lorenzi, Oscar Camara, and Maxime Sermesant

Model of Left Ventricular Contraction: Validation Criteria
and Boundary Conditions.................................... 294
Aditya V. S. Ponnaluri, Ilya A. Verzhbinsky, Jeff D. Eldredge,
Alan Garfinkel, Daniel B. Ennis, and Luigi E. Perotti

End-Diastolic and End-Systolic LV Morphology in the Presence
of Cardiovascular Risk Factors: A UK Biobank Study............... 304
Kathleen Gilbert, Avan Suinesiaputra, Stefan Neubauer,
Stefan Piechnik, Nay Aung, Steffen E. Petersen, and Alistair Young

Solution to the Unknown Boundary Tractions in Myocardial Material
Parameter Estimations 313
Anastasia Nasopoulou, David A. Nordsletten, Steven A. Niederer,
and Pablo Lamata

**Advanced Cardiac Image Analysis Tools for Diagnostic
and Interventions**

Fully Automated Electrophysiological Model Personalisation Framework
from CT Imaging 325
Nicolas Cedilnik, Josselin Duchateau, Frédéric Sacher, Pierre Jaïs,
Hubert Cochet, and Maxime Sermesant

Validation of Equilibrated Warping—Image Registration with Mechanical
Regularization—On 3D Ultrasound Images 334
Lik Chuan Lee and Martin Genet

Ventricle Surface Reconstruction from Cardiac MR Slices
Using Deep Learning....................................... 342
Hao Xu, Ernesto Zacur, Jurgen E. Schneider, and Vicente Grau

FR-Net: Joint Reconstruction and Segmentation in Compressed Sensing
Cardiac MRI   352
    Qiaoying Huang, Dong Yang, Jingru Yi, Leon Axel,
    and Dimitris Metaxas

SMOD - Data Augmentation Based on Statistical Models of Deformation
to Enhance Segmentation in 2D Cine Cardiac MRI   361
    Jorge Corral Acero, Ernesto Zacur, Hao Xu, Rina Ariga,
    Alfonso Bueno-Orovio, Pablo Lamata, and Vicente Grau

Comparing Subjects with Reference Populations - A Visualization Toolkit
for the Analysis of Aortic Anatomy and Pressure Distribution   370
    Sahar Karimkeshteh, Lilli Kaufhold, Sarah Nordmeyer, Lina Jarmatz,
    Andreas Harloff, and Anja Hennemuth

Model-Based Indices of Early-Stage Cardiovascular Failure
and Its Therapeutic Management in Fontan Patients.   379
    Bram Ruijsink, Konrad Zugaj, Kuberan Pushparajah,
    and Radomír Chabiniok

3D Coronary Vessel Tree Tracking in X-Ray Projections   388
    Emmanuelle Poulain, Grégoire Malandain, and Régis Vaillant

Interactive-Automatic Segmentation and Modelling of the Mitral Valve   397
    Patrick Carnahan, Olivia Ginty, John Moore, Andras Lasso,
    Matthew A. Jolley, Christian Herz, Mehdi Eskandari,
    Daniel Bainbridge, and Terry M. Peters

Cardiac Displacement Tracking with Data Assimilation Combining
a Biomechanical Model and an Automatic Contour Detection   405
    Radomír Chabiniok, Gautier Bureau, Alexandra Groth,
    Jaroslav Tintera, Jürgen Weese, Dominique Chapelle,
    and Philippe Moireau

An Adversarial Network Architecture Using 2D U-Net Models
for Segmentation of Left Ventricle from Cine Cardiac MRI   415
    Roshan Reddy Upendra, Shusil Dangi, and Cristian A. Linte

Analysis of Three-Chamber View Tagged Cine MRI in Patients
with Suspected Hypertrophic Cardiomyopathy   425
    Mikael Kanski, Teodora Chitiboi, Lennart Tautz, Anja Hennemuth,
    Dan Halpern, Mark V. Sherrid, and Leon Axel

Author Index   433

Electrophysiology: Mapping and Biophysical Modelling

Transcriptomic Approaches to Modelling Long Term Changes in Human Cardiac Electrophysiology

Furkan Bayraktar[1], Alan P. Benson[2], Arun V. Holden[2(✉)],
and Eleftheria Pervolaraki[2]

[1] Biomedical Engineering, Afyon Kocatepe University,
Afyonkarahisar, Afyon, Turkey
[2] School of Biomedical Sciences, University of Leeds, Leeds LS2 9JT, UK
a.v.holden@leeds.ac.uk

Abstract. Slow changes in the activity of the heart occur with time scales from days through to decades, and may in part result from changes in cardiomyocyte properties. The cellular mechanisms of the cardiomyocyte action potential have time scales from < ms to hundreds of ms. Although the quantitative dynamic relations between mRNA transcription, protein synthesis, trafficking, recycling, and membrane protein activity are unclear, mRNA-Seq can be used to inform parameters in cell excitation equations. We use such transcriptomic data from a non-human primate to scale maximal conductances in the O'Hara-Rudy (2011) family of human ventricular cell models, and to predict diurnal changes in human ventricular action potential durations. These are related to circadian changes in the incidence of sudden cardiac deaths. Transcriptomic analysis of human fetal hearts between 9 and 16 weeks gestational age is beginning to be used to inform ventricular cell and tissue models of the electrophysiology of the developing fetal heart.

Keywords: Transcriptomics · Circadian · Fetal

1 Introduction

1.1 Cardiac Excitation Systems

Normal sinus rhythm in the human is robust, and persists from 21 days after fertilization until death, maybe 100 years later. During the lifetime there can be $\sim 10^9$ beats, each triggered by a propagating excitation wave, at a rate from 40 to 180 beats/min. Noninvasive monitoring of activity of the heart, say by an ECG, show beat to beat fluctuations in the intervals of the cardiac cycles, of the order of tens of ms; accelerations and decelerations of heart rate, lasting $10^1 - 10^2$ s; and longer diurnal, monthly, developmental and ageing changes.

A simple computational model of propagating activity in the heart is a nonlinear diffusion system that represents cardiac tissue as mono-domain excitable medium. The nonlinearity is produced by the total membrane ionic current density I_{tot} (μA/μF), given by appropriate excitation equations of the form

© Springer Nature Switzerland AG 2019
Y. Coudière et al. (Eds.): FIMH 2019, LNCS 11504, pp. 3–10, 2019.
https://doi.org/10.1007/978-3-030-21949-9_1

$$dV/dt = -I_{tot}/C_m, \tag{1}$$

$$I_{tot} = I_{ion} + I_{pump} + I_{exchanger} = f(\boldsymbol{v}(t), \boldsymbol{\mu}),$$

where V is the membrane potential (mV), t is the time (ms), $I_{ion} + I_{pump} + I_{exchanger}$ are the ionic, pump and exchanger current densities and C_m the specific membrane capacitance (µF). The equations for these currents have been obtained from extensive series of voltage clamp experiments on isolated myocytes, and patch clamp experiments on recombinant ionic channels expressed in cells. The $\boldsymbol{v}(t)$ are dynamic variables that change during the solution of the equation, *e.g.* membrane potential, activation and inactivation variables, and some intracellular ionic concentrations. The $\boldsymbol{\mu}$ are parameters, some are constants that define the cell type. Some parameters can be changed, to model changes in experimental conditions (*e.g.* temperature, extracellular ionic concentrations).

We assume that changes in cardiac electrical activity over 10^4–10^9 s in an individual lifetime may be produced by changes in protein expression and can be modelled by changes in the cell population and spatial distributions of these parameters over time. For a human cardiac cell models, the parameters are estimated from experiments on human cardiac cells and tissue, obtained *ex vivo* during surgery, or from healthy human donor hearts [1].

1.2   Channel Expression

In an individual cell, a given specific ion channel density corresponds to a number of functioning channels, each of which may be composed of several protein subunits and associated regulatory proteins. The number of functioning channels is in a dynamic balance with silent channels, and the total channel number increased by trafficking and insertion to the membrane, and decreased by internalization, recycling and degradation. Proteins newly synthesized by ribosomes on the endoplasmic reticulum can be transported by endosomes to the membrane. The dynamics for the alpha subunit of the I_{Ks} channel are schematized in Fig. 1, with α, β the rates of formation and degradation of the mRNA and protein $K_{v7.1}$. The overall amount of protein depends on the rate of translation and degradation [2]: any given steady state level of protein ($\alpha_{mRNA}.\alpha_{Kv7.1}/\beta_{mRNA}.\beta_{Kv7.1}$) could be produced by differing combinations of transcription [mRNA h^{-1}] and translation rates [protein mRNA^{-1} h^{-1})]. As well as synthesis and degradation reactions, there are intracellular transport processes. A compartmental model would be a system of ordinary differential equations with largely unknown rate coefficients. It is clear that the total amount of mRNA for the protein subunit, the total amount of the protein subunit, and the ionic maximal conductance G_{Ks} need not be linearly or even simply related, that there will delays between mRNA, protein and changes in maximal conductance, and that there could be oscillatory or complex dynamics.

However, simultaneous measurement by quantitative mass spectrometry of absolute mRNA and protein levels and half-lives in mouse fibroblasts by pulse labelling has shown a correlation between absolute mRNA and protein levels in the same samples [3]. This partially justifies approximating the dynamics of Fig. 1 by a simple linear

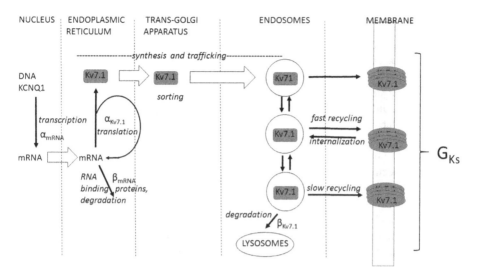

Fig. 1. Schematic of the dynamics of the protein alpha subunit Kv7.1 of the I_{Ks} channel. Each channel is a tetramer and the four pore-forming alpha subunits are associated with modulatory KCNE1 beta subunits.

filter, that would give a phase lag and a linear relation between mRNA and maximal conductance.

There are two ways transcriptomics can lead to estimates of the parameters in the cell and tissue excitation equations.

First, where the excitation mechanisms are well described, the transcriptomic data can be taken as quantitative, and used to scale the values of maximal conductances exchanger, release, and uptake processes of a cell model, as in [4, 5] for the human sinoatrial node (SAN). The assumption that mRNA expression can be used to scale maximal conductances was justified by numerical solutions reproducing SAN action potential characteristics.

Second, where there are experimental cell electrophysiology recordings, but not a voltage clamp based mechanistic model for excitation, the transcriptomic data can be treated as qualitative, to indicate what channel systems are expressed, and the electrophysiological data used to constrain the possible parameter values, as in [6], where the electro-chemical (membrane potential and intracellular $[Ca^{++}]$) activity of single uterine myocytes is reconstructed. Transcriptomic analysis provides a qualitative list of what mRNAs are expressed [7]. The kinetics of each ion channel or transporter system is modelled separately, usually from data obtained from patch clamp experiments on an expression system. The assumption is that the kinetics of the channel or transporter in the expression system used are approximately the same as in the normal cell, even though accessory subunits and anchoring proteins found in myocytes are not present. Experimentally observed $\{V(t), Ca^{++}(t)\}$ constrain the space of conductances/densities that are consistent with the observed electrochemistry, and so the cell is modelled not by a specific set of conductances, but by any member of the space of ionophore densities that is consistent with the observed $\{V(t), Ca^{++}(t)\}$.

1.3   Circadian Rhythms

Long term, slow diurnal or developmental changes in cardiac activity over 10^4–10^9 s will have associated changes in cardiac myocyte properties and behavior, but longitudinal or population transcriptomic and cellular electrophysiological studies over these time scales are impractical. Computational modelling of human cell and tissue electrophysiology, informed by the limited available transcriptomic data, may suggest possible cellular electrophysiological changes that could provide the mechanisms for these changes in activity. Here we approach the modeling of circadian (10^4 s) and fetal developmental (10^6 s) from 9 to 16 weeks gestational age (WGA).

Cardiac electrophysiological variables – heart rate, ECG intervals and their variability, and the onset of arrhythmia, all show diurnal rhythms [8]. The timing of sudden cardiac deaths, obtained from out of hospital and in-hospital data [9, 10], and of the onset of ventricular fibrillation, obtained from ICD data downloads [11], both show a diurnal rhythm - see Fig. 2a. The time is local clock time. The four-fold change in sudden cardiac death rates has a rise time of ~6 h and a fall time of ~18 h. The rhythms for sudden cardiac death, onset of ventricular tachycardia and fibrillation are mirrored by the diurnal changes in RR interval [12, 13] – see Fig. 2b. These diurnal rhythms in heart rate and SCD could be driven by external factors or autonomic activity [14], or could be generated by circadian changes in ion channel expression within the heart [8] that are entrained, *via* autonomic and humoral factors, by the suprachiasmatic nucleus that provides synchronization to the daylight [15].

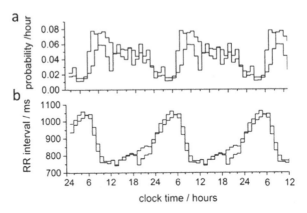

Fig. 2. Circadian changes in cardiac events, averaged over 10^3–10^4 subjects and several years. (a) Replot of SCD data from Berlin, n = 2406 [9]; Massachusetts, n = 2203 [10]. (b) Average RR interval for normal adult subjects, replot from [12, 13].

Diurnal changes in heart rate will produce rate dependent changes in ventricular action potential duration, dispersion, and variability that provide an obvious possible explanation for diurnal changes in SCD. Any intrinsic diurnal changes in myocyte membrane channel expression could directly modulate the risk of SCD.

2   Methods

An extensive diurnal transcriptome atlas of the transcriptome of different tissues for the primate *Papio anubis* (baboon) on a 12 h light-12 h dark cycle has been published and archived in NCBI's GEO (GSE98965) [16]. 12 animals were used, to give 12 time points every 2 h *i.e.* one animal/time point. 15% of the cardiac transcriptome products shows a diurnal rhythm, with the times of peak expression clustered in a temporal window <6 h wide, and periodicity identified by meta2d of Metacycle [17]. Transcriptomics data was extracted for the α-subunits proteins that form part of the membrane ionic channels for the inward currents I_{Na} (SCN5A), $I_{Ca,L}$ (CACNA1C) and outward currents I_{to} (KCND3, KCNA4), I_{Kur} (KCNA5), I_{Kr} (KCNH2), I_{Ks} (KCNQ1) and I_{K1} (KCNJ2/12/4). $I_{K,ACh}$ (KCNJ3/5), $I_{K,ATP}$ (KCNJ11). Only changes in KCNA4 and KCNQ1 are statistically significant (Metacycle integrated $p < 0.05$). The gene level read counts are normalized.

Fetal developmental changes in mRNA for cardiac ion channel protein were extracted from the data [18] deposited in European Genome-Phenome Archive (EGA) which is hosted at the EBI and the CRG, (E-MTAB-7031) for time points 9, 12 and 16 WGA, with each time point being the average from 3 hearts.

The percentage change in mRNA were used to scale the ionic conductances in the O'Hara-Rudy [19] model for human healthy ventricular cells. Equations for isolated cell models were solved using a forward Euler method with a time step of $\Delta t = 0.05$ ms; ion channel gating equations were solved using the Rush-Larsen scheme. The action potential duration at 90% repolarization (APD_{90}) restitution was measured using a standard protocol, with 100 S1 pacing beats at a cycle length of 1,000 ms to establish periodicity, followed by a single premature stimulus (S2) at progressively shorter S1–S2 intervals.

3   Results

Circadian Changes. The normalized individual gene level read counts are plotted for KCNA4 and KCNQ1 in Fig. 3. The Zeitgeber time (ZT) relates to the 24 h light–dark cycle, with light on at ZT = 0. Each 2 h wide column over 24 h in the histogram is based on one animal, and the observation that there are compact peaks, at different times for the two mRNAs, suggests that the modulation is related to ZT and not the individual animal. When the G_{to} and G_{Kr} are scaled by the normalized channel expression data of Fig. 3 there is a diurnal modulation of APD_{90} and its restitution. This is plotted in Fig. 4 over 48 h, with S1–S2 intervals from 200 to 1000 ms, with APD_{90} colour-coded over for each 2 h time window of the day. Light on (ZT0) is identified as the average local clock time of dawn over the year (6 *a.m.*), and the diurnal pattern of sudden cardiac death of Fig. 2(a) is superimposed. There appears to be a coincidence between the diurnal pattern of APD_{90} and the timing of SCD's.

Developmental Changes. The development of ventricular structure, intercellular coupling, and transcriptome in the human fetus has been mapped from 9 to 20 WGA by DT-MRI [20, 21], RNA-Seq [18] and qPCR [20], and related to noninvasive electro-physiology [21].

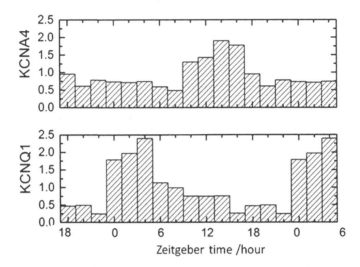

Fig. 3. Diurnal changes over 36 h in cardiac ion channel expression in baboons that had been entrained on a 24 h (12 h light, 12 h dark) cycle.

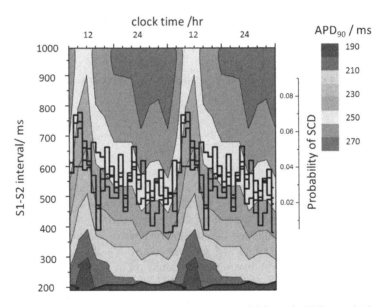

Fig. 4. Circadian changes in ventricular endocardial cell model dynamic APD_{90} restitution curves, computed using data in Fig. 3. The sudden cardiac death rates from Fig. 2 are superimposed, with ZT0 (light on) taken as 06.00 *a.m.* APD_{90} is colour-coded.

The organization of the ventricular wall, quantified by fractional anisotropy, apparent diffusion coefficients and fiber paths lengths extracted by tractography, increases linearly with gestational age, with the helical organisation, and transmural change in fiber helix angle characteristic of the adult mammalian heart developed by 20 weeks. RNAseq showed no significant differential expression of Cx40 or Cx43 between 9 and 16 WGA [18], while both Cx40 and Cx43 expression relative to GAPDH obtained by Western blotting increased significantly [20]. The increase in organisation correlates with an increase in Connexin expression [20]. In tissue models these increase both propagation velocity and its anisotropy.

Differential expression of genes coding ion channel subunits has been extracted from the data of [18]; of the 806 genes differentially expressed between 9 and 16 (WGA) only 20 relate to cardiac excitation with potassium voltage-gated channel (KCNJ2, KCNJ2-AS1, KCTD9, KCNJ8), sodium voltage-gated channel (SCN2B, SCN7A), gap junction protein alpha (GJA1, GJA3) and ryanodine receptor 2 (RYR2), all upregulated; and potassium voltage-gated channel and modifiers (KCNAB3, KCNA1, KCNG1, KCNQ2); calcium voltage-gated channels (CACNG4, CACNG7, CACNA1G/H/S), potassium calcium activated channel (KCNN1) and Ryanodine receptor RYR2 all downregulated.

The ORd model is parameterized for the healthy adult, and there are no well-founded, experimentally based models of human fetal myocardial cells whose parameters could be scaled by the transcriptomic data. In [21, 22] conductance parameters of an adult endocardial myocyte model were scaled so their APD_{90} restitution curves reproduce 35 week gestational age fetal QT restitution curve. The transcriptomic data would give an increase in maximal rate of rise of action potential, and increase in its conduction velocity and duration, but quantitative values are speculative in the absence of fetal cell and tissue electrophysiological data. However, transcriptomics of *ex-vivo* or *postmortem* material could be applied to the ORd system to model ageing in the adult.

4   Conclusions

Transcriptomic data provides a route to parameterizing human cardiac excitation equations to reproduce slow developmental changes, seen over time scales of days to decades, time scales that are not accessible for cell and tissue electrophysiology. What is needed is a clearer, quantitative description of the dynamics of transcription and membrane protein synthesis: the first order approximation of a linear relation between mRNA level and functional membrane ionophore activity is oversimplified, even though it leads to plausible behaviors.

Acknowledgements.   F.B. was supported by an ERAMUS + traineeship.

References

1. Holzem, K.M., Madden, E.J., Efimov, I.R.: Human cardiac systems electrophysiology and arrhythmogenesis. Europace **16**(Suppl 4), iv77–iv85 (2014)
2. Curran, J., Mohler, P.J.: Alternative paradigms for channelopathies. Ann. Rev. Physiol. **77**, 505–524 (2015)

3. Schwanhäusser, B., et al.: Global quantification of mammalian gene expression control. Nature **473**, 337–342 (2011)
4. Chandler, N.J., et al.: Molecular architecture of the human sinus node: insights into the function of the cardiac pacemaker. Circulation **119**(12), 1562–1575 (2009)
5. Boyett, M.R.: 'And the beat goes on' The cardiac conduction system: the wiring system of the heart. Exp. Physiol. **94**(10), 1035–1049 (2009)
6. Atia, J., et al.: Reconstruction of cell surface densities of ion pumps, exchangers, and channels from mRNA expression, conductance kinetics, whole-cell calcium, and current-clamp voltage recordings. PloS Comput. Biol. **12**(4), e1004828 (2016)
7. Chan, Y., et al.: Assessment of myometrial transcriptome changes associated with spontaneous human labour by high-throughput RNA-seq. Exp. Physiol. **99**(3), 510–524 (2014)
8. Black, N., et al.: Circadian rhythm of cardiac electrophysiology, arrhythmogenesis, and the underlying mechanisms Heart Rhythm **16**(2), 298–307 (2018)
9. Arntz, H.-R., et al.: Diurnal, weekly and seasonal variation of sudden death. Eur. Heart J. **21**(4), 315–320 (2000)
10. Willich, S.N., et al.: Circadian variation in the incidence of sudden cardiac death in the framingham heart study population. Am. J. Cardiol. **60**(10), 803–804 (1987)
11. Tofler, G.H., et al.: Morning peak in ventricular tachyarrhythmias detected by time of implantable cardioverter/defibrillator therapy. Circulation **92**, 1203–1208 (1995)
12. Gang, Y., et al.: Circadian variation of the ST interval in patients with sudden cardiac death after myocardial infarction. Am. J. Cardiol. **81**, 950–956 (1998)
13. Bonnemeier, H., et al.: Circadian profile of cardiac autonomic nervous modulation in healthy subjects: differing effects of aging and gender on heart rate variability. J. Cardiovasc. Electrophysiol. **14**, 791–799 (2003)
14. West, A.C., et al.: Misalignment with the external light environment drives metabolic and cardiac dysfunction. Nature Commun. **18**, 417 (2017). https://doi.org/10.1038/s41467-017-00462-2
15. Crnko, S., et al.: Circadian rhythms and the molecular clock in cardiovascular biology and disease. Nature Rev. Cardiol. (2019). https://doi.org/10.1038/s41569-019-0167-4
16. Mure, L.S., et al.: Diurnal transcriptome atlas of a primate across major neural and peripheral tissues. Science **359**, eaao0318 (2018)
17. Wu, G., et al.: MetaCycle: an integrated R package to evaluate periodicity in large scale data. Bioinformatics **32**, 3351–3353 (2016)
18. Pervolaraki, E., et al.: The developmental transcriptome of the human heart. Sci Rep. **8**(1), 15362 (2018)
19. O'Hara, T., Virag, L., Varro, A., Rudy, Y.: Simulation of the undiseased human cardiac ventricular action potential. PLoS Comput. Biol. **7**, e1002061 (2011)
20. Pervolaraki, E., Dachtler, J., Anderson, R.A., Holden, A.V.: Ventricular myocardium development and the role of connexins in the human fetal heart. Sci Rep. **7**(1), 12272 (2017)
21. Pervolaraki, E., et al.: Antenatal architecture and activity of the human heart. Interface Focus. **3**(2), 20120065 (2013). https://doi.org/10.1098/rsfs.2012.0065
22. Pervolaraki, E., Hodgson, S., Holden, A.V., Benson, A.P.: Towards computational modelling of the human foetal electrocardiogram: normal sinus rhythm and congenital heart block. Europace **16**(5), 758–765 (2014). https://doi.org/10.1093/europace/eut377

Virtual Catheter Ablation of Target Areas Identified from Image-Based Models of Atrial Fibrillation

Aditi Roy[1(✉)], Marta Varela[1], Henry Chubb[1], Robert S. MacLeod[2],
Jules Hancox[3], Tobias Schaeffter[4], Mark O'Neill[1], and Oleg Aslanidi[1]

[1] School of Biomedical Engineering and Imaging Sciences,
King's College London, London, UK
`aditi.roy@kcl.ac.uk`
[2] Bioengineering Department, University of Utah, Salt Lake City, USA
[3] School of Physiology, Pharmacology and Neuroscience,
University of Bristol, Bristol, UK
[4] Physikalisch-Technische Bundesanstalt, Berlin, Germany

Abstract. Catheter Ablation (CA) is an effective strategy for rhythm control in atrial fibrillation (AF) patients. However, success rate remains suboptimal in chronic AF patients, where targets for optimal ablation are unknown. Recent clinical evidence suggests an association of atrial fibrosis with locations of re-entrant drivers (RDs) sustaining AF. However, the knowledge of optimal ablation locations based on patient-specific fibrosis distribution is lacking. The aim of this study is to provide a proof-of-concept method to (1) predict patient-specific ablation targets from 3D models of fibrotic atria and (2) perform virtual ablation.

Left atrial (LA) geometry and fibrosis distribution of a persistent AF patient was obtained from MR imaging data. AF simulations were performed by initiating RDs at 12 different locations in the LA model. The tip of the meandering RDs was tracked in all simulations to identify atrial wall regions with the highest probability of harbouring RDs – target areas (TAs). Finally, virtual ablation was performed based on the knowledge of TAs to identify strategies that eliminate RDs.

Our simulations showed that the TAs are typically located at specific regions within the fibrotic patches where RDs stabilize. Ablation strategies that connect these TAs to the nearest pulmonary vein (PV) or the mitral valve can both terminate the existing RD and reduce the inducibility of new RDs, thus preventing AF.

Keywords: Atrial fibrillation · Fibrosis · Image-based modelling · Ablation

1 Introduction

Catheter ablation (CA) is a first-line treatment for atrial fibrillation (AF), the most common sustained cardiac arrhythmia [14]. CA is particularly effective in paroxysmal AF patients, where isolation of the pulmonary veins (PVI) – believed to be prone to

© Springer Nature Switzerland AG 2019
Y. Coudière et al. (Eds.): FIMH 2019, LNCS 11504, pp. 11–19, 2019.
https://doi.org/10.1007/978-3-030-21949-9_2

generating ectopic beats and harbouring re-entrant drivers (RDs) – is a proven strategy [8]. However, PVI has high post-ablation AF recurrence rates [1], and CA success rates also remain suboptimal in patients with chronic forms of AF [10, 16]. This is due to empirical nature of the procedure and lack of mechanistic knowledge of optimal ablation sites and strategies in patients whose atria is altered by AF-induced structural remodelling.

Increasing evidence is suggestive of a link between atrial fibrosis and the decrease in success rates of CA procedure [12]. Fibrosis a product of AF-induced structural remodelling of the atria, which results in increased deposition of extracellular collagen and can be identified in patients using late gadolinium enhanced (LGE) MR imaging. Fibrosis burden has been applied as a predictor of post-ablation AF recurrence [15], and a recent clinical study has directly correlated patient-specific LGE areas with locations of RDs recoded using electrocardiography [2]. Moreover, ablation strategies that target low-voltage areas (believed to be a surrogate of fibrosis) in addition to PVI have shown increased success rates [9, 11]. Modelling studies have suggested that regions with a high heterogeneity in fibrotic density (such as border zones between fibrosis and healthy tissue, BZ) are common anchoring sites for RDs sustaining AF [13, 19]. CA strategies targeting the fibrosis BZ may therefore improve the procedure outcome.

The aim of the study is to develop an image-based computational method to predict the RD locations – potential CA targets – from patient-specific fibrosis distribution, and to simulate ablation in these predicted locations and evaluate the success of this method.

2   Methods

2.1   Image Processing

The patient geometry and fibrosis distribution were reconstructed from LGE MRI of a persistent AF patient using our general workflow [17] (Fig. 1A, B). The MR images were acquired after written informed consent using cardiac and respiratory gating at a spatial resolution of $1.3 \times 1.3 \times 1.3$ mm^3. The LA geometry was manually segmented from the images using MITK. The endocardial wall was identified by segmenting the blood pool. The epicardial wall was obtained by dilating the endocardial wall by 3 mm.

The patient-specific fibrosis distribution was constructed by analysing the image intensity ratio (IIR), computed by dividing individual voxel intensities by the mean blood pool intensity of the LGE images. Voxels with IIR > 1.24 were classified as dense fibrosis and the IIR < 1.08 as healthy tissue, while the regions surrounding the dense fibrosis was labelled as the BZ with IIR values between 1.08 and 1.24 [17].

2.2   Atrial Electrophysiology Model

All the simulations were performed with the Fenton Karma (FK) cell model modified to accurately match the restitution properties of AF-remodelled atrial cells [7]. The monodomain equation with no-flux boundary conditions was solved using forward

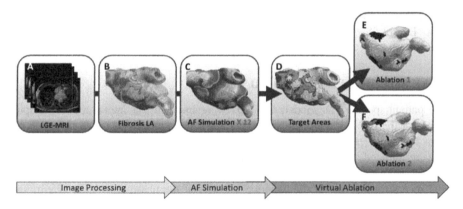

Fig. 1. The workflow for reconstruction of the LA geometry and fibrosis distribution from MRI data (A–B) to develop a patient-specific LA model (C), predict the locations of RDs (D) from AF simulations and then perform virtual ablations (E and F) on these predicted target areas (TA).

Euler and centred finite-difference schemes with a temporal and spatial resolution of 0.005 ms and 0.3 mm. Our model was isotropic, therefore a diffusion coefficient, D of 0.1 mm^2ms^{-1} was chosen to produce atrial conduction velocity of 0.60 ms^{-1} typical of AF.

The macroscopic effect of fibrosis on slowing atrial conduction [5] was modelled by progressively decreasing the diffusive coupling by reducing D within segmented fibrotic tissue region. In the current study, D was set to 0.1 mm^2/ms in the healthy

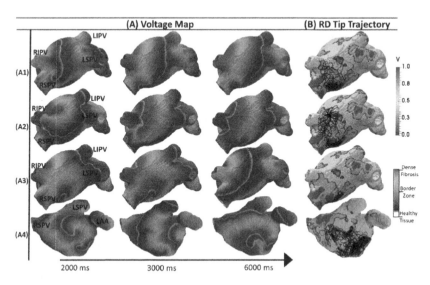

Fig. 2. The dynamics of RDs in the LA model with patient-specific fibrosis distribution. The images show snapshots of colour coded atrial voltage maps (A) for RDs initiated at 4 different locations (A1–A4) over time 6 s of simulations (same applies to in Figs. 4 and 5 below), with their respective RD tip trajectories (black) overlaid on the fibrosis distribution (B).

tissue (IIR < 1.08) and 0.017 mm^2/ms in dense fibrotic tissue (IIR > 1.24); in the fibrotic BZ (1.08 < IIR < 1.24) D had intermediate values calculated via linear interpolation [17].

2.3   AF Simulation Protocol and RD Tracking

In the patient-specific LA model, the RDs were initiated via interaction of ectopic pacing near the LIPV and a plane wave (Fig. 1C). The former was paced at a BCL of 130 ms for 7 to 13 times before introducing the latter to facilitate RDs formation. The RDs were initiated at 12 different locations by varying the direction and the timing of the plane wave across the LA wall. In each simulation, we tracked the RD tip for a period of 6 s by identifying the dynamic location of the phase-singularity (PS) point [4]. The trajectories were then used to construct a map of RD probability by recording the number of times each voxel was visited by the RD tip over the entire simulation.

2.4   Target Areas and Virtual Ablation

The RD tip probability maps obtained from 12 simulations (RDs initiated from different locations) were combined to create the overall RD probability map. The voxels which had a probability higher than 80% were marked as the patient-specific TA (Fig. 1D).

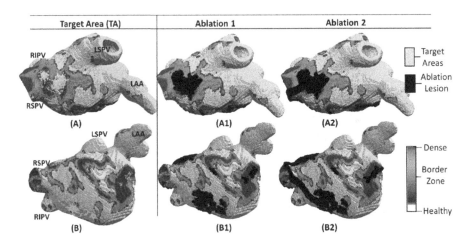

Fig. 3. Patient-specific ablation targets. TAs (yellow) predicted from the image-based LA model are presented from two views, anterior (A) and floor (B). The two ablation strategies (black lesions) are based on the predicted the TAs: (1) ablation lesions only on the TAs (A1 and B1) and (2) additional linear lesions connecting the TAs to the PV and MV (A2 and B2).

The ablation lesions were implemented using two strategies: (1) within the identified TAs (Fig. 1E) and (2) within the TAs plus linear lesions joining them to nearest anatomical boundary, i.e. either the PV or the mitral valve (MV) (Fig. 1F). For both strategies, we tested if these lesions (a) prevented AF inducibility and (b) enabled the

RD termination. In case (a), the protocol used for RD initiation was repeated in the presence of the ablation lesions to check inducibility; in case (b) the lesions were applied after 6 s of the simulation to analyse if these lesions terminated the existing RDs. The CA lesions were modelled as one or more transmural regions of unexcitable tissue. They were cylindrical with a diameter of 2.4 mm to account for the catheter tip shape.

3 Results

The fibrosis distribution in the patient-specific LA model is shown in Fig. 2B. The fibrosis burden was 29% and therefore the patient was classified as Utah 3 category. In the LA model simulations, the interaction of a plane wave and ectopic pacing resulted in the formation of either 1 or 2 RDs per simulation. The RDs stabilized at different region of the atrial wall depending on their initiation location and the surrounding fibrosis distribution (Fig. 2). In fact, across all 12 simulations of this model, a minimum of 1 RD in every case stabilized at the slow conducting fibrotic or BZ regions.

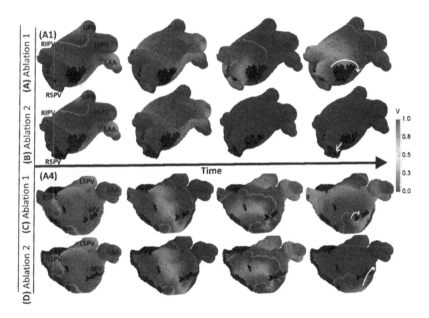

Fig. 4. AF inducibility following two ablation strategies. The RDs were initiated following ablation strategy 1 in both cases (A and C). However, following ablation strategy 2, where the TAs were connected to the PV or MV, the RD either terminated (B) or anchored around the MV (D).

The dynamic behaviour of the RD tip was influenced by the shape and distribution of the fibrotic patches across the LA wall. After their formation, the RDs initially meandered to specific regions of a fibrotic patch where they variously (i) stabilized for the remaining time, (ii) drifted to other patches or (iii) destabilized and terminated. In summary, the highly heterogeneous fibrosis patterns led to the RDs anchoring at specific regions and hence generating a non-uniform distribution of these anchoring locations.

Hence, by simulating RDs at multiple locations in the LA model and constructing the RD tip probability map, we identified patient-specific regions where the RDs were more likely to stabilize – the TAs (Fig. 3A–B). In this model, 68% of TAs were in fibrotic regions, 48% of which were in the dense fibrotic regions and 20% in the BZ. As the volume of the TAs (2%) is substantially smaller than the total fibrosis (29%), targeting these regions in CA procedures may help increase the treatment efficiency and decrease the scar tissue formation and its impact on the LA mechanical function.

Virtual ablation lesions in strategy 1 were only applied in the TAs and in strategy 2, additional linear lesions were applied connecting the TAs to the nearest atrial boundary. For both these strategies 1 and 2, the inducibility and termination of RDs were assessed. The respective voltage dynamics in the LA model after the application of CA lesions using each strategy are shown in Fig. 4 (inducibility) and Fig. 5 (termination).

In the inducibility test (Fig. 4), the virtual ablation performed using strategy 1 did not prevent the RDs from being initiated in any of the 12 simulations. Instead, in majority of cases (10/12), the newly-generated RDs anchored to the ablation scar irrespective of the location where they were initiated (Fig. 4A and C). In the remaining 2/12 cases, RDs were located outside the TAs, and hence were induced as before CA.

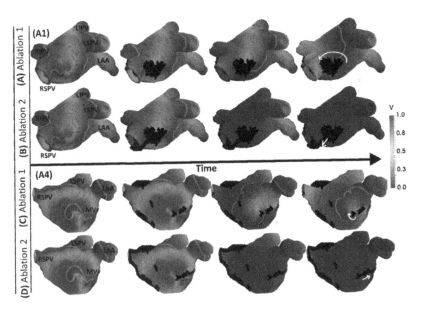

Fig. 5. The termination using two ablation strategies. The RDs did not terminate following ablation strategy 1 in either case (A and C). However, the RD terminated following ablation strategy 2 in both cases, where the TAs where connected to the PV (B) and the MV (D).

However, the virtual ablation performed using strategy 2 resulted in failure to induce RDs in the majority (8/12) of the simulation. The linear CA lesions connecting lesions in the TA to the PV (Fig. 4B), prevented the inducibility of these RDs. However, in 2/12 cases where the TA lesions were connected to the MV with linear lesions, the RD anchored to the MV opening instead (Fig. 5D). In the two remaining cases, where the RDs were located outside the TAs, they were induced as before CA.

In the termination test (Fig. 5), the virtual ablation was performed in 10/12 cases, as in the remaining 2 cases RDs self-terminated before 6 s. In these simulations, CA lesions applied using strategy 1 did not terminate the RDs in any of the 10 cases. In fact, the RDs anchored to the ablation scar in 6/10 cases (Fig. 5A and C). In the remaining 4 cases, the RDs were located outside the TAs; 3 anchored to PVs and 1 stabilized at the LAA. However, in the virtual ablation using strategy 2, the linear lesions connecting the TAs to the PV resulted in the RD termination in 7/10 cases (Fig. 4B and D). In the rest 3/10 cases, RDs were outside the TAs and remained as before.

4   Discussion

We presented an image-based computational method for the prediction of CA targets in chronic AF patients with large amount of fibrosis, which can be applied in the clinic. In this study, we considered a persistent AF patient from Utah 3 category. Our findings suggest that the RDs are localized at specific regions of fibrotic tissue which could be targeted to optimize CA procedure. These results agree with computational studies that provided first evidence for the RD anchoring to atrial fibrotic regions [13, 19].

Moreover, virtual ablations were performed on the identified TAs with the aim of either preventing the RD inducibility or terminating the existing RDs. We tested two strategies: (1) ablation only in the TAs and (2) in addition to ablating the TAs, application of linear lesions joining the TAs to the nearest boundary (PV or MV). Our simulation results demonstrated clear superiority of the second strategy. These results could explain why linear lesions in addition to PVI have higher success rate compared to PVI alone in chronic AF patients [6]. The linear lesions are known to modify atrial substrate and compartmentalize the atria into smaller regions, which prevents the RD formation. However, applying too many linear lesions could compromise the mechanical function of the atria. Therefore, by computationally identifying the patient-specific TAs we can choose the number and locations of the linear lesions to optimise the CA procedure.

The main limitation of the study is the absence of patient-specific data on electrophysiological remodelling and atrial anisotropy, which both could influence the RD behaviour [3, 18]. However, the non-invasive collection of such data from a patient is limited by current electroanatomical mapping (EAM) and imaging techniques, whereas our aim was to design a framework which could be implemented in a clinical setting with ease. In future, we aim to validate the model predictions with EAM patient data.

5   Conclusions

Simulations of the patient-specific LA model showed that RDs sustaining AF were localized at specific regions of fibrotic tissue (TAs) which potentially could be targeted to optimize CA. Performing virtual ablations on the TAs and connecting them to the nearest PV or MV has superior anti-fibrillatory effect compared to ablating TAs alone.

Acknowledgements. This work was supported by the British Heart Foundation [PG/15/8/31138], the EPSRC [EP/L015226/1, EP/R511559/1] and the Wellcome Trust/EPSRC Centre for Medical Engineering [WT 203148/Z/16/Z].

References

1. Calkins, H., et al.: 2017 HRS/EHRA/ECAS/APHRS/SOLAECE expert consensus statement on catheter and surgical ablation of atrial fibrillation. Hear. Rhythm. **14**(10), e275–e444 (2017). https://doi.org/10.1016/J.HRTHM.2017.05.012
2. Cochet, H., et al.: Relationship between fibrosis detected on late gadolinium-enhanced cardiac magnetic resonance and re-entrant activity assessed with electrocardiographic imaging in human persistent atrial fibrillation. JACC Clin. Electrophysiol. **4**(1), 17–29 (2018). https://doi.org/10.1016/j.jacep.2017.07.019
3. Colman, M.A., et al.: Evolution and pharmacological modulation of the arrhythmogenic wave dynamics in canine pulmonary vein model. Europace **16**(3), 416–423 (2014)
4. Fenton, F., et al.: Vortex dynamics in three-dimensional continuous myocardium with fiber rotation: Filament instability and fibrillation. Chaos An Interdiscip. J. Nonlinear Sci. **8**, 20–47 (1998). https://doi.org/10.1063/1.166311. May 2016
5. Fukumoto, K., et al.: Association of left atrial local conduction velocity with late gadolinium enhancement on cardiac magnetic resonance in patients with atrial fibrillation. Circ. Arrhythmia Electrophysiol. **9**(3), e002897 (2016)
6. Gaita, F., et al.: Long-term clinical results of 2 different ablation strategies in patients with paroxysmal and persistent atrial fibrillation. Circ. Arrhythmia Electrophysiol. **1**(4), 269–275 (2008). https://doi.org/10.1161/CIRCEP.108.774885
7. Goodman, A.M., et al.: A membrane model of electrically remodelled atrial myocardium derived from in vivo measurements. Europace **7**(Suppl. 2), 135–145 (2005)
8. Hocini, M., et al.: Techniques for curative treatment of atrial fibrillation. J. Cardiovasc. Electrophysiol. **15**(12), 1467–1471 (2004)
9. Jadidi, A.S., et al.: Ablation of persistent atrial fibrillation targeting low-voltage areas with selective activation characteristics. Circ. Arrhythmia Electrophysiol. **9**(3), e002962 (2016). https://doi.org/10.1161/CIRCEP.115.002962
10. Kanagaratnam, L., et al.: Empirical pulmonary vein isolation in patients with chronic atrial fibrillation using a three-dimensional nonfluoroscopic mapping system: long-term follow-up. Pacing Clin. Electrophysiol. **24**(12), 1774–1779 (2001)
11. Kottkamp, H., et al.: Box isolation of fibrotic areas (BIFA): a patient-tailored substrate modification approach for ablation of atrial fibrillation. J. Cardiovasc. Electrophysiol. **27**(1), 22–30 (2016). https://doi.org/10.1111/jce.12870
12. Marrouche, N.F., et al.: Association of atrial tissue fibrosis identified by delayed enhancement MRI and atrial fibrillation catheter ablation: the DECAAF study. JAMA, J. Am. Med. Assoc. **311**(5), 498–506 (2014). https://doi.org/10.1001/jama.2014.3

13. Morgan, R., et al.: Slow conduction in the border zones of patchy fibrosis stabilizes the drivers for atrial fibrillation: insights from multi-scale human atrial modeling. Front. Physiol. **7**, 1–15 (2016). https://doi.org/10.3389/fphys.2016.00474

14. Nattel, S.: New ideas about atrial fibrillation 50 years on. Nature **415**(6868), 219–226 (2002). https://doi.org/10.1038/415219a

15. Okamura, T., et al.: Diagnosis of cochleovestibular neurovascular compression syndrome: a scoring system based on five clinical characteristics. Neurol. Surg. **45**(2), 117–125 (2017). https://doi.org/10.1001/jama.2014.3

16. Oral, H., et al.: Pulmonary vein isolation for paroxysmal and persistent atrial fibrillation. Circulation **105**(9), 1077–1081 (2002). https://doi.org/10.1161/hc0902.104712

17. Roy, A., et al.: Image-based computational evaluation of the effects of atrial wall thickness and fibrosis on re-entrant drivers for atrial fibrillation. Front. Physiol. **9**, 1352 (2018). https://doi.org/10.3389/fphys.2018.01352

18. Varela, M., et al.: Atrial heterogeneity generates re-entrant substrate during atrial fibrillation and anti-arrhythmic drug action: mechanistic insights from canine atrial models. PLoS Comput. Biol. **12**(12), e1005245 (2016)

19. Zahid, S., et al.: Patient-derived models link re-entrant driver localization in atrial fibrillation to fibrosis spatial pattern. Cardiovasc. Res. **110**(3), 443–454 (2016)

Deep Learning Formulation of ECGI for Data-Driven Integration of Spatiotemporal Correlations and Imaging Information

Tania Bacoyannis[1]([✉]), Julian Krebs[1], Nicolas Cedilnik[1,2], Hubert Cochet[2], and Maxime Sermesant[1]

[1] Inria, Université Côte d'Azur, Nice, France
`tania-marina.bacoyannis@inria.fr`
[2] Liryc Institute, Bordeaux, France

Abstract. The challenge of non-invasive Electrocardiographic Imaging (ECGI) is to re-create the electrical activity of the heart using body surface potentials. Specifically, there are numerical difficulties due to the ill-posed nature of the problem. We propose a novel method based on Conditional Variational Autoencoders using Deep generative Neural Networks to overcome this challenge. By conditioning the electrical activity on heart shape and electrical potentials, our model is able to generate activation maps with good accuracy on simulated data (mean square error, MSE = 0.095). This method differs from other formulations because it naturally takes into account spatio-temporal correlations as well as the imaging substrate through convolutions and conditioning. We believe these features can help improving ECGI results.

Keywords: ECGI · Deep learning · Simulation · Generative model

1 Introduction

Electrocardiographic Imaging (ECGI) has been an active research area for decades. Important progress was achieved but there are still challenges in robustness and accuracy due to the ill-posedness of the classical formulation.

In the last few years, deep learning (DL) based methods have been used to solve inverse problems, e.g. in medical image reconstruction [7]. Autoencoders are popular for these problems, as they are specifically designed to reconstruct high dimensional data in an unsupervised fashion. Autoencoders learn an identity function in order to reconstruct the input image after having first encoded it in a latent representation and then decoded it to the original input. Such methods have been recently introduced into the ECGI problem to regularize the temporal information while processing it [4].

© Springer Nature Switzerland AG 2019
Y. Coudière et al. (Eds.): FIMH 2019, LNCS 11504, pp. 20–28, 2019.
https://doi.org/10.1007/978-3-030-21949-9_3

In this manuscript, we propose to reformulate the whole ECGI problem as a conditional variational autoencoder based on convolutional neural networks. This has four main advantages:

- Spatio-temporal correlations: the convolutional model learns interactions in space and time between signals, while most ECGI methods solve each time step independently.
- Imaging substrate: the correlation between the substrate from imaging and the signals is also learned, therefore we can seamlessly integrate any 3D image information in ECGI, while this is still difficult in the classical formulation.
- Data-driven regularisation: using a generative model from a low dimensional space should ensure smooth variations between similar cases; this could alleviate the ill-posedness problem.
- Fast computations: once trained, DL methods are very fast to evaluate.

In order to achieve this, we use Cartesian coordinates for all the data (space and time) to leverage the power of convolutional neural networks.

2 Context

2.1 Electrocardiographic Imaging

ECGI is a non invasive modality which aims to better understand the electrical activity of the heart, both quantitatively and qualitatively. ECGI allows the visualisation of the electrical potential distribution of the electrical wave on the heart surface from body surface potentials (BSP). It requires medical imaging to obtain geometrical information and methods to solve the inverse problem.

Forward Problem. This refers to the estimation of the ECG data from cardiac data. The two classical numerical approaches for this are the Boundary Element Method (BEM), based on surfaces, and the Finite Element Method (FEM), where the 3D torso model is approximated by small volume elements. Both propagate the epicardial action potentials to the body surface with chosen boundary conditions, e.g., null current across the body surface. There are also methods using a dipole formulation, assuming that the torso domain is homogenous and infinite [3]. In [11], the authors demonstrated that noninvasive ECGI reconstruction does not require first order approximations for torso heterogeneities.

Inverse Problem. This allows to reconstruct cardiac electrical activity using BSP. The classical approach is based on a transfer matrix between epicardial potentials and the torso potentials. To compute this transfer matrix different approaches can be used such as BEM [2] or FEM [13]. However, inverting it is ill-posed [13]. Therefore, different formulations and regularization methods were proposed, see for instance the publications of the ECGI consortium[1].

---

[1] http://www.ecg-imaging.org.

2.2   Deep Learning

Variational Autoencoders. Variational Autoencoders (VAEs) are powerful probabilistic generative models [9]. VAEs consist of two connected networks: an encoder and a decoder. The former takes an input x and compresses it into a low-dimensional latent representation, which variables are denoted z, with a distribution $P(z)$, often assumed to be a centred isotropic multivariate Gaussian $P(z) = N(z : 0, I)$. The latter takes z as input and reconstructs the data from the generative distribution $P_\Theta(x|z)$. This likelihood distribution $P_\Theta(x|z)$ is learned in the decoder neural network with parameters Θ. The resulting generative process induces the distribution $P_\Theta(x) = E_{P(z)}P_\Theta(x|z)$. The term Variational in Variational Autoencoder refers to variational inference or variational Bayes. Due to the intractability of the true posterior distribution $P(z|x)$, the posterior is approximated by learning the probability $Q_\Phi(z|x)$ in the encoder neural network with parameters Φ (Fig. 1).

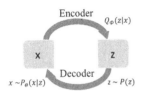

A VAE is trained in order to minimize the Kullback-Leibler (KL) divergence between the variational probability $Q_\Phi(z|x)$ and the true posterior distribution $P(z|x)$. As this is intractable, it can be reformulated as the evidence lower bound (ELBO) of the log marginalized data likelihood $P_\Theta(x)$ [9]:

Fig. 1. VAE model [9]

$$logP_\Theta\left(x|z\right) - D_{KL}\left(Q_\Phi\left(z|x\right) \| P\left(z\right)\right) \leq logP_\Theta\left(x\right)$$

The VAE loss function is then defined by a reconstruction term and the KL divergence between variational and prior probability:

$$L_{VAE}\left(\Theta, \Phi\right) = -logP_\Theta\left(x|z\right) + D_{KL}\left(Q_\Phi\left(z|x\right) \| P\left(z\right)\right)$$

By minimizing the loss function, the lower bound of the probability of generating real data samples is maximised. To compute the gradient of the variational lower bound, we use the reparameterization trick with respect to VAE [9]. Commonly, $Q_\Phi\left(z|x\right)$ is Gaussian with a diagonal covariance matrix: $z \sim Q_\Phi\left(z|x\right) = N\left(z : \mu, \sigma\right)$ where $z = \mu + \sigma \odot \epsilon$ and $\epsilon \sim N\left(0, I\right)$.

Conditional Variational Autoencoders. Conditional VAEs (CVAEs) [10] are an extension of VAEs [9] where the latent variables and the data are both conditioned on additional random variables c. The encoder of the CVAE is not only conditioned on the data x but also on the conditioning data c which results in the variational probability: $Q_\Phi\left(z|x, c\right)$. Respectively, the decoder is also conditioned on the conditioning data c. Thus, the generative distribution becomes: $P_\Theta\left(z|x, c\right)$. The CVAE is trained to maximize the conditional log-likelihood where the ELBO is:

$$L_{CVAE}\left(\Theta, \Phi\right) = -logP_\Theta\left(x|z, c\right) + D_{KL}\left(Q_\Phi\left(z|x, c\right) \| P\left(z\right)\right) \leq logP_\Theta\left(x|c\right)$$

β-Variational Autoencoders. β-VAEs are an evolution of VAEs which intends to discover disentangled latent factors. In [8], it is shown that this model achieves similar disentanglement performance compared with VAEs, both quantitatively and qualitatively. β-VAEs and VAEs have the same goals: maximize the probability of generating real data and minimize the distance between the real and estimated posterior distributions (smaller than a constraint ε). In this approach, the prior is an isotropic Gaussian $P(z) = N(0, I)$.

The β-VAE lower bound is defined as:

$$L_{\beta\text{-VAE}}(\Theta, \Phi, \beta) = -log P_\Theta(x|z) + \beta D_{KL}(Q_\Phi(z|x) \parallel P(z))$$

with β a Lagrangian multiplier hyperparameter under the Karush-Kuhn-Tucker conditions.

If $\beta = 1$, β-VAEs correspond to the original VAEs [9]. If $\beta > 1$, β-VAEs apply a stronger constraint on the latent bottleneck and so limit the capacity of z. It allows to learn the most efficient representation of the data. However, if β is too big, β-VAE learns an entangled latent representation because of its excessive capacity in the latent z bottleneck. Same remarks if β is too small, it will have too little capacity. To sum up, $\beta > 1$ is capital to achieve good disentanglement.

3   Methods

3.1   ECGI Forward Problem: Data Simulation

We simulated cardiac activation maps and BSP data using the Eikonal Model directly on a Cartesian grid from image segmentation [1]. The Eikonal model is a fast generic model of wave front propagation. Its inputs are the myocardial wall mask, a local conduction velocity v for each voxel x of the given wall mask and the pacing zone.

$$v(x) \parallel \nabla T(x) \parallel = 1$$

with $T(x)$ the local activation time in x. It is solved using The Fast Marching Method (FMM) [12].

BSP were generated using the dipole formulation [6] associating each activation time from the Eikonal model with an action potential signal from the Mitchell Schaeffer model. 100 torso electrodes were positioned on a 10×10 Cartesian grid in front of the heart.

In total, 120 activation maps and the corresponding BSP were simulated using a patient image segmentation from CT images. In order to accelerate computation and ease memory requirements, we currently present 2D results, where we separated the 3D simulated data in 2D slices.

3.2   Models Structure and Training

The different models described below were trained on 80% of the data, tested on the remaining 20% and have been implemented using Keras[2].

VAE for BSP. It follows the architecture presented in Fig. 2, with convolutional layers in order to extract spatiotemporal correlations.

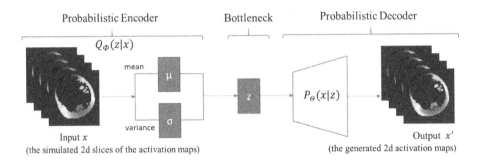

Fig. 2. Variational Autoencoder architecture

The encoder consists of two convolutional layers followed by one dense layer. The bottleneck layers (μ, σ, z) are fully-connected. The decoder consists of a fully-connected and three transposed convolutional layers. For all convolutional layers except the output one: strides are set to 2, 16 filters are applied, the kernel size is 3 and the activation functions are *ReLU*. In the last layer, one filter is applied with a kernel size of 3 and a sigmoid activation function. The latent code size is set to 6. Mean squared error (MSE) is used to calculate the reconstruction loss. The optimization is performed with the NADAM solver with a batch size of 32.

VAE for Activation Maps. The encoder architecture is the same as the previous one. However, network parameters and activations differ: the convolution kernel size is 3 (even in the last layer) and all activations are *tanh*. The latent code size is set to 6. MSE is used to calculate the reconstruction loss. Optimization is performed with the RMSPROP solver with a batch size of 25.

β-CVAE for Activation Maps from BSP and Images. Our conditional autoencoder aims to generate heart activation maps using two conditions: the simulated ECG signals and the shape of the patient's heart. This is a novel formulation of ECGI using DL in order to learn the influence of cardiac shape/structure.

The architecture of our conditioned generative model (encoder) and our conditioned variational approximation (decoder) is described in Fig. 3. Empirically, we found that the best value for Lagrangian multiplier hyperparameter β in the loss function was 5. The ADAM optimiser was used with 10^{-4} as learning rate and a batch size of 1. Latent space is set to 16.

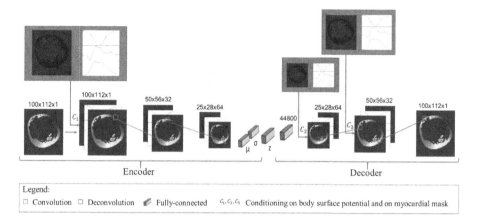

Fig. 3. CVAE architecture for ECGI with imaging data (C_1: The first convolution layer in the encoder was conditioned by concatenating the input data with the BSP mapping signals and the myocardial mask. C_2, C_3: The deconvolution layers in the decoder were conditioned by concatenating each layer's output with sub-sampled versions of the BSP mapping signals and of the mask).

4   Results

4.1   Evaluation of Body Surface Potentials VAE

Figure 4 shows two examples of input BSP mapping signals and their corresponding generated signals by our model.

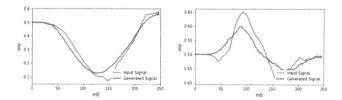

Fig. 4. Evaluation of input (red) and output (blue) signals from VAE for 2 different electrodes. (Color figure online)

The performance of our model was evaluated measuring 5 metrics between the input BSP signals and their corresponding generated signals. We found a

correlation of 0.95, a mean difference of 3.345%, an abscisse area difference of
−2.152% and a maximum amplitude difference of −3.468%; 90.854% of the time,
the sign of first peak was the same in the input and in the output.

4.2   Evaluation of Activation Maps VAE

Using our VAE model developed to generate the electrical activation of the heart
(Fig. 5), we obtained a MSE of 0.018 for reconstruction accuracy.

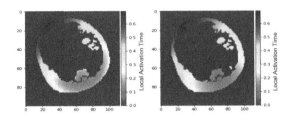

Fig. 5. Predicted (left) and simulated (right) activation maps with the Activation Maps
VAE model.

4.3   Evaluation of β-CVAE

β-CVAE is a probabilistic generative method, so we can generate several proba-
ble solutions for a given input. We generated ten activation maps per prediction
in order to evaluate the accuracy and stability of our method. The MSE metric
was 0.095 over all the tests. Figure 6 shows an example of (a) an input activation
map, (b) its corresponding generated mean activation map by our β-CVAE proposed method, (c) standard deviation map for 10 predictions, and (d) the error
map.

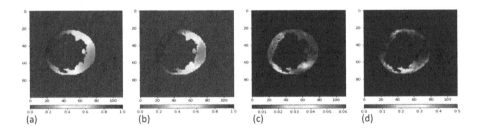

Fig. 6. (a) Simulated and (b) predicted mean activation maps for proposed deep learning based ECGI, (c) Standard deviation map calculated over 10 predictions, (d) error
map, difference between predicted and simulated activation maps.

In Fig. 6 (c) and (d), small values imply that the reconstruction performs
well (small error and small standard deviation), while large values mean that

the reconstruction is suffering. We can observe that the areas with the highest standard deviation are close to areas with the highest error, therefore the probabilistic aspect of our method can help in quantifying the uncertainty in the predictions.

5 Discussion and Conclusion

As a direct application of the Forward Problem of ECGI, electrical activation and potentials of the heart were simulated using a patient CT-scan image and the Eikonal Model. The first step of our work was to understand and choose the best hyperparameters of the Variational Autoencoders to generate potential or activation maps. Finally, we proposed a novel method based on Conditional β Variational Autoencoder able to solve ECGI inverse problem in 2D. This generative probabilistic model learns geometrical and spatio-temporal information and enables to generate the corresponding activation map of the specific heart. We showed that these generated electrical activities were very similar to the simulated ones. The presented method will now be generalised to 3D and evaluated on clinical data. Theoretically there is no impediment to extend our 2D model to 3D, as all the convolution operators have a 3D version. However, it will require to simulate more data to train the model, as we will have more hyperparameters to optimise. We will also explore transfer learning to apply it on clinical data [5].

Acknowledgements. The research leading to these results has received European funding from the ERC starting grant ECSTATIC (715093) and French funding from the National Research Agency grant IHU LIRYC (ANR-10-IAHU-04).

References

1. Cedilnik, N., et al.: Fast personalized electrophysiological models from CT images for ventricular tachycardia ablation planning. EP-Europace **20**, November 2018
2. Chamorro-Servent, J., Dubois, R., Potse, M., Coudière, Y.: Improving the spatial solution of electrocardiographic imaging: a new regularization parameter choice technique for the tikhonov method. In: Pop, M., Wright, G.A. (eds.) FIMH 2017. LNCS, vol. 10263, pp. 289–300. Springer, Cham (2017). https://doi.org/10.1007/978-3-319-59448-4_28
3. Chávez, C.E., Zemzemi, N., Coudière, Y., Alonso-Atienza, F., Álvarez, D.: Inverse problem of electrocardiography: estimating the location of cardiac Ischemia in a 3D realistic geometry. In: van Assen, H., Bovendeerd, P., Delhaas, T. (eds.) FIMH 2015. LNCS, vol. 9126, pp. 393–401. Springer, Cham (2015). https://doi.org/10.1007/978-3-319-20309-6_45
4. Ghimire, S., Dhamala, J., Gyawali, P.K., Sapp, J.L., Horacek, M., Wang, L.: Generative modeling and inverse imaging of cardiac transmembrane potential. In: Frangi, A.F., Schnabel, J.A., Davatzikos, C., Alberola-López, C., Fichtinger, G. (eds.) MICCAI 2018. LNCS, vol. 11071, pp. 508–516. Springer, Cham (2018). https://doi.org/10.1007/978-3-030-00934-2_57

5. Giffard-Roisin, S., et al.: Transfer learning from simulations on a reference anatomy for ECGI in personalised cardiac resynchronization therapy. IEEE Trans. Biomed. Eng. **20** (2018)
6. Giffard-Roisin, S., et al.: Non-invasive personalisation of a cardiac electrophysiology model from body surface potential mapping. IEEE Trans. Biomed. Eng. **64**(9), 2206–2218 (2017)
7. Hammernik, K., et al.: Learning a variational network for reconstruction of accelerated MRI data. Magn. Reson. Med. **79**(6), 3055–3071 (2018)
8. Higgins, I., et al.: β-vae: Learning basic visual concepts with a constrained variational framework. In: International Conference on Learning Representations (2017)
9. Kingma, D.P., Welling, M.: Auto-encoding variational bayes. In: Proceedings of the International Conference on Learning Representations (ICLR) (2014)
10. Kingma, D.P., Mohamed, S., Jimenez Rezende, D., Welling, M.: Semi-supervised learning with deep generative models. In: Ghahramani, Z., Welling, M., Cortes, C., Lawrence, N.D., Weinberger, K.Q. (eds.) Advances in Neural Information Processing Systems, vol. 27, pp. 3581–3589. Curran Associates, Inc. (2014)
11. Ramanathan, C., Rudy, Y.: Electrocardiographic imaging: effect of torso inhomogeneities on noninvasive reconstruction of epicardial potentials, electrograms, and isochrones. J. Cardiovasc. Electrophysiol. **12**, 241–252 (2001)
12. Sermesant, M., Coudière, Y., Moreau-Villéger, V., Rhode, K.S., Hill, D.L.G., Razavi, R.S.: A fast-marching approach to cardiac electrophysiology simulation for XMR interventional imaging. In: Duncan, J.S., Gerig, G. (eds.) MICCAI 2005. LNCS, vol. 3750, pp. 607–615. Springer, Heidelberg (2005). https://doi.org/10.1007/11566489_75
13. Zemzemi, N., et al.: Effect of the torso conductivity heterogeneities on the ECGI inverse problem solution. In: Computing in Cardiology, Nice, France, September 2015

Spatial Downsampling of Surface Sources in the Forward Problem of Electrocardiography

Steffen Schuler[1]([✉]), Jess D. Tate[2], Thom F. Oostendorp[3],
Robert S. MacLeod[2], and Olaf Dössel[1]

[1] Institute of Biomedical Engineering, KIT, Karlsruhe, Germany
publications@ibt.kit.edu
[2] SCI Institute, University of Utah, Salt Lake City, USA
[3] Donders Centre for Neuroscience, Radboud University, Nijmegen, Netherlands

Abstract. The boundary element method is widely used to solve the forward problem of electrocardiography, i.e. to calculate the body surface potentials (BSP) caused by the heart's electrical activity. This requires discretization of boundary surfaces between compartments of a torso model. Often, the resolution of the surface bounding the heart is chosen above 1 mm, which can lead to spikes in resulting BSPs. We demonstrate that this artifact is caused by discontinuous propagation of the wavefront on coarse meshes and can be avoided by blurring cardiac sources before spatial downsampling. We evaluate different blurring methods and show that Laplacian blurring reduces the BSP error 5-fold for both transmembrane voltages and extracellular potentials downsampled to 3 different resolutions. We suggest a method to find the optimal blurring parameter without having to compute BSPs using a fine mesh.

Keywords: Boundary element method · Spatial downsampling · Transmembrane voltages · Extracellular potentials · Body surface potentials

1 Introduction

It is well known that a spatial resolution below 1 mm is required for mono- or bidomain simulations of cardiac excitation propagation in order to capture the sharp upstroke of the transmembrane voltage (TMV) during depolarization or the corresponding downslope of the extracellular potential (EP). For solving the forward problem of electrocardiography, i.e. to calculate body surface potentials (BSP) from a set of cardiac sources, often much coarser resolutions are used to reduce the computational load. However, direct mapping of sources from a fine mesh onto a coarse one causes the activation to abruptly "jump" from one node to the other, which can lead to jagged time courses of BSPs. Using a volumetric finite element model (FEM), Potse et al. showed that translating transmembrane currents from a fine to a coarse mesh using regional summation of currents rather than "simple injection" reduces such artifacts in simulated electrograms [3]. Here, we aim to develop a similar method to spatially downsample surface sources used in forward calculations with the boundary element method (BEM). We hypothesize that, although high spatial frequencies associated with depolarization cannot be represented on a coarse mesh, low frequency components are sufficient to obtain a good approximation of BSPs, as the torso itself has

© Springer Nature Switzerland AG 2019
Y. Coudière et al. (Eds.): FIMH 2019, LNCS 11504, pp. 29–36, 2019.
https://doi.org/10.1007/978-3-030-21949-9_4

a strong smoothing effect. Following the sampling theorem, we suggest to blur the surface sources before downsampling in order to represent the low frequency components correctly. We evaluate the method for the two most common surface source models: TMVs and EPs.

2   Methods

2.1   Simulation Setup

For evaluating different downsampling strategies, we considered four levels of ventricular mesh resolution. Starting from the coarsest surface mesh consisting of approximately equilateral triangles (resolution 0, Fig. 1, left), three finer meshes were created through 1-, 2- and 4-fold linear subdivision (Table 1). This ensures that the geometry stays exactly the same across different resolutions and changes in BSPs are exclusively related to the spatial sampling density.

An excitation originating at the center of the right ventricular septum was simulated using the bidomain model and the Ten Tusscher et al. ionic model. To this end, we tetrahedralized the surface mesh at resolution 4 and added 2 mm of blood on either side of the myocardium. Epi- and endocardial fiber angles with respect to the circumferential direction were chosen to be ±60° and intra- and extracellular anisotropy ratios were set to 9 and 3, respectively. Myocardial conductivities were adapted to get a global conduction velocity of about 1 m/s.

For forward calculations, we used a homogeneous torso model (Fig. 1, right) and BEM with point collocation weighting and linear basis functions [4]. Transfer matrices were computed for both TMVs and EPs at all four ventricular mesh resolutions. In order to get the same BSPs for both source models at the finest resolution and thus to make the effect of downsampling as comparable as possible, we did not use the EPs from the bidomain simulation, but performed an FEM forward calculation of volumetric TMVs using a fully isotropic torso model that is equivalent to the BEM torso model and extracted the resulting EPs at the myocardial surface.

Table 1. Ventricular mesh resolutions.

Resolution (level of subdivision)	0	1	2	4
Edge length (mean ± SD in mm)	12.42 ± 1.00	6.21 ± 0.50	3.10 ± 0.25	0.78 ± 0.06
Number of nodes	578	2,306	9,218	147,458

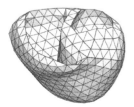 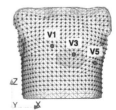 

Fig. 1. Left: Ventricular mesh at resolution 0. Right: Torso mesh with 1000 electrodes (black dots) and positions of Wilson leads V1, V3 and V5 (red spheres). (Color figure online)

2.2 Blurring Methods

For reducing high spatial frequencies before downsampling, we considered three spatial and, for comparison, two temporal blurring methods, which are shortly described in the following.

Circular Box Blur: The values x_i of all nodes i within a radius r, measured along the surface, are averaged:

$$\tilde{x}_i = \frac{1}{N} \sum_j x_j \quad \text{with} \quad N = \sum_j 1, \quad d_{ij} \leq r, \tag{1}$$

where d_{ij} is the geodesic distance between nodes i and j computed using the fast marching method.

Gaussian Blur: The values are distance-weighted using a Gaussian with standard deviation σ_S that is truncated at 30 mm and summed up:

$$\tilde{x}_i = \frac{1}{N} \sum_j x_j \, e^{-\frac{1}{2}\left(\frac{d_{ij}}{\sigma_S}\right)^2} \quad \text{with} \quad N = \sum_j e^{-\frac{1}{2}\left(\frac{d_{ij}}{\sigma_S}\right)^2}, \quad d_{ij} \leq 30\,\text{mm}. \tag{2}$$

Laplacian Blur: The sum of second derivatives along the surface is minimized:

$$\tilde{\mathbf{X}} = \arg\min_{\tilde{\mathbf{X}}} \|\mathbf{X} - \tilde{\mathbf{X}}\|_F^2 + \lambda_S \|\mathbf{L}\tilde{\mathbf{X}}\|_F^2 = (\mathbf{I} + \lambda_S \mathbf{L}^\mathsf{T}\mathbf{L})^{-1}\mathbf{X}, \tag{3}$$

where $\mathbf{L}$ is an approximation of the surface Laplacian operator [2]. $\mathbf{I}$ is an identity matrix and $\mathbf{X}$ contains nodes along rows and time steps along columns, thus enabling spatial blurring for all time steps simultaneously.

Temporal Gaussian Blur: Gaussian-weighted average along time:

$$\tilde{x}(t) = \frac{1}{N} \sum_{\tau=-50\,\text{ms}}^{50\,\text{ms}} x(t - \tau) \, e^{-\frac{1}{2}\left(\frac{\tau}{\sigma_T}\right)^2} \quad \text{with} \quad N = \sum_{\tau=-50\,\text{ms}}^{50\,\text{ms}} e^{-\frac{1}{2}\left(\frac{\tau}{\sigma_T}\right)^2}. \tag{4}$$

The signal $x(t)$ is padded before filtering by repeating its border elements.

Temporal Laplacian Blur: The second temporal derivative is minimized:

$$\tilde{\mathbf{X}} = \left((\mathbf{I} + \lambda_T \mathbf{D}^\mathsf{T}\mathbf{D})^{-1}\mathbf{X}^\mathsf{T}\right)^\mathsf{T} = \mathbf{X}(\mathbf{I} + \lambda_T \mathbf{D}^\mathsf{T}\mathbf{D})^{-1}, \tag{5}$$

where the rows of $\mathbf{D}$ contain finite difference coefficients of the second derivative.

2.3 Interpolation Methods

To transfer signals from coarse to fine meshes, we considered three interpolation methods. These were needed to find the optimal blurring parameter, for downsampling in a least-squares sense and to blur signals that are only available on a coarse mesh (see Sects. 3.2 and 3.3).

Linear Interpolation: Values are interpolated within triangles by weighting the values at the vertices with the barycentric coordinates of the target point.

Laplacian Interpolation: Values at unknown mesh points are interpolated by minimizing the Laplacian at all mesh points (method B in [2]).

Fig. 2. Left: Streamlines along the gradient of ATs. Right: Original and blurred TMVs along these streamlines for $t = 80$ ms. The ventricles are clipped for better visualization.

Wave Equation Based Interpolation: Interpolation according to [1]:

– Signals of all nodes are aligned in time by shifting them by their negative activation times (AT).
– Time-aligned signals are interpolated in space using Laplacian interpolation.
– Signals are shifted in time by their Laplacian-interpolated activation times.

2.4 Streamline Analysis

To analyze the effect of blurring on the spatial course of signals in the direction perpendicular to the wavefront, we computed streamlines along the gradient of ATs (Fig. 2, left). ATs were defined as the time of maximum slope of TMVs. Streamlines were started for all nodes, whose AT lay within the 40 central time steps of activation (61...100 ms) and stopped at ± 30 mm to minimize border effects. TMVs and EPs of the time step corresponding to the AT of the starting point were then extracted at intervals of 0.375 mm along these streamlines (Fig. 2, right). Finally, signals along all 68k streamlines and corresponding AT-aligned time signals were averaged for visualization.

2.5 Error Measures

To quantify errors between two signals x and $\hat{x}$, we used the root-mean-square error (RMSE), the relative RMSE (rRMSE) and the mean absolute error (MAE). For BSP errors, we always used the BSPs at the finest resolution as truth $\hat{x}$ and computed the errors across all 1000 electrodes (Fig. 1) and/or all 500 time steps.

$$\text{RMSE} = \sqrt{\frac{\sum_{i=1}^{n}(x_i - \hat{x}_i)^2}{n}}, \quad \text{rRMSE} = \sqrt{\frac{\sum_{i=1}^{n}(x_i - \hat{x}_i)^2}{\sum_{i=1}^{n}\hat{x}_i^2}},$$

$$\text{MAE} = \sqrt{\frac{\sum_{i=1}^{n}|x_i - \hat{x}_i|}{n}}.$$

3   Results

3.1 Comparison of Blurring Methods

Before actually downsampling the sources, the isolated effect of blurring on spatial and temporal courses of source signals shall be analyzed and the blurring method causing the least error in resulting BSPs shall be identified. Therefore, we applied all blurring methods at four different blurring levels and performed forward calculations at resolution 4. To make sure that blurring with the different methods is approximately equally effective in suppressing high spatial frequencies, comparable blurring parameters had to be used. We found that the following empirical relations led to comparable blurring strengths:

$$\sigma_S = \{1, 2, 4, 8\}\,\text{mm}, \quad r = 2\,\sigma_S, \quad \lambda_S = 2\,\sigma_S^4, \quad \sigma_T = 1.25\,\sigma_S, \quad \lambda_T = 0.5\,\sigma_T^4.$$

Figure 3 shows the average spatial signals along streamlines and, superimposed in dotted lines, the average temporal signals. It can be seen that TMVs have a very similar spatial and temporal course and that this spatio-temporal congruency largely persists during blurring. For EPs, spatial and temporal courses differ much more, even before blurring, but again spatial blurring also leads to temporal blurring and vice versa. However, the blurring effect is always slightly stronger in the domain in which the blurring has been applied. As expected, the transition between regions strongly affected by the filter and regions that are not is smoother with the Gaussian blur than with the circular box blur, yet without producing any ringing artifacts. The Laplacian blur, however, produces mild ringing artifacts, as it seeks to minimize the curvature of the signal. TMVs for the second largest blurring level are depicted in Fig. 4. While spatial and temporal blurs have a very similar effect when the wave runs "freely", there are substantial differences in areas of breakthrough and wave collision. Figure 5 demonstrates that most details in BSP time courses can be preserved with spatial blurring, while they are lost with temporal blurring. BSP errors for all blurring methods and levels are listed in Table 2. Temporal methods generally cause much larger errors than their spatial counterparts, with the Laplacian blur clearly performing best.

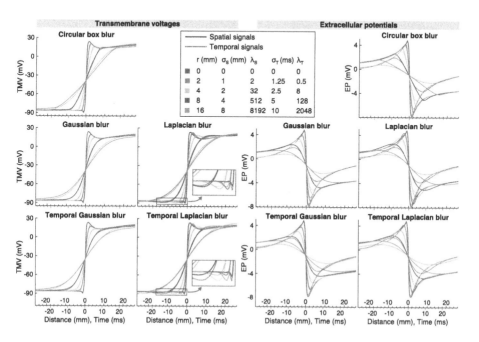

Fig. 3. Effect of blurring on average TMVs and EPs along streamlines and along time.

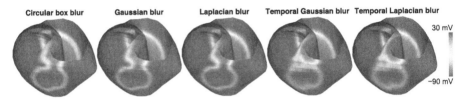

Fig. 4. Effect of blurring at the second largest level on TMVs for $t = 80$ ms.

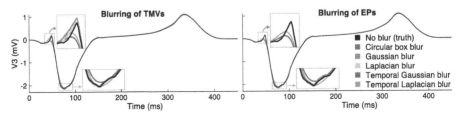

Fig. 5. Effect of blurring TMVs and EPs at the second largest level on lead V3.

Table 2. rRMSE (%) of BSPs for all blurring methods and levels (best   worst).

Source model	Transmembrane voltages				Extracellular potentials			
Blurring level	1	2	3	4	1	2	3	4
Circular box blur	0.25	0.93	3.13	10.61	0.41	1.42	4.82	18.39
Gaussian blur	0.26	0.90	3.05	10.22	0.39	1.34	4.72	17.88
Laplacian blur	0.11	0.49	1.53	6.44	0.29	1.04	2.03	7.10
Temporal Gaussian blur	0.82	2.33	6.35	15.72	0.83	2.34	6.35	15.72
Temporal Laplacian blur	0.52	1.65	4.86	13.93	0.53	1.65	4.87	13.93

3.2 Comparison of Downsampling Methods

Having identified Laplacian blurring as the method of choice for retaining BSPs, we proceeded with its application for downsampling. Figure 6 shows BSP time courses obtained by direct downsampling, i.e. extracting the signals from the fine mesh without any blurring, and by downsampling after Laplacian blurring on the fine mesh. For both TMVs and EPs, spikes caused by the discontinuous movement of the wavefront on coarser meshes are eliminated with blurring and the RMSE of BSPs is markedly reduced.

To determine the blurring parameter λ_S, the downsampled signal was Laplacian interpolated onto the original mesh, yielding a reconstructed signal x_r. λ_S was then found by minimizing the sum of MAEs between x_r and the original signal x_o and between x_r and the signal x_b after blurring, but before downsampling:

$$\lambda_S = \arg\min_{\lambda_S} \left\{ \text{MAE}\big(x_r(\lambda_S), x_o\big) + \text{MAE}\big(x_r(\lambda_S), x_b(\lambda_S)\big) \right\} \quad (6)$$

Figure 7 shows that the minimum of this objective function (bottom row) coincides well with the region of smallest BSP error (top row) for both TMVs and EPs and for all resolutions. We used golden-section search over a log-spaced parameter space across 7 decades to solve Eq. (6). On average, this required 10 iterations.

For comparison, we applied one more downsampling approach: Given a matrix **M** that interpolates from a coarse onto a fine mesh, downsampling in a least-squares (LS) sense with respect to $\mathbf{M}$ can be expressed as $\mathbf{X} = (\mathbf{M}^T\mathbf{M})^{-1}\mathbf{M}^T\mathbf{X}_o$, where $\mathbf{X}$ is the downsampled and $\mathbf{X}_o$ the original signal. The method "LS lin interp" uses linear and "LS Lap interp" uses Laplacian interpolation.

Figure 8 depicts TMVs before and after downsampling to resolution 1. Direct downsampling leads to a very poor representation of the wavefront's original shape, whereas it is rendered much better with the other methods. While least-squares downsampling retains more sharpness than Laplacian blurring, it also produces more ringing artifacts, which are most severe for LS Lap interp.

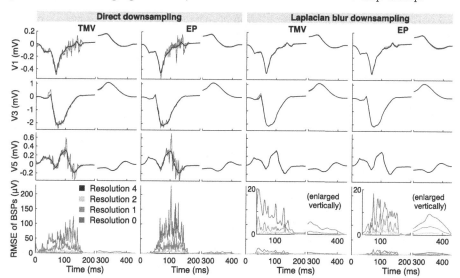

Fig. 6. Direct vs. Laplacian blur downsampling with λ_S determined using Eq. (6).

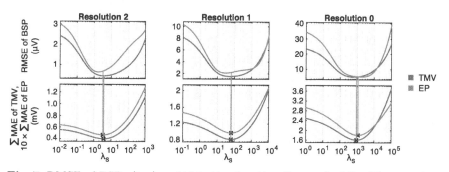

Fig. 7. RMSE of BSPs (top) and objective function (bottom) of Eq. (6) across λ_S.

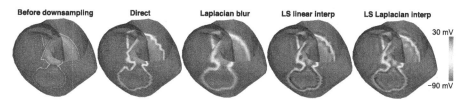

Fig. 8. TMVs before and after downsampling to resolution 1 using different methods.

Table 3 lists the rRMSE for all resolutions. For TMVs, LS lin interp performs best, although Lap blur still reduces the error 5-fold compared to direct downsampling. For EPs, results with LS methods are to the contrary. LS lin interp is worst, while Lap blur is best, again yielding a 5-fold improvement.

Table 3. rRMSE (%) of BSPs for different downsampling methods (best   worst).

Source model	Transmembrane voltages				Extracellular potentials			
Downsampling method	Direct	Lap blur	LS lin interp	LS Lap interp	Direct	Lap blur	LS lin interp	LS Lap interp
Resolution 2	2.44	0.43	0.20	0.33	2.93	0.72	0.98	0.77
Resolution 1	8.70	1.46	0.87	1.21	10.58	2.32	3.65	2.98
Resolution 0	25.05	4.82	2.19	3.66	34.96	5.19	10.67	6.24
Avg. improvement	1	5.61	11.21	7.12	1	5.13	3.06	4.32

3.3 Blurring of Coarsely Sampled Sources

Sometimes source signals are only available coarsely sampled. While Laplacian interpolation from resolution 0 to 4 cannot reduce BSP errors, wave eq. based interpolation can reduce the error for EPs and even more for TMVs (Table 4). This improvement largely persists for subsequent Laplacian blur downsampling.

Table 4. rRMSE (%) of BSPs for sources provided at resolution 0 (best   worst).

Source model	Transmembrane voltages				Extracellular potentials			
Interpolation	—	Lap	Wave eq	Wave eq	—	Lap	Wave eq	Wave eq
Downsampling	—	—	—	Lap blur	—	—	—	Lap blur
Forward resolution	0	4	4	0	0	4	4	0
rRMSE	25.05	25.31	6.32	8.06	34.96	34.79	14.72	16.07
Improvement	1	0.99	3.96	3.11	1	1.00	2.38	2.18

4 Conclusion

We showed that Laplacian blurring before downsampling of surface sources can greatly reduce errors in BSPs. We developed a method to optimize the blurring parameter. Although LS downsampling works even better for TMVs, Laplacian blurring gives more consistent results across source models and produces less artifacts in source signals. Using a different numerical scheme for BEM forward calculations, e.g. Galerkin weighting, might slightly change quantitative results.

References

1. Ni, Q., MacLeod, R.S., Lux, R.L., Taccardi, B.: A novel interpolation method for electric potential fields in the heart. Ann. Biomed. Eng. **26**(4), 597–607 (1998)
2. Oostendorp, T.F., van Oosterom, A., Huiskamp, G.: Interpolation on a triangulated 3d surface. J. Comput. Phys. **80**(2), 331–343 (1989)
3. Potse, M., Kuijpers, N.H.: Simulation of fractionated electrograms at low spatial resolution in large-scale heart models. In: Comput. Cardiol. pp. 849–852. IEEE (2010)
4. Stenroos, M., Mäntynen, V., Nenonen, J.: A matlab library for solving quasi-static volume conduction problems using the boundary element method. Comput. Methods Programs Biomed. **88**(3), 256–263 (2007)

GRÖMeR: A Pipeline for Geodesic Refinement of Mesh Registration

Jake A. Bergquist[1,2,3](✉) ⓘ, Wilson W. Good[1,2,3] ⓘ, Brian Zenger[1,2,3] ⓘ,
Jess D. Tate[1,2,3] ⓘ, and Robert S. MacLeod[1,2,3] ⓘ

[1] University of Utah, Salt Lake City, UT 84112, USA
jbergquist@sci.utah.edu
[2] Scientific Computing and Imaging (SCI) Institute, Salt Lake City, UT 84112, USA
[3] Nora Eccles Harrison Cardiovascular Research and Training Institute (CVRTI),
Salt Lake City, UT 84112, USA

Abstract. The electrical signals produced by the heart can be used to assess cardiac health and diagnose adverse pathologies. Experiments on large mammals provide essential sources of these signals through measurements of up to 1000 simultaneous, distributed locations throughout the heart and torso. To perform accurate spatial analysis of the resulting electrical recordings, researchers must register the locations of each electrode, typically by defining correspondence points from post-experiment, three-dimensional imaging, and directly measured surface electrodes. Often, due to the practical limitations of the experimental situation, only a subset of the electrode locations can be measured, from which the rest must be estimated. We have developed a pipeline, GRÖMeR, that can perform registration of cardiac surface electrode arrays given a limited correspondence point set. This pipeline accounts for global deformations and uses a modified iterative closest points algorithm followed by a geodesically constrained radial basis deformation to calculate a smooth, correspondence-driven registration. To assess the performance of this pipeline, we generated a series of target geometries and limited correspondence patterns based on experimental scenarios. We found that the best performing correspondence pattern required only 20, approximately uniformly distributed points over the epicardial surface of the heart. This study demonstrated the GRÖMeR pipeline to be an accurate and effective way to register cardiac sock electrode arrays from limited correspondence points.

Keywords: Registration · Surface meshes · Geodesic

1 Introduction

The detection and diagnosis of electrical abnormalities of the heart continues to be a topic of active research despite over a century of study [5]. The electrical signals produced by the heart are used extensively to assess cardiac health and

ⓒ Springer Nature Switzerland AG 2019
Y. Coudière et al. (Eds.): FIMH 2019, LNCS 11504, pp. 37–45, 2019.
https://doi.org/10.1007/978-3-030-21949-9_5

diagnose adverse pathologies [11]. However, many clinical applications are limited by a lack of mechanistic understanding. To better understand these mechanisms, *i.e.*, how electrical signals relate to specific pathologies, researchers use *in situ* experiments with high-density electrode arrays to measure up to 1000 individual signals directly from the hearts and torsos of large mammals [2,3]. One example of these high-density arrays is the Utah High Density Epicardial Sock (UHDES), a flexible mesh that is stitched with 240–490 evenly spaced electrodes and stretched over the ventricles of the heart [1,3,8]. The signals are then analyzed in space and time to visualize electrical changes occurring during experimental manipulation of the heart [3,6]. To include space in this analysis, researchers must know the locations of each electrode on the epicardial surface. However, in practical experimental conditions, often only a subset of the electrode locations can be measured directly due to limitations of the *in situ* preparations, *e.g.*, physical access to the electrodes, heart motion, and limited range of motion of mechanical digitizers. This limited subset of electrodes must then be used to estimate the locations of the entire array using surface registration of a preexisting geometrical model or one derived from subsequent imaging. The resulting registration problem is challenging because of the limited subset of electrode locations recorded, the non-rigid nature of the sock placement, and variations in cardiac anatomy between the experiment and any subsequent imaging.

Many examples of similar surface-to-surface registration problems exist, in which one must align a template model to a sub-sampled and distorted specific shape. Various registration approaches have been employed to address such problems, such as rigid or affine iterative closest points (ICP), thin plate splines, model based, Procrustes, and combinations of these methods [3,4,7,9,10]. Each approach and the resulting software pipelines have limitations, often with small numbers of correspondence points and the ambiguities of non-rigid deformation [2,3]. We developed a registration pipeline, GRÖMeR, that performs global deformations of an epicardial sock electrode array, calculates a rigid registration informed by the correspondence points, and then non-rigidly deforms the points into their final locations according to the known positions of correspondence-point electrodes. The result is a smooth, non-rigid deformation of the sock dictated by the correspondence points that results in an accurately registered electrode array.

The results of this study showed the GRÖMeR pipeline was able to register subject-specific and template UHDES arrays using a limited number of correspondence points. Among the correspondence point patterns considered, we found the most accurate registration across all target geometry scenarios with a relatively small number of measured correspondence points (20) distributed approximately uniformly over the heart. Similarly, we found that correspondence patterns that included both anterior and posterior coverage showed improved registration accuracy over posterior or anterior coverage alone.

2 Methods

Registration Pipeline: The GRÖMeR pipeline registers a template geometry to a target geometry using three inputs: (1) a template geometry, (2) correspondence points from the target geometry measured in the experiment coordinate system, and (3) a constraining surface derived from post-experiment imaging. We assume that the target geometry exists on the surface of the constraining surface. The first step in GRÖMeR uses the positioning of the correspondence points to calculate a deformation along the long axis of the template geometry. This deformation is then applied to the template geometry to estimate the longitudinal stretch of the sock, which varies greatly across experiments. The longitudinally deformed template geometry is then rigidly registered to the constraining surface using a modified iterative closest points (ICP) algorithm [10]. This modified algorithm ensures that known correspondence points are included in registration calculations. The rigidly registered template is then snapped to the constraining surface by a nearest-node proximity. This snapped template geometry is deformed along the constraining surface as follows. For each correspondence pair, a Euclidean vector is constructed to move each correspondence point on the template geometry to the target geometry. This Euclidean vector is projected onto the constraining surface to calculate a geodesic path. An effect weighting (v) is then calculated using a radial basis function (Eq. 1) for each correspondence point and applied to each non-correspondence point on the template geometry, where d is the Euclidean distance between the correspondence point and a non-correspondence point, and r is the radial basis factor that determines the area of effect for a correspondence point. For this study, we selected $r = 50$ based on empirical assessment. Each non-correspondence point is then assigned a movement vector based on the sums of all correspondence point movement vectors and the applied radial basis effect. If the non-correspondence movement vector is normal to the local tangent space, that point remains static. Otherwise, the movement vector is projected onto the constraining surface to calculate a geodesic path for each non-correspondence point. Finally, the non-correspondence points are relaxed to maintain nearly average edge lengths across the registered geometry. The final result is a registered geometry, dictated by correspondence points and the constraining surface.

$$v = e^{-(\frac{d}{r})^3} \tag{1}$$

Generation of Geometries: An essential goal of this study was to test the GRÖMeR pipeline under realistic conditions. To do so we evaluated it using a phantom 3D-printed heart, from which we generated target geometries to which we tried to register a template geometry. The template geometry was created by fitting the Utah High Density Epicardial Sock (UHDES) onto a plastic mold constructed from a heart harvested from a past experiment. All 247 electrode locations for this particular UHDES were measured using a mechanical digitizer (MicroScribe), and these electrode positions formed an open, cup-shaped surface, which was triangulated manually. For the constraining surface, a heart surface geometry was created according to Burton *et al.* [3] using a heart from an *in situ* preparation. Briefly, post-experiment the heart was excised and imaged with a seven Tesla MRI (Bruker BIOSPEC 70/30, Billerica, MA) using FISP (fast imaging with steady-state precession) and FLASH (fast low angle shot) imaging sequences. Utilizing images from both sequences, the heart volume was segmented using Seg3D open-source software (www.seg3d.org), and a heart surface mesh was extracted using the SCIRun problem solving environment (www.sci.utah.edu/cibc-software/scirun). To simulate an experimental placement of the UHDES, we 3D printed the constraining surface as a phantom heart in PLA filament (Lulzbot Taz 3D printer) and then placed the UHDES on the 3D printed heart as in an *in situ* preparation. The locations of each electrode was again digitized (Fig. 1) and used as one target geometry. We then generated five additional target geometries by perturbing the phantom sock geometry as follows: uneven placement, in which posterior nodes were moved uniformly 20 mm toward the apex (Fig. 1b); the Dilated or Contracted cases, in which all nodes (and the constraining surface) were dilated from or eroded toward, respectively, the center of the sock by 10 mm (Fig. 1c); and the Rotated cases, in which nodes were rotated about the long axis of the sock by 30° clockwise or counterclockwise (Fig. 1d). These target geometries were used as ground truths that we sought to recreate based on a limited sampling (Fig. 1).

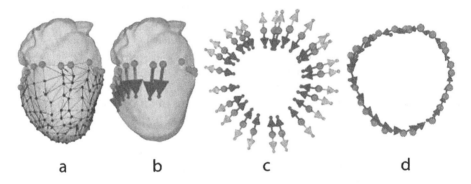

Fig. 1. Generation of target geometries. Digitized target geometry (black) on the constraining heart surface (gray) (a). Deformations used to generate the other target geometries are demonstrated on the top ring of nodes (red). Each node is an electrode. Uneven sock placement, in which the anterior electrodes are set and posterior electrodes are moved toward the apex by 20 mm (b). View from the base of the heart: electrodes and the heart mesh are either eroded (blue) or dilated (green) by 10 mm (c). Nodes are rotated by 30 degrees clockwise or counterclockwise. Counterclockwise shown in (d). (Color figure online)

Exploration of Correspondence Selection: A key parameter of any correspondence-based registration scheme is the choice and availability of correspondence points. We selected four patterns of correspondence points that mimicked realistic scenarios in our experiments: an *Anterior* group, *Posterior* group, evenly spaced *Vertical Strips*, and a *Uniformly Distributed* group over the entire target geometry. For each pattern, we selected either 8 or 20 correspondence points.

The Anterior group simulated the limited access situation common during *in situ* experimental preparations. The posterior group simulated similarly limited access on the back of the heart. The vertical strips and distributed points were selected to explore the improvements possible through larger regions of correspondence points. Each correspondence point pattern was tested for each target geometry.

Analysis: The accuracy of the resulting registration was based on statistics of the Euclidean distances between each electrode of the registered and target geometries measured in millimeters (localization error). We also computed the percent of electrodes for which the localization error was below a threshold of 5 mm, termed the "percent of nodes below threshold" (PNBT). The mean, median, and standard deviation of the localization error and the PNBT were compared across the various target geometries and correspondence patterns.

3   Results

Registration Results: The accuracy achieved using the GRÖMeR pipeline depended on the locations and the number of correspondence points. The results

Table 1. Numerical results from registration scenarios given 8 correspondence points. For each registration scenarios, the mean and median localization error (mm) are shown as well as the percent of nodes below the threshold localization error of 5 mm. All errors are listed as ± one standard deviation.

Target		Anterior	Posterior	Distributed	Vertical strips
Basal	Mean	6.52 ± 3.3	13.87 ± 9.09	4.13 ± 2.08	4.45 ± 2.18
	Median	6.55 ± 3.3	12.92 ± 9.09	4.03 ± 2.08	4.49 ± 2.18
	PNBT	33.6%	21.05%	66.8%	59.51%
Dilated	Mean	10.92 ± 5.3	19.02 ± 12.26	6.36 ± 2.97	7.15 ± 3.16
	Median	10.73 ± 5.3	17.49 ± 12.26	6.51 ± 2.97	7.15 ± 3.16
	PNBT	13.77%	14.98%	32.79%	23.48%
Contracted	Mean	3.46 ± 1.94	7.28 ± 4.26	2.8 ± 1.52	3.14 ± 1.56
	Median	3.25 ± 1.94	7.07 ± 4.26	2.7 ± 1.52	3.22 ± 1.56
	PNBT	76.92%	36.44%	91.5%	88.26%
Uneven	Mean	9.05 ± 5.28	15.62 ± 9.18	3.87 ± 2.16	5.32 ± 3.03
	Median	8.34 ± 5.28	15.58 ± 9.18	3.78 ± 2.16	5.1 ± 3.03
	PNBT	21.86%	16.19%	74.49%	47.77%
Rotated +30°	Mean	6.19 ± 3.24	10.49 ± 6.32	3.51 ± 1.95	4.02 ± 1.88
	Median	6.01 ± 3.24	9.87 ± 6.32	3.45 ± 1.95	4.14 ± 1.88
	PNBT	34.82%	24.29%	79.76%	67.21%
Rotated −30°	Mean	6.99 ± 4.74	12.71 ± 8.3	4.18 ± 2.24	4.49 ± 2.16
	Median	6.2 ± 4.74	11.41 ± 8.3	4.09 ± 2.24	4.5 ± 2.16
	PNBT	44.13%	21.46%	63.56%	58.3%

summarized in Table 1 for sets of 8 correspondence points and Table 2 for sets of 20 correspondence points show the statistical results for each correspondence pattern and target geometry scenario.

The best performing correspondence pattern across many of the target geometries was the Uniformly Distributed correspondence pattern, closely followed by the Vertical Strips pattern. Within the target geometry perturbations, the dilated and uneven target geometries produced the highest localization error. The Distributed Correspondence pattern was the most robust to the target geometry variations according to all three metrics shown for both numbers of correspondence points. The Distributed pattern with only 8 points was able to outperform 20 correspondence points in both Anterior and Posterior groupings. Figure 2 shows an example, comparing the Distributed-8 correspondence to the Anterior-20 for the uneven target geometry. Figure 2b shows the per node localization error mapped onto each node of the registered sock geometry, with higher error in red and lower error in blue. The maximum localization error was 8.56 mm for the Distributed-8 correspondence set, and 20.15 mm for the Anterior-20 set. Figure 2c shows the nodes above the 5 mm threshold in red and the nodes below the threshold in blue.

Table 2. Numerical results from registration scenarios given 20 correspondence points. Otherwise identical layout as Table 1

Target		Anterior	Posterior	Distributed	Vertical strips
Basal	Mean	6.12 ± 3.42	13.53 ± 9.95	3.25 ± 1.93	3.26 ± 1.93
	Median	6.06 ± 3.42	11.96 ± 9.95	3.1 ± 1.93	3.19 ± 1.93
	PNBT	38.87%	24.29%	80.57%	82.59%
Dilated	Mean	9.95 ± 5.66	18.77 ± 13.9	4.79 ± 2.71	4.8 ± 2.87
	Median	10.06 ± 5.66	16.75 ± 13.9	4.9 ± 2.71	4.86 ± 2.87
	PNBT	19.84%	18.62%	50.61%	53.44%
Contracted	Mean	3.17 ± 1.94	7.82 ± 5.38	2.33 ± 1.43	2.51 ± 1.52
	Median	3.13 ± 1.94	7.18 ± 5.38	2.2 ± 1.43	2.52 ± 1.52
	PNBT	81.38%	34.41%	95.55%	93.52%
Uneven	Mean	8.03 ± 5.27	16.37 ± 12.24	3.39 ± 2.43	3.26 ± 2.34
	Median	7.12 ± 5.27	14.18 ± 12.24	3.14 ± 2.43	2.94 ± 2.34
	PNBT	26.72%	21.05%	77.33%	80.97%
Rotated +30°	Mean	6.89 ± 4.77	10.37 ± 7.04	2.94 ± 1.79	3.15 ± 1.85
	Median	6.17 ± 4.77	9.22 ± 7.04	2.84 ± 1.79	3.02 ± 1.85
	PNBT	36.84%	27.94%	85.02%	85.02%
Rotated −30°	Mean	6.42 ± 4.49	13.29 ± 9.91	3.42 ± 1.97	3.15 ± 1.85
	Median	6.14 ± 4.49	11.07 ± 9.91	3.41 ± 1.97	3.03 ± 1.85
	PNBT	42.91%	24.29%	78.95%	82.59%

4 Discussion and Conclusions

In this study, we sought to develop a pipeline that could perform surface-to-surface registration based on a limited and often asymmetrically distributed set of correspondence points. In order to perform such a registration, we leveraged the availability of a high-resolution heart surface derived from MRI imaging over which the target sock geometry was stretched and over which our registration was constrained. We applied this novel registration approach to a cardiac sock electrode array, and tested it with a phantom heart model that we used to generate target geometries, which we then perturbed in realistic ways. We were able to achieve a high degree of fidelity using as few as 8 well-placed correspondence points. By selecting different correspondence point patterns, we were able to explore the robustness of the GRÖMeR pipeline under realistic sampling scenarios.

For all correspondence point patterns, contracting the target geometry and constraining mesh resulted in the most accurate registrations (highest PNBT and

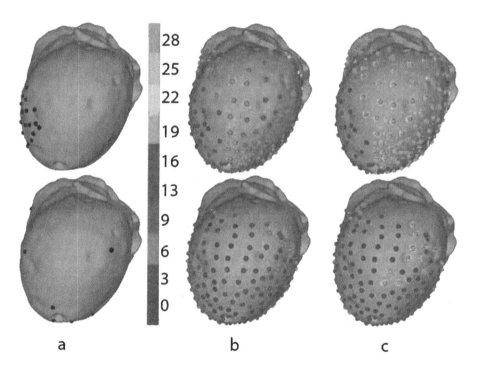

Fig. 2. Registration comparison between the Anterior 20 points (top) and the Distributed-8 points (bottom). The anterior surface of the constraining heart surface was oriented to the left and the posterior to the right. The correspondence points used are shown in black (a). Localization error (in millimeters) was mapped onto the registered electrodes (b) with high error in red and low error in blue. Electrodes with localization error below 5 mm were colored blue and with localization error at or above the same threshold were colored red (c). (Color figure online)

lowest localization errors), whereas dilating has the opposite effect. We consider that the localization error is a metric of the average electrode distance from the target geometry, whereas the PNBT is a metric of how well the registration performed overall compared to a predefined threshold. Therefore these results were expected, given that when the target geometry was dilated, the overall spacing of electrodes had to increase. Thus, error was magnified on this larger surface, and with the same threshold fewer nodes fell below the error threshold. The opposite is seen on the contracted target geometry. In these cases, the localization error provides a better interpretation for how the different correspondence patterns performed when compared to other target geometries in which the average electrode spacing is not eroded or dilated.

When examining the other target geometry perturbations, the uneven target geometry also showed high localization error and low PNBT. The local deformations were different on the anterior and the posterior sides and the non-uniform nature of the uneven perturbation likely contributed to the difficulty in registration. Correspondence patterns that did not cover locally deformed regions struggled to reconstruct these deformations. However, correspondence patterns that covered at least some part of these local deformation resulted in improved registrations. Tables 1 and 2 document these findings as does Fig. 2, which shows how a correspondence pattern that provides sampling closer to these local deformations resulted in better registration, as is expected.

Of the correspondence point patterns, the Posterior performed poorly in all cases. A possible explanation is that the posterior surface of the heart and therefore the target geometry is, generally, smooth and flat. The anterior surface by contrast, particularly along the junction between the left and right ventricles, has a more complex shape, which is captured by the sock geometry. Thus, correspondence patterns that had at least some sampling in the anterior performed better than those that did not, because without anterior sampling it was difficult to accurately reconstruct the anterior geometry. Correspondence patterns with sampling over both the anterior and posterior were superior, because these patterns could capture local deformations across the majority of the sock geometry.

We also compared the performance of correspondence points spread evenly over the entire sock geometry rather than unevenly constrained to one side. An evenly distributed, or four strips of correspondence points across the target geometry, resulted in improved registration results compared to the Anterior and Posterior patterns. This result is particularly apparent in Fig. 2, in which the Distributed-8 correspondence pattern resulted in a superior registration accuracy than did the Anterior-20 correspondence pattern. The increase in PNBT of 44.77% and decrease in mean localization error of 4.16 mm demonstrates the benefit of even a modest number of correspondence points distributed across the target geometry and highlights that simply increasing the number of correspondence point measurements on the anterior does not prove superior to a more even coverage. The Vertical strips pattern behaved similarly to the distributed, resulting in superior registration as compared to the anterior and posterior patterns, likely due to the increased correspondence coverage of the Distributed and

Vertical strips patterns. In some cases, the Vertical strips 20 pattern resulted in a slightly higher PNBT than even the Distributed 20 pattern. However, at the 8 correspondence points level, these two patterns perform similarly. The results of this analysis highlight the motivation to collect more distributed correspondence points to get accurate registration of the complete sock electrode array. In the future we plan to address these technical limitations of our experimental preparation to achieve a correspondence distributions with errors that approache those of the ideal distributed pattern.

This study has demonstrated the novel GRÖMeR pipeline to be an accurate and effective way to register surface electrode arrays from limited correspondence points. Using the GRÖMeR pipeline, we were also able to suggest ways to improve the registration process by evenly distributing and perhaps increasing the number of recorded correspondence points. In the future, we will use this tool to register our cardiac sock electrodes and test ideal correspondence points to register the UHDES array.

References

1. Aras, K., Burton, B., Swenson, D., MacLeod, R.: Sensitivity of epicardial electrical markers to acute ischemia detection. J. Electrocardiol. **47**(6), 836–841 (2014). https://doi.org/10.1016/j.jelectrocard.2014.08.014
2. Bear, L.R., et al.: Forward problem of electrocardiography: is it solved? Circ. Arrhythmia Electrophysiol. **8**(3), 677–684 (2015). https://doi.org/10.1161/CIRCEP.114.001573
3. Burton, B.M., Aras, K.K., Good, W.W., Tate, J.D., Zenger, B., MacLeod, S.: Image-based modeling of acute myocardial ischemia using experimentally derived ischemic zone source representations. J. Electrocardiol. **51**(4), 725–733 (2018). https://doi.org/10.1016/j.jelectrocard.2018.05.005
4. Chen, Y., Zhao, J., Deng, Q., Duan, F.: 3D craniofacial registration using thin-plate spline transform and cylindrical surface projection. PLoS ONE **12**(10), 1–19 (2017). https://doi.org/10.1371/journal.pone.0185567
5. Einthoven, W.: Le telecardiogramme. Arch. Int. de Physiol. **4**, 132–164 (1906)
6. Rodenhauser, A., et al.: PFEIFER: preprocessing framework for electrograms intermittently fiducialized from experimental recordings. J. Open Source Softw. **3**, 472 (2018). https://doi.org/10.21105/joss.00472
7. Sablatnig, R., Kampel, M.: Model-based registration of front- and backviews of rotationally symmetric objects. Comput. Vis. Image Underst. **87**(1–3), 90–103 (2002). https://doi.org/10.1006/cviu.2002.0985
8. Shome, S., Lux, R., Punske, B., MacLeod, R.: Ischemic preconditioning protects against arrhythmogenesis through maintenance of both active as well as passive electrical properties in ischemic canine hearts. J. Electrocardiol. **40**(4), S5–S6 (2007). https://doi.org/10.1016/j.jelectrocard.2007.06.012
9. Tate, J., et al.: Reducing error in ECG forward simulations withimproved source sampling. Frontiers Physiol. **9**, 1304 (2018). https://doi.org/10.3389/fphys.2018.01304
10. Zhang, Z.: Iterative point matching for registration of free-form curves and surfaces. Int. J. Comput. Vis. **13**(2), 119–152 (1994). https://doi.org/10.1007/BF01427149
11. Zipes, D., Jalife, J., Stevenson, W. (eds.): Cardiac Electrophysiology, From Cell to Bedside, 7th edn. Elsevier, Philadelphia (2014)

A Numerical Method for the Optimal Adjustment of Parameters in Ionic Models Accounting for Restitution Properties

Jacob Pearce-Lance[1], Mihaela Pop[2], and Yves Bourgault[1(✉)]

[1] Department of Mathematics and Statistics, University of Ottawa, Ottawa, Canada
{jpear019,ybourg}@uottawa.ca
[2] Sunnybrook Research Institute, University of Toronto, Toronto, Canada
mihaela.pop@utoronto.ca

Abstract. We developed new numerical methods to optimally adjust the parameters in cardiac electrophysiology models, using optimal control and non-differentiable optimization methods. We define an optimal control problem to adjust parameters in single-cell models so that the trans-membrane potential predicted by a model fits in a least-square (LS) sense the potential recorded over time. To account for restitution properties, this LS function measures the discrepancy between predictions and experiments for a cell paced at various heart rates (HR) of increasing frequency. The methodology is used to adjust parameters in the Mitchell-Schaeffer model to unscaled non-smoothed experimental recording of the trans-membrane potential obtained in pig heart using optical fluorescence imaging based on voltage-sensitive dye, and simultaneously identify scaling factors for the experimental data. The methodology is validated by adjusting the model for multiple heart beats at a single HR. The fit for a single HR is excellent (LS function = 0.0065–0.02). The methodology is applied to adjust the MS model to multiple heart beats at three different HR. It is observed that the fit remains good when the range of HR is moderately large (LS function = 0.052), while a larger HR gap is more challenging (LS function = 0.17).

Keywords: Ionic models · Restitution · Optimal control · Non-differentiable optimization · Parameter identification

1 Introduction

Several ionic models are available to describe the evolution of the electrical potential across cardiac cell membranes. These models usually read as a system of coupled highly nonlinear differential equations with many adjustable parameters. The adjustment of parameters becomes increasingly important to be able to

Supported by the National Science and Engineering Council of Canada.

Y. Coudière et al. (Eds.): FIMH 2019, LNCS 11504, pp. 46–54, 2019.
https://doi.org/10.1007/978-3-030-21949-9_6

personalize these models using medical data (see for instance [10,11]), to compare models with each other in the best possible way or to represent the more complex dynamical behavior of cardiac cells such as restitution properties. It is not easy to study the combined effect of varying the parameters and the literature is usually not too explicit on the way the parameters are adjusted in ionic models.

Parameter adjustment is possible with simpler ionic models using asymptotic formula connecting the parameters with the phase durations [6,11,12]. Few attempts have been made to address the adjustment of the ionic model parameters using fully nonlinear models. We are aware of the recent paper [2] where a genetic algorithm was used to build a cell-specific cardiac electrophysiology model and [5] where simulated annealing is used to compare two ionic models. Genetic algorithms were also used to adjust conductances in nonlinear models [13], then a larger set of parameters in [3]. Direct recording of action potentials (AP) or membrane resistance were used to build least-square functions. In [13], AP recorded at various frequencies were included in the least-square function to eventually match restitution properties.

In [7], we introduced numerical methods to optimally adjust the parameters in ionic models. Our method is based on the numerical solution of an optimal control problem with a least-square objective function to fit the main features of the cardiac AP, e.g. the action potential duration (APD), the depolarization time (DT), recovery time (RT), etc. We then provided a second function to fit the trans-membrane potential predicted by the model to experimental recording on a single cell. The goal of the current paper is to show that this methodology naturally leads to parameter identification to match restitution properties in fully nonlinear ionic models directly from simple potential recordings in a cell or heart at multiple frequencies. No assumption will be made on the amplitude of the recorded potential (a limitation of indirect measurement techniques), consequently scaling factors for the data will have to be identified together with the model parameters.

We will illustrate the efficiency of the method for the Mitchell-Schaeffer model [6], which is a simple two variables ionic model with a limited set of parameters. Our methodology is not limited to this model. Numerical results are presented, in particular model fitting to experimental AP measurements obtained through an optical fluorescence imaging technique.

2   Mathematical Models

2.1   Mitchell-Schaeffer Model

As one particular example where the proposed parameter identification technique can be applied, we consider the Mitchell-Schaeffer (MS) two-variable model [6]. This model describes the dynamics of the trans-membrane potential u in the myocardium and a gating variable v representing in a lumped way the opening and closing of ionic channels controlling the passage of ions across the cell membranes. Here we consider the 0D model for a single cell (no space dependence of the variables u and v).

48   J. Pearce-Lance et al.

The dependent variables $u = u(t)$ and $v = v(t)$, $t > 0$, are solutions of:

$$\frac{du}{dt} = f(u,v) + I_{stim}(t), \quad \text{with} \quad f(u,v) = \frac{1}{\tau_{in}}vu^2(1-u) - \frac{1}{\tau_{out}}u, \quad (1)$$

$$\frac{dv}{dt} = g(u,v), \quad \text{with} \quad g(u,v) = \begin{cases} \frac{1-v}{\tau_{open}}, & \text{if} \quad u < u_{gate}, \\ \frac{-v}{\tau_{close}}, & \text{if} \quad u \geq u_{gate}. \end{cases} \quad (2)$$

The trans-membrane current $f(u,v)$ is the sum of the gated inward current $vu^2(1-u)/\tau_{in}$ with time scale τ_{in} that tends to depolarize the cardiac cell and the ungated current $-u/\tau_{out}$ that tends to repolarize the cardiac cell with time scale τ_{out}. Finally, I_{stim} represent an external current produced by a stimulation electrode. The dynamics of the gating variable v depends on the threshold potential u_{gate} for the initiation of an AP, and on two time constants, τ_{open} and τ_{close}, respectively controlling the opening and closing of the gate. We set $\tau = [\tau_{in}, \tau_{out}, \tau_{open}, \tau_{close}]$ to simplify notations. The functions f and g depend on the parameter τ. Equations (1)–(2) require initial conditions $u(0) = u_0$ and $v(0) = v_0$, where $u_0, v_0 \in [0,1]$ are given. In the MS model, the variables u and v are non-dimensional, while the time t is in ms. Consequently, the parameters τ are in ms, and the source terms f and g are in ms^{-1}.

Since the model needs to be periodically stimulated in order to account for restitution properties, I_{stim} can then be written as

$$I_{stim}(t) = \begin{cases} A & \text{if } t \in [n \cdot BCL, n \cdot BCL + \Delta t], \ n \in \mathbb{N} \\ 0 & \text{otherwise}, \end{cases} \quad (3)$$

where BCL, or Basic Cycle Length, is the delay between stimulations, Δt is the duration of the stimulations and A is their amplitude. As stated in [6], the amplitude A depends on the model parameters and indirectly on BCL:

$$A = \frac{1}{\Delta t}\left(\frac{1}{2} - \sqrt{\frac{1}{4} - \frac{\tau_{in}}{\tau_{out}v^*}}\right) \cdot (1+\beta) \quad (4)$$

where v^* is the value of v at the time of stimulation, $\beta > 0$ is a small "safety factor" set to guarantee depolarization.

2.2   Optimal Control Problem

We introduce a control problem to fit in the least-square (LS) sense the trans-membrane potential $u = u(t)$ from the model to a recorded potential $\tilde{u} = \tilde{u}(t)$, $t \in [0,T]$, measured experimentally, normalized in a specific way and rescaled using a scaling factor $s \in (0,1)$.

Define $\tilde{J} = \tilde{J}(\tau, s)$

$$\tilde{J}(\tau, s) = \frac{\int_0^T |u(t,\tau) - s\,\tilde{u}(t)|^2\, dt}{\int_0^T |s\,\tilde{u}(t)|^2\, dt},$$

where u and v are solution of (1)–(2) with parameters τ and I_{stim} is adjusted to match the BCL of $\tilde{u}$. For the solution of the MS model to reach a stable cyclical response from beat to beat, we disregard the first five AP in the evaluation of the cost function $\tilde{J}$. The parameter u_{gate} is intentionally left out from the parameter identification as a senstivity analysis showed that u_{gate} has little impact on the phase durations or the shape of the AP [8].

Now consider multiple different $\tilde{u}_i$, $i = 1, \ldots, N$, obtained by stimulating the same heart or cardiac cell at different frequencies and find (τ^*, S^*) minimizing

$$J(\tau, S) = \sum_{i=1}^{N} \tilde{J}(\tau, s_i) \tag{5}$$

where $S = (s_1, \ldots, s_N)$ are the scaling factors for each $\tilde{u}_i$. The connection between the parameter τ and the objective function J is occurring through the dependance of the solutions (u, v) on the parameters τ.

3   Numerical Methods

The equations (1)–(2) are solved using the function `ode45` in `Octave`, which implements an explicit Dormand-Prince method of order 4.

The function g is discontinuous in u, which leads to a lack of regularity of the solution (u, v) of (1)–(2) and consequently of the function $J = J(\tau, S)$. The derivatives of J with respect to τ may not be well defined. Work done in [7] showed that numerical derivatives do not converge when the increments in τ are reduced. To avoid the computation of the sensitivities and the gradient of J with respect to τ, we use non-differentiable optimization methods [1]. The Compass Search method described in Algorithm 1 is taken from [4]. This method can be changed in a few ways to improve performance, but the principle behind the method is maintained for all modified versions.

Algorithm 1 (Compass Search: CS) *Given a function f, an initial guess $x \in \mathbb{R}^n$, an initial step-size $\delta > 0$, a contraction factor $c \in (0, 1)$ and stopping criteria, the following is applied:*
while *the stopping criteria are not met*

   let $D = \{p\,\delta\,e_i \mid p = -1, +1$ and e_i is an element of the standard basis for $\mathbb{R}^n\}$
   let $f(x^*) = \min\limits_{d \in D} f(x + d)$
   if *$f(x^*) < f(x)$, **then***
       set $x \leftarrow x^*$
   else
       set $\delta \leftarrow c\,\delta$

The algorithm repeats the process of constructing D and moving x, if possible, as long as the stopping criteria are not met. Options for stopping criteria include $f(x) \leq ftol$ for a chosen tolerance $ftol$ on the value of the cost function,

$\delta \le \delta tol$ for a chosen tolerance δtol on the step-size, and maximal numbers of iterations and function evaluations.

The optimization methods used introduce a sensitivity to the initial guess of the parameters and scaling factors. If convergence is reached but the value of the least-square function J is not small for the final iterate, the minimum is likely to be local only and new initial guesses must be attempted in the hope of getting a better fit. We used our experience with the fitted model as well as trial and error to find good initial guesses.

4 Data Acquisition

For parameterization of mathematical models, AP waves were recorded using voltage-based optical fluorescence imaging, as described in [9]. The fluorescence dye (di4-ANEPPS) and uncoupler to block contraction (2, 3 BDM) were injected into the coronary circulation of a healthy explanted swine heart perfused by a Langendorff system. The optical dye was excited with green light ($\sim$530 nm) while the emitted epicardial signals were filtered ($>$610 nm) and captured by a high-speed CCD camera (MICAM02, BrainVision Inc. Japan) at 256 frames/second (Fig. 1). The field of view was 184×124 pixels (12×10 cm), yielding an $\sim$0.7 mm spatial resolution. We stimulated the heart at several frequencies to study the restitution properties (see Table 1 below). The relative change in fluorescence signal intensity ($\Delta F/F$) recorded at each pixel, gives directly the AP waves. For model fitting, we used the AP waves recorded at one pixel selected from an area in the left ventricle (LV) where tissue was homogeneously illuminated, and also both fluorescence signal and tissue perfusion were homogeneous.

5 Numerical Results

5.1 Preparation of the Experimental Data

The potential measured with fluorescence imaging are recorded as signal intensity, consequently these must be normalized between 0 and 1 before fitting with the MS model. We worked directly on the raw data. In order to normalize the data, averaged extrema must be calculated. The average values are considered, since taking the absolute minimum or maximum of a set of noisy values would not make sense when trying to normalize the recorded potential. The average minimum was found by taking the average of the values during a single diastolic interval. Since the data is very noisy, determining which values were to be considered to be in the diastolic interval presented some challenges, so this was done manually. A representative diastolic interval was identified by looking at the graph of raw data and noting the approximate times at which the diastolic interval starts and ends. A similar logic was used when finding the averaged maximum. The maximum was taken manually to be approximately the value at which each potential peaks. Once the average extrema are found, the data

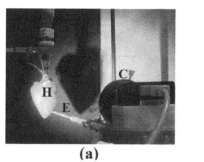 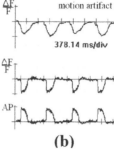

(a)   **(b)**

Fig. 1. (a) Snapshot of the optical experiment to record epicardial AP wave propagation using a fast CCD camera (C), where the pig heart (H) was stimulated via an electrode (E). (b) Examples of waves recorded at one pixel in the heart without the uncoupler (top) as well as after the uncoupler (bottom) was injected. Note that the inverse of the relative loss of fluorescence signal $\Delta F/F$ (arbitrary units) gives the AP. The waves were displayed with BV-Ana software (BrainVision, Japan). (Color figure online)

is normalized, by subtracting the minimum and dividing by the amplitude, to obtain a normalized potential $\tilde{u}_i$ between 0 and 1. Also, when considering the problem of matching the MS model to the data, it is convenient for the data to start with an upstroke, so the first part of most of the data sets is ignored (data acquisition does not always start in sync with the stimulation, it can start mid heartbeat). Finally, the aforementioned scaling $s_i\tilde{u}$ further adjusts the data so that the peaks of the potential have the same values as the peaks in the model. This is required since the peak trans-membrane potential u predicted by the MS model gets smaller as BCL is reduced (the usual mechanism for restitution).

5.2   Single Pacing Frequency

To validate our approach, we verify that our method is efficient at fitting data sets individually (i.e $N = 1$ in (5)). For each of the cases given in Table 1 (presented in order of increasing frequencies), we pace the MS model according to (3)–(4) with $(u_0, v_0) = (0, 1)$, $\Delta t = 2\,\mathrm{ms}$, $\beta = 0.25$ and BCL calculated from column 2 of Table 1. Figure 2 shows the measured potential and the trans-membrane potential from the fitted MS model for dataset 5. The potential predicted by the model matches the noisy data very well, which is confirmed by the small value of the least-square function J in Table 1. The fit is good for all datasets, with datasets 3 and 4 showing a slightly larger deviation. This shows that the parameter identification method is capable of identifying parameters τ simultaneously with a scaling factor.

Table 1. Statistics of single dataset fitting

Dataset	HR (bpm/Hz)	Final τ	Scaling factor s	J value
2	59.17/0.99	$[0.358, 9.24, 153.98, 177.95]$	0.96779	$8.62e^{-3}$
1	62.37/1.04	$[0.595, 10.83, 211.39, 198.44]$	0.91763	$8.06e^{-3}$
3	72.29/1.20	$[0.632, 6.51, 287.48, 325.83]$	0.85545	$1.20e^{-2}$
4	74.81/1.25	$[0.448, 5.31, 207.84, 202.22]$	0.87435	$2.05e^{-2}$
5	104.35/1.74	$[0.584, 8.86, 81.41, 168.39]$	0.91417	$6.52e^{-3}$
6	145.63/2.43	$[0.700, 7.61, 48.16, 207.68]$	0.86944	$8.65e^{-3}$

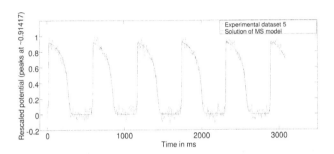

Fig. 2. Result of fitting dataset 5 individually

Table 2. Statistics of multiple datasets fitting

Datasets	Final τ	Scaling factors S	J value
$2 - 1 - 3$	$[0.500, 9.36, 538.54, 244.59]$	$[0.890, 0.913, 0.923]$	$5.23e^{-2}$
$4 - 5 - 6$	$[0.651, 6.32, 53.66, 245.33]$	$[0.877, 0.857, 0.872]$	$6.09e^{-2}$
$2 - 4 - 5$	$[0.405, 8.04, 612.26, 239.70]$	$[0.889, 1.06, 0.834]$	$1.68e^{-1}$
$2 - 1 - 6$	$[0.535, 9.06, 251.15, 224.97]$	$[0.906, 0.936, 0.822]$	$7.02e^{-2}$
$1 - 3 - 6$	$[0.621, 9.04, 183.05, 219.20]$	$[0.884, 0.927, 0.827]$	$7.72e^{-2}$

5.3   Multiple Pacing Frequency

We next illustrate how problem (5) is capable of adjusting parameters for restitution properties. We use different combinations of three datasets among six, as
given in Table 2. The stimulation current and initial conditions for the MS model
are set as in Sect. 5.2. Figure 3 shows the experimental and fitted potentials for
combinations that give the best and an average-quality fit, respectively. This is
confirmed by the values of J in Table 2. We note that a comparison of single and
multiple dataset fitting can be made by dividing the values of J by the number
N of datasets used. As expected, multiple dataset fitting is more difficult.

The value v^* of the gating variable at pacing time and the maximal potential reached for each AP predicted by the MS model decrease with increasing
frequency. This is easily seen for case 6, which has the largest frequency, by com-

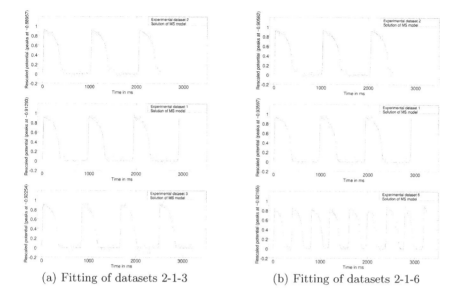

(a) Fitting of datasets 2-1-3   (b) Fitting of datasets 2-1-6

Fig. 3. Best (left) and average-quality (right) fitting of multiple datasets

paring the amplitude of the potential with the other cases on Fig. 3. Restitution is thus well represented by the model, at least qualitatively and to some extent quantitatively. When the quality of fit deteriorates, the MS model still predicts decreasing v^* and maximal potential, only with a small mismatch of the APD for either low or high frequencies. We should recall that the optical measurements are taken at a specific location on a heart, not on an isolated cell. This may explain the difficulties at matching a cell ionic model uniformly across all frequencies.

6   Conclusions and Perspectives

We provided a new framework for fitting electrophysiology model parameters based on control theory in order to adjust the dynamic response of the model. The only required data is indirect measurement of the potential (e.g. obtained through fluorescence imaging) for N pacing frequencies. As opposed to direct measures (e.g. obtained through clamping), the scale of our data had to be fitted simultaneously with the model parameters. The control problem is solved by a non-differentiable optimization method, hence differentiability of the ionic model is not required and complex dynamic response functions can be included in the cost function (e.g. restitution curve). The compass search method used is a simple alternative to genetic methods, requiring a reasonable number of function evaluations (200 to 500) as opposed to 10^4 for genetic methods [13]. The least-square function (5) gave a good fit of the trans-membrane potential predicted by the model to the potential recorded over time for N frequencies. However, the fit for this latter is as good as a given ionic model can represent the data. The applicability to more complex models will have to be investigated. We know from previous work [7] that models with 8–10 parameters can be identified for non-dynamic response (e.g. phase durations of a single AP). We are convinced that such models will be identifiable in the dynamic case with our new method.

References

1. Chong, E.K.P., Zak, S.H.: An Introduction to Optimization, 3rd edn. Wiley, Hoboken (2008)
2. W. Groenendaal et al.: Cell-specific cardiac electrophysiology models. PLOS Comput. Biol. 22 pages (2015)
3. Kaur, J., Nygren, A., Vigmond, E.J.: Fitting membrane resistance along with action potential shape in cardiac myocytes improves convergence: application of a multi-objective parallel genetic algorithm. PLOS One **9**(9), 1–10 (2014)
4. Kolda, T.G., Lewis, R.M., Torczon, V.: Optimization by direct search: new perspectives on some classical and modern methods. SIAM Rev. **45**(3), 385–482 (2003)
5. Lombardo, D.M., Fenton, F.H., Narayan, S.M., Rappel, W.-J.: Comparison of detailed and simplified models of human atrial myocytes to recapitulate patient specific properties. PLOS Comput. Biol. 15 pages (2016)
6. Mitchell, C.C., Schaeffer, D.G.: A two-current model for the dynamics of cardiac membrane. Bull. Math. Biol. **65**(5), 767–793 (2003)
7. Pongui Ngoma, D.V., Bourgault, Y., Pop, M., Nkounkou, H.: Adjustment of parameters in ionic models using optimal control problems. In: Pop, M., Wright, G.A. (eds.) FIMH 2017. LNCS, vol. 10263, pp. 322–332. Springer, Cham (2017). https://doi.org/10.1007/978-3-319-59448-4_31
8. Pongui-Ngoma, D.V., Bourgault, Y., Nkounkou, H.: Parameter identification for a non-differentiable ionic model used in cardiac electrophysiology. Appl. Math. Sci. **9**(150), 7483–7507 (2015)
9. Pop, M., et al.: Fusion of optical imaging and MRI for the evaluation and adjustment of macroscopic models of cardiac electrophysiology: a feasibility study. Med Image Anal. **13**(2), 370–80 (2009)
10. Relan, J., Pop, M., Delingette, H., Wright, G., Ayache, N., Sermesant, M.: Personalization of a cardiac electrophysiology model using optical mapping and MRI for prediction of changes with pacing. IEEE Trans. Biomed. Eng. **10**(10), 11 pages (2011)
11. Relan, J., et al.: Coupled personalization of cardiac electrophysiology models for prediction of ischaemic ventricular tachycardia. Interface Focus **1**, 396–407 (2011)
12. Rioux, M., Bourgault, Y.: A predictive method allowing the use of a single ionic model in numerical cardiac electrophysiology. ESAIM: Math. Modell. Numer. Anal. **47**, 987–1016 (2013)
13. Syed, Z., Vigmond, E., Nattel, S., Leon, L.J.: Atrial cell action potential parameter fitting using genetic algorithms. Med. Biol. Eng. Comput. **43**, 561–571 (2005)

EP-Net: Learning Cardiac Electrophysiology Models for Physiology-Based Constraints in Data-Driven Predictions

Ibrahim Ayed[1], Nicolas Cedilnik[2,3]([✉]), Patrick Gallinari[1], and Maxime Sermesant[2]

[1] Sorbonne University, LIP6, Paris, France
[2] Université Côte d'Azur, Epione Research Project, Inria, Sophia Antipolis, France
nicolas.cedilnik@inria.fr
[3] Liryc Institute, Bordeaux, France

Abstract. Cardiac electrophysiology (EP) models achieved good progress in simulating cardiac electrical activity. However numerical issues and computational times hamper clinical applicability of such models. Moreover, personalisation can still be challenging and model errors can be difficult to overcome. On the other hand, deep learning methods achieved impressive results but suffer from robustness issues in healthcare due to their lack of physiological knowledge. We propose a novel approach which is based on deep learning in order to replace numerical integration of partial differential equations. This has the advantage to directly learn spatio-temporal correlations, which increases stability. Moreover, once trained, solutions are very fast to compute. We present first results in state estimation based on few measurements and evaluate the forecasting power of the trained network. The proposed method performed very well on this preliminary evaluation. It opens up possibilities towards data-driven personalisation, to overcome model error by learning from the data.

Keywords: Electrophysiology · Deep learning · Simulation

1 Introduction

Mathematical modelling of the cardiac cell has been an active research area for the last decades. Cardiac electrophysiology models can accurately reproduce cardiac cells electrical behaviour. This provides a mathematical framework to simulate cardiac activity, however it remains challenging to achieve in 3D due to numerical issues and computational time. These difficulties become all the more apparent when trying to personalise such model by matching patient data. This matching can also be hampered by the approximations of the model.

© Springer Nature Switzerland AG 2019
Y. Coudière et al. (Eds.): FIMH 2019, LNCS 11504, pp. 55–63, 2019.
https://doi.org/10.1007/978-3-030-21949-9_7

The last decade has seen huge progress in data-driven approaches, with deep learning achieving impressive results in numerous tasks. It has been particularly successful in image and signal processing, where spatio-temporal correlations can be learned. However it suffers from robustness issues in healthcare due to its lack of physiological knowledge. More recently, physics-based learning proposed to use machine learning in order to solve physics equations [5,7,9]. This has the potential to alleviate some numerical and computational time issues, and also to provide a framework to overcome model error. Additionally, it is a way to introduce physics-based prior knowledge in learning methods.

In this manuscript, we present a method to learn the dynamics from a database of cardiac electrophysiology simulations and use the trained network to forecast the behaviour of unseen simulations. A state estimation method is coupled with a forecasting network to produce predictions for a few iterations.

These preliminary results are promising in terms of learning the dynamical behaviour of cardiac electrophysiology models.

2 Electrophysiological Modelling

Cardiac electrophysiology modelling is a very active research area, with a large variety in terms of model complexity. In this manuscript, we used the Mitchell-Schaeffer model [8] which is a two variables model that has been successfully used in patient-specific modelling [11]. The variable v represents the transmembrane potential while the "gating" variable h controls the repolarisation:

$$\partial_t v = \Delta v + \frac{hv^2(1-v)}{\tau_{in}} - \frac{v}{\tau_{out}} + J_{stim} \tag{1}$$

$$\partial_t h = \begin{cases} \frac{1-v}{\tau_{open}} & \text{if } v < v_{gate} \\ \frac{v}{\tau_{close}} & \text{if } v > v_{gate} \end{cases} \tag{2}$$

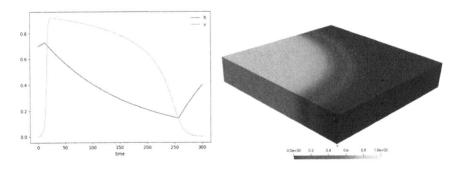

Fig. 1. Cardiac electrophysiology model simulation. (Left) v and h values along time for a given point of the domain, (Right) spatial propagation of v at a given time point.

In order to use convolutional neural networks, it is much more efficient to work on a Cartesian grid. To this end, we used a Lattice Boltzmann method to solve the EP model [10].

The domain is a slab of cardiac tissue of size $25 \times 25 \times 5 \, \text{mm}^3$, which we will denote Ω, discretised in voxels of $1 \, \text{mm}^3$ (Fig. 1). To initiate the propagation (J_{stim}), a current of 1 normalised transmembrane potential unit was applied on a single voxel for $10 \, \text{ms}$. The Mitchell-Schaeffer model parameters are taken from the original paper, except h_0 which was set to 0.7. The simulation was conducted for $300 \, \text{ms}$, and stored every ms with a discrete time step of $0.1 \, \text{ms}$. The simulation database was created by stimulating the domain from every voxel. We therefore have 3 125 simulations in the database. Let V and H be the complete solutions in space and time of these equations.

3 Learning Method

In this section, we reformulate the problem we want to solve with learning and outline the features of the learning model we used.

3.1 Formulation

Combining Eqs. (1) and (2), the studied dynamical system can be written in the form:

$$\frac{dX_t}{dt} = F(X_t) \tag{3}$$

where X is a spatio-temporal three-dimensional vector field over the domain $\Omega \subset \mathbb{R}^3$. In this particular case, we have:

$$X = \begin{pmatrix} V \\ H \end{pmatrix}$$

In other words, for any given time t and point of the domain $x \in \Omega$, we have $X_t(x)$, a two-dimensional vector.

In practice, while the value of V can be measured, H is a hidden variable which is difficult to estimate. We model this fact by adding to our system an observation operator $\mathcal{H}$ defined in our case by:

$$\mathcal{H}(X) = V \tag{4}$$

which makes our model a partially observed one, as only some parts of the full state can be directly measured.

It is important to note that $\mathcal{H}$ can be used to model more general observation operators which can be any transformation of the state. Obviously, the more information is lost through the observation process, the harder it will be to reconstruct the dynamics of the system. Moreover, for practical purposes, we also require $\mathcal{H}$ to be differentiable. This might seem as a strong hypothesis but, in practice, the operators which are used can generally be smoothed without a significant departure from reality.

The general system that we model can thus be written:

$$
\begin{cases}
X_0 \\
\dfrac{dX_t}{dt} = F(X_t) \\
Y_t = \mathcal{H}(X_t)
\end{cases}
\tag{5}
$$

3.2   Approach

In the system (5), starting from observations Y, we need to estimate both the initial state and the dynamics it should follow. Obviously, from one observation alone, it is often impossible to estimate the full state but hopefully this becomes possible for most operators $\mathcal{H}$ when a long enough sequence of observations is used.

Our idea is to use an operator g_θ to estimate X_0 from a sequence of past observations $Y_{-k} = (Y_{-k+1}, ..., Y_0)$ of length k and a second operator F_θ to model the ODE governing the dynamics of X. Thus, θ is a variable summarizing the parameterization of both the dynamics and the initial state.

This means that the system can be written as:

$$
\begin{cases}
X_0 = g_\theta(Y_{-k}) \\
\dfrac{dX_t}{dt} = F_\theta(X_t) \\
Y_t = \mathcal{H}(X_t)
\end{cases}
\tag{6}
$$

We assume that F and g are parameterized so that the above system has a unique solution which will be denoted X^θ.

Defining a cost functional:

$$
\mathcal{J}(Y, \widetilde{Y}) = \int_0^T \|Y_t - \widetilde{Y}_t\|^2 dt
\tag{7}
$$

we can then frame the statistical learning problem as:

$$
\begin{aligned}
&\underset{\theta}{\text{minimize}} \ \ \mathbb{E}_{Y \in \text{Dataset}} \mathcal{J}(Y, \mathcal{H}(X)) \\
&\text{subject to} \ \ \frac{dX_t}{dt} = F_\theta(X_t), \\
&\qquad\qquad\quad X_0 = g_\theta(Y_{-k})
\end{aligned}
\tag{8}
$$

This optimization problem is generally a difficult one and cannot be solved exactly for most choices of parameterization families for F and g. This makes necessary the use of gradient descent algorithms for which it is necessary to estimate the derivative $w.r.t.$ θ of $\mathbb{E}_{Y \in \text{Dataset}} \left[\mathcal{J}(Y, \mathcal{X}^\theta) \right]$ which justifies the use of differentiable parameterizations.

In our case, starting from a random initialization of θ, we follow the steps:

1. Solve the forward state Eq. (6) to find X^θ with an explicit differentiable solver;
2. Get the gradient of $\theta \rightarrow \mathbb{E}_{Y \in \text{Dataset}} \left[\mathcal{J}(Y, \mathcal{X}^\theta) \right]$ with automatic differentiation tools;
3. Update θ in the steepest descent direction.

In our case, we parameterize g and F as neural networks so that the corresponding automatic differentiation tools can be used. The simplest example for explicit solvers is a Euler scheme but more complicated numerical methods can also be used. As shown in [1], this framework can also be extended to the modelling of stiff equations by using implicit solvers through the more general adjoint method.

4 Experiments

In this section we present the experiments we undertook. Even though we are still at a preliminary stage regarding the choice of architectures and hyperparameters, the results are already encouraging.

4.1 Implementation Details

Operator g is implemented as a three-dimensional U-Net inspired from [3] with 25 filters at the initial stage, Fig. 2 shows the general structure of this architecture. F is implemented as a Residual Network [4] with three-dimensional convolutions, two downsampling initial layers, three intermediary blocks (Fig. 3 shows the structure of one) and an inner dimension of 24. In order to solve the forward equation, we use an explicit Euler scheme.

Input: 3D image

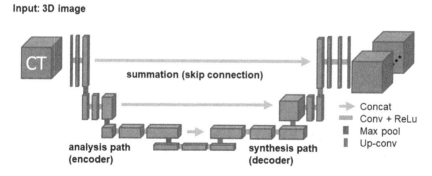

Fig. 2. Figure illustrating the general form of the U-Net architecture. Illustration taken with a slight modification from [12].

To construct a training dataset, we have randomly selected 500 simulations, temporally downsampled three times. A validation set of 300 simulations and a test set of 200 were also randomly selected from the remaining simulations and

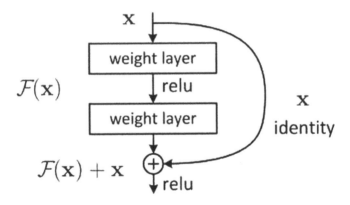

Fig. 3. Figure showing a typical residual block for the ResNet. Illustration taken from [4].

downsampled likewise. In particular, this means that test and training simulations have different initial conditions so that the model is tested with data it has never seen. Only the V variable was used and we have taken $k = 4$ and a training horizon of 8 time-steps.

The optimization over θ uses the ADAM optimizer [6] with a learning rate of 10^{-3}. We also use exponential scheduled sampling [2] with parameter 0.9999 during training and start with a reweighted orthogonal initialization for the parameters of F.

Table 1. Relative mean-squared error of normalised transmembrane potential per time-step for different forecasting horizons.

Time horizon K	2 (6 ms)	5 (15 ms)	8 (24 ms)	11 (33 ms)	14 (42 ms)	17 (51 ms)
Mean Squared Error	0.005	0.012	0.048	0.060	0.14	0.58

4.2 Results

We present here results on the forecast over 8 time frames after assimilating the first 4 frames (see Fig. 4). We can observe very good agreement with the ground truth on this forecast, which represents an important part of cardiac dynamics within this virtual slab of tissue, from early depolarisation to full depolarisation.

Table 1 shows MSE results for our algorithm for different forecasting horizons: for a horizon K, this means that we feed the model with three initial measurements then let it predict K steps forward without any additional information. We can see that up to a reasonable number of steps, here corresponding to 51 ms, our model does reasonably well which is very encouraging given that we are still at an early stage of experimenting with it. Figure 4 visually confirms those numerical results.

Ground
Truth

Network

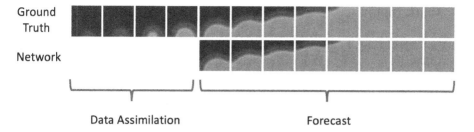

Data Assimilation   Forecast

Fig. 4. Transmembrane potential ground-truth (top) and forecasted (bottom) for one slice of the tissue slab.

It has to be noted that for such cardiac model, predicting the propagation between time 10 and 60 ms on such a slab of tissue represents the whole depolarisation wave, which is the most important part of cardiac electrophysiology dynamics.

Moreover, it is important to notice that, while the model takes a few hours to train, once this is done the inference is very quick and does not require any recalibration.

5   Discussion

We have presented a learning model able to learn the considered dynamics and thus showing promise regarding the automated learning of cardiac electrophysiology models. Moreover, while we only presented results which are completely unsupervised regarding the full state of the studied system, which makes us unable to reconstruct H for example, it is possible with our approach to add in stronger physical priors by pre-training the model over fully observed states, adding some information about the equation into F. The model we present is very versatile and can thus be used with different types of dynamics, degrees of supervision... which could open the way for many other applications where quick simulation is needed.

However, this is still a first step and we need to experiment with our model in more challenging settings *i.e.*, with a more realistic H which is not a simple and dense linear projection in real-world applications. Another shortcoming of our model is the fact that uncertainty is not taken into account, at input as well as at output level, while it is an important feature of many real-world applications especially in healthcare. This should be an essential direction of research for our future work. Finally, it is also important to note that, while Mean Squared Error was chosen for training as well as for testing, it is not necessarily the best metric for all applications: depending on the use case, it might be useful to design a different one, more adapted to the problem.

6   Conclusion

We proposed in this manuscript an approach to learn the dynamics of a cardiac electrophysiology model along with a data assimilation procedure, so that limited measurements can be used to estimate an initial state and predict the future evolution of the electrophysiology model.

In clinical practice, it is challenging to completely map cardiac activation, especially in case of arrhythmia. This method could be used to complete in space and time limited clinical measurements.

Such coupling between simulation and learning opens up many possibilities, especially in order to make such approach more clinically applicable and patient-specific. It has the potential to tackle both computational as well as robustness issues that appear when using computer models or machine learning separately.

Acknowledgements. The research leading to these results has received French funding from the National Research Agency grant IHU LIRYC (ANR-10-IAHU-04).

References

1. Ayed, I., de Bezenac, E., Pajot, A., Brajard, J., Gallinari, P.: Learning dynamical systems from partial observations (2019). http://arxiv.org/abs/1902.11136
2. Bengio, S., Vinyals, O., Jaitly, N., Shazeer, N.: Scheduled sampling for sequence prediction with recurrent neural networks. In: Advances in Neural Information Processing Systems 28 (NIPS 2015) (2015)
3. Çiçek, Ö., Abdulkadir, A., Lienkamp, S.S., Brox, T., Ronneberger, O.: 3D U-Net: learning dense volumetric segmentation from sparse annotation. In: Ourselin, S., Joskowicz, L., Sabuncu, M.R., Unal, G., Wells, W. (eds.) MICCAI 2016. LNCS, vol. 9901, pp. 424–432. Springer, Cham (2016). https://doi.org/10.1007/978-3-319-46723-8_49
4. He, K., Zhang, X., Ren, S., Sun, J.: Deep residual learning for image recognition. In: 2016 IEEE Conference on Computer Vision and Pattern Recognition (CVPR) (2015)
5. Herzog, S., Wörgötter, F., Parlitz, U.: Data-driven modeling and prediction of complex spatio-temporal dynamics in excitable media. Front. Appl. Math. Stat. **4** (2018). https://doi.org/10.3389/fams.2018.00060
6. Kingma, D.P., Ba, J.: Adam: a method for stochastic optimization. In: Proceedings of International Conference for Learning Representations, San Diego (2015)
7. Long, Z., Lu, Y., Ma, X., Dong, B.: PDE-Net: learning PDEs from data. In: International Conference on Machine Learning, pp. 3208–3216, July 2018. http://proceedings.mlr.press/v80/long18a.html
8. Mitchell, C.C., Schaeffer, D.G.: A two-current model for the dynamics of cardiac membrane. Bull. Math. Biol. **65**(5), 767–793 (2003)
9. Raissi, M., Karniadakis, G.E.: Hidden physics models: machine learning of nonlinear partial differential equations. J. Comput. Phys. **357**, 125–141 (2018). https://doi.org/10.1016/j.jcp.2017.11.039, http://www.sciencedirect.com/science/article/pii/S0021999117309014

10. Rapaka, S., et al.: LBM-EP: Lattice-Boltzmann method for fast cardiac electro-physiology simulation from 3D images. In: Ayache, N., Delingette, H., Golland, P., Mori, K. (eds.) MICCAI 2012. LNCS, vol. 7511, pp. 33–40. Springer, Heidelberg (2012). https://doi.org/10.1007/978-3-642-33418-4_5

11. Relan, J., et al.: Coupled personalization of cardiac electrophysiology models for prediction of ischaemic ventricular tachycardia. Interface Focus **1**(3), 396–407 (2011)

12. Roth, H.R., et al.: Deep learning and its application to medical image segmentation. Med. Imaging Technol. **36**(2), 63–71 (2018)

Pipeline to Build and Test Robust 3D T1 Mapping-Based Heart Models for EP Interventions: Preliminary Results

Mengyuan Li[1,2], Maxime Sermesant[3], Sebastian Ferguson[1],
Fumin Guo[1,2], Jen Barry[1], Xiuling Qi[1], Peter Lin[1,2], Matthew Ng[1,2],
Graham Wright[1,2], and Mihaela Pop[1,2(✉)]

[1] Sunnybrook Research Institute, Toronto, Canada
mihaela.pop@utoronto.ca
[2] Medical Biophysics, University of Toronto, Toronto, Canada
[3] Inria - Epione Group, Sophia Antipolis, France

Abstract. Computational models are powerful tools in electrophysiology (EP), helping us understand and predict arrhythmia associated with heart attack (i.e., myocardial infarction), a major cause of sudden cardiac death. Our broad aim is to combine novel scar imaging methods with fast computational models to enable accurate predictions of electrical wave propagation, and then to test these models in preclinical frameworks prior to clinical translation. In this work we used n = 3 swine with chronic infarct, which underwent MR followed by conventional x-ray guided electro-anatomical EP mapping. For scar imaging, we employed our T1-mapping MR method based on multi-contrast late enhancement (MCLE) at 1×1 mm in-plane resolution and 5 mm slice thickness. Next, we used the MCLE images as input to a fuzzy-logic algorithm and segmented the infarcted area into two zones: infarct core IC (dense fibrosis) and grey-zone, GZ (i.e., arrhythmia substrate). We further built 3D heart models from the stack of segmented 2D MCLE images, integrating tissue zones (healthy, IC and GZ) into detailed tetrahedral heart meshes (~ 1.5 mm element size). Finally, we investigated the accuracy of model predictions by comparing measured maps of activation times (i.e., depolarization times) with simulated maps obtained by employing a macroscopic formalism and reaction-diffusion equations. We obtained an acceptable small mean absolute error between the simulated and measured depolarization times (~ 12 ms, in average). Future work will focus on refining MR imaging resolution and use the models to guide ablation procedures.

Keywords: Cardiac MRI · T1-mapping · Modelling · Electrophysiology

1 Introduction

Malignant arrhythmia due to high heart rates (e.g. fast ventricular tachycardia, VT) is an important cause of sudden cardiac death in patients with structural disease such as myocardial infarction [1]. During VT, an aberrant electrical wave anchors around

© Springer Nature Switzerland AG 2019
Y. Coudière et al. (Eds.): FIMH 2019, LNCS 11504, pp. 64–72, 2019.
https://doi.org/10.1007/978-3-030-21949-9_8

electrically inert infarct scars, propagating through viable *isthmuses* of slow conductivity (i.e., channels formed by a mixture of viable myocytes and collagen fibrils). For assessment of electrical function and risk of arrhythmia, these patients undergo an invasive x-ray guided catheter-based electrophysiology (EP) study, during which a map of electrical signals are recorded and used to detect the isthmuses within or at the periphery of infarcted areas, followed by an even more invasive test of arrhythmia inducibility. Once identified, the isthmuses (i.e., VT foci) become targets for VT ablation. Unfortunately, the success rate of ablations is currently lower and VT reoccur in >50% of patients [1]. The ablation failure is often due to inadequate substrate identification in electrical maps which are limited to catheter-based invasive recordings acquired only on the surface of the heart.

Owing to its excellent soft-tissue contrast and lack of radiation, MR imaging has found an increasing role in clinical exams prior to EP interventional studies. Non-invasive MRI is now routinely used to provide structural information regarding infarct location and transmural extent, as well as functional information (i.e., wall motion). Typically, late gadolinium enhancement (LGE) imaging is used for scar imaging [2, 3], where the VT substrate is identified as a *'grey zone'* (GZ) due to its intermediate signal intensity (SI) between scar and healthy tissue [4]. Unfortunately, this method has several drawbacks, particularly missing subtle sub-endocardial GZ. An alternative is to use T1-mapping MR methods, which provide superior contrast and sensitivity in scar/GZ detection compared to LGE images [5].

However, the key limitations due to the surface-derived EP maps and to VT test invasiveness still need to be addressed. To overcome them, *computational modelling* tools can be used [6] in combination with scar information from MR imaging to build 3D heart models coupled with numerical methods. These models can simulate in silico the abnormal propagation of electrical waves and predict VT inducibility. We previously built such 3D heart models from ex vivo diffusion tensor MR images of explanted pig hearts with chronic infarction, and then validated the simulated activation maps and VT test outcome vs. EP measurements recorded by conventional x-ray guided electro-anatomical systems [7]. A next logical step is to develop a similar framework using accurate in vivo MR methods.

In this work we propose a novel pipeline to build preclinical 3D heart models from T1-mapping images and to test them using data obtained from in vivo x-ray guided EP studies from a pig model of chronic infarction. Specifically, here we employ our robust scar/GZ segmentation method which was validated against histology [8], and also correct the motion-induced errors during multi-phase T1 image acquisition. Lastly, we compare measured activation times with those simulated using fast computational models. A simplified diagram illustrating various components of the T1-based pipeline is shown in Fig. 1.

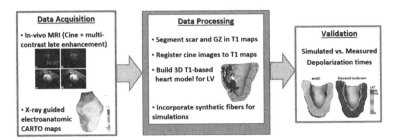

Fig. 1. Diagram for the pipeline to build and test 3D T1-mapping models (including: MR-EP data acquisition, image processing and model validation).

2   Materials and Methods

2.1   Animal Preparation

All in vivo animal studies (i.e., infarct creation, MR imaging and EP procedures) were approved by our Sunnybrook Research Institute. In this work we included results from n = 3 MR-EP studies performed in chronically infarcted pigs. For infarction, a major coronary artery was occluded under x-ray by a balloon catheter for 90 min, followed by reperfusion. The pigs were allowed to heal for ∼5 weeks prior to MR-EP studies. By this time point, fibrosis had replaced dead myocytes in the infarct core (IC), whereas a mixture of viable and collagen fibrils was found in the peri-infarct (i.e., GZ), as confirmed by collagen-sensitive histological stains.

2.2   Image Analysis Pipeline

Step (1) Data Acquisition: MR Imaging and EP Studies
All MRI studies were performed using a 1.5T GE SignaExcite MR scanner. For heart anatomy we used a 2D Cine SSFP sequence. For scar imaging we used a T1-mapping method based on a 2D multi-contrast late enhancement (MCLE) method. The MCLE images were acquired over one R-R interval about 20 min following the injection of Gd-based contrast agent, resulting in 20 phases (images) per cardiac cycle at different inversion times (one of them nulling the signal from blood and healthy tissue). Both Cine MR images and MCLE images were acquired at 1 × 1 × 5 mm spatial resolution. For the EP study we used an x-ray guided electro-anatomical CARTO system (Biosense). The LV endocardial maps (∼120 points/map) were acquired via a catheter introduced in the LV cavity, in sinus rhythm and/or pacing conditions. In the case of pacing at CL = 500 ms, a second (pacing) catheter was inserted in the RV and positioned at the apex, while the mapping catheter recorded the intracardiac electrograms from the LV-endo. The MR images helped guide the mapping catheter such that denser EP points (∼2–3 mm apart) were acquired from within the infarct area.

Step (2) Data Processing: Image Analysis and 3D Heart Model Building
Segmentation: The MCLE images were used to extract steady-state (SS) and T1* maps (Fig. 2a), which were used as an input to a robust fuzzy-logic segmentation algorithm (Matlab). The resulting clusters of healthy, GZ, dense scar (IC) and blood pixels

(Fig. 2b), were used to generate tissue parametric maps (Fig. 2c), which were compared to tissue histological maps using collagen-sensitive stains (Fig. 2d).

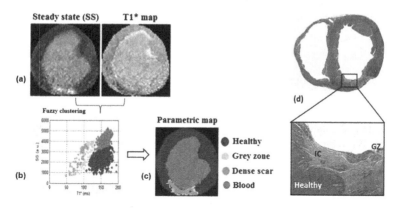

Fig. 2. Segmented MCLE image (a–c). Corresponding histology stain (Masson's Trichrome): IC (dense collagen) in green, GZ as a mixture of green and red, and healthy tissue in dark red (d). (Color figure online)

<u>Image Registration:</u> Prior to building 3D heart models from stacks of 2D segmented MCLE images, we performed a motion correction step to align all segmented MCLE images in the same cardiac phase. To do so, we selected a diastole phase in the Cine sequence, and registered segmented MCLE and Cine images for this phase as in [9]. The MCLE-cine image alignment was initialized using a block-matching-based rigid registration approach [10] followed by a deformable registration refinement step. The deformable registration approach employed [11]: a self similarity context descriptor for image similarity measurements; optical flow as a transformation model; and a convex optimization to derive the optimal solution.

<u>Mesh and Fibers Integration:</u> Our generated 3D MCLE-based LV meshes were constructed using CGAL libraries (www.cgal.org) from the stacks of segmented 2D T1-SS images. These meshes were of sufficiently high density (i.e., between 150–200 K tetrahedral elements, with mean element size 1.3–1.5 mm) to adequately simulate the wave propagation. All 3D heart meshes integrated synthetic fiber directions, which were generated using rule-based methods that obey analytical equations [12].

Step (3) Computational Modelling and Validation Tests

The 3D MCLE-based heart models were further used to simulate the electrical wave propagation through the heart using a mono-domain macroscopic formalism with reaction-diffusion equations. This model proposed by Aliev-Panfilov model [13, 14] solves for the action potential (V) and recovery term (r), and was implemented by Inria researchers [15]:

$$\frac{\partial V}{\partial t} = \nabla \cdot (D\nabla V) - kV(V - a)(V - 1) - rV \tag{1}$$

$$\frac{\partial r}{\partial t} = -\left(\varepsilon + \frac{\mu_1 r}{\mu_2 + V}\right)(kV(V - a - 1) + r) \tag{2}$$

where a tunes the action potential duration and k corresponds to the recovery phase. This fiber directions are accounted for via the diffusion tensor D, where d is the 'bulk' electrical conductivity of tissue. A reduced value of d results in a low conduction velocity (c) of wave:

$$c = \sqrt{2 \cdot k \cdot d}(0.5 - a) \tag{3}$$

In this work we personalized only the key model parameter 'd' corresponding to the tissue electrical conductivity, because we compared only the simulated and measured maps of depolarization times, whereas the other parameters were taken from previous studies [7, 15]. We calibrated $d\_healthy$ in the healthy zone by employing a calibration curve [16]. We then assigned $d\_GZ = 0.5*d\_healthy$ to the slow-conductive GZ, and $d\_scar = 0$ to the non-conductive scar, as in [7, 17].

For all Finite Element simulations, we used a 4,096(1x)MB machine with an Intel® Core™ i3-2310M processor, 640 GB HD, NVIDIA® GeForce® 315M graphic adapter. Typically, it took <5 min to simulate 200 ms of the cardiac cycle on a mesh of about 150K elements (~ 1.5 mm element size).

Lastly, we compared the simulated and measured depolarization time maps. For quantitative comparisons, we projected the point-based measured endocardial maps onto the LV surfaces of each mesh, and then we interpolated these maps using our image visualization platform, Vurtigo (www.vurtigo.ca).

3   Results

Figure 3 shows results from the MCLE-to-Cine MR image registration step, before (*left*) and after (*right*) motion correction of the myocardial wall seen in a longitudinal view through the heart segmented in Fig. 2. The registration process was automated, requiring approximatively 3 min per heart. Note the improved alignment of the myocardial contours with Dice coefficients (%) of $\sim 82\%$, which resulted in smooth endocardial/epicardial surfaces of the 3D LV model.

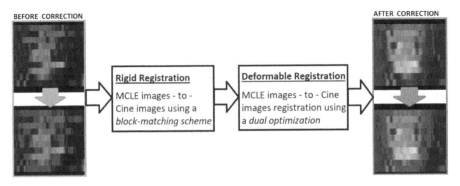

Fig. 3. Motion correction performed through the MCLE-to-Cine image registration, resulting in smoothly aligned endocardial/epicardial surfaces

Next, exemplary results from the construction of a 3D LV model for one infarcted pig heart are presented in Fig. 4. From the stack of 2D segmented MCLE images (registered to the Cine images as described above), we generated an interpolated 3D anatomical LV model using CGAL libraries. The model integrated the three classes of tissue (i.e., healthy zone, GZ and infarct core, IC) determined from the T1-SS mapping images (Fig. 4a). In addition, shown are also the tetrahedral mesh (Fig. 4b) as well as synthetic fibers rotating from −70° to +70° (from endocardium to epicardium) integrated into the 3D LV mesh by assigning the fiber directions to each vertex (Fig. 4c).

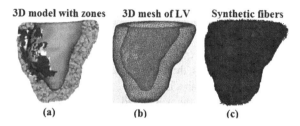

3D model with zones	3D mesh of LV	Synthetic fibers
(a)	(b)	(c)

Fig. 4. Results from building a T1-based model for an infarcted pig heart: (a) 3D LV anatomic model integrating the three MCLE-defined tissue zones (GZ area is in white and the IC area is in black); (b) corresponding mesh ($\sim$150K elements); and (c) synthetic fibers.

Figure 5 shows LV models integrating various fiber inclinations. Based on our previous work [7, 17] and comparisons with EP data in healthy hearts, our results suggest that an angle inclination range of [−70°/+70°] produces the closest pattern and smallest error between simulated and measured endocardial activation maps.

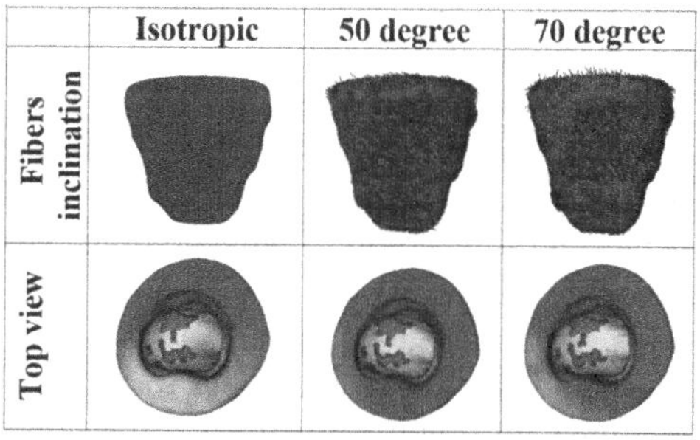

	Isotropic	50 degree	70 degree
Fibers inclination			
Top view			

Fig. 5. Simulated activation times for LV modes with different fibers inclinations

Figure 6a shows the calibration curve for the speed of wave vs. *d_healthy* (bulk conductivity) used for model personalization. Figure 6b shows simulated depolarization times (left) and experimental maps (*right*), with LAT being the local activation time (*blue* corresponds to late depolarization times). The scar (IC) was assigned $d = 0$. The simulations on this LV mesh were performed in $\sim$4 min.

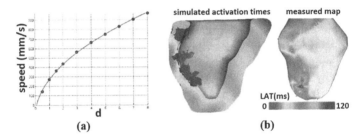

Fig. 6. Results: calibration curve for speed vs *d* conductivity (a); and comparison between the simulated and measured depolarization maps in an infarcted heart (b).

Overall, there was a very good qualitative correspondence between computed and measured activation patterns in all three infarcted hearts. Furthermore, quantitative comparison led to an acceptable absolute error (~ 12 ms mean values, in average among the three hearts) between the simulated depolarization times vs. measured depolarization maps recorded from the endocardial surface. The largest differences were observed at the periphery of infarct, which was expected due to the relatively sparse EP points compared to the MR-derived mesh density. All quantitative comparisons were performed using custom codes developed in Matlab.

4   Discussion and Future Work

Innovative biomedical technologies using exquisite cardiac MR scar imaging methods in combination with predictive image-based computer models can used as powerful non-invasive diagnostic and treatment-planning tools in the clinical EP lab. To sum up, in this work we proposed a novel image analysis pipeline to augment the information from conventional electro-anatomical EP studies with 3D electro-physiology simulations using high-resolution T1-mapping-based computer models. Such models could supplement important information that is currently lacking in EP maps due to the sparse point-base recordings, typically limited to the LV endocardial surface. Our T1-mapping MCLE method for scar imaging was recently validated using quantitative histopathology [8], giving us confidence that our 3D anatomical models integrating three zones: scar, healthy tissue and GZ are sufficiently accurate.

The preclinical results in this work suggest that macroscopic theoretical models can provide rapid (<5 min) simulation results for depolarization times on relatively dense MCLE-derived LV meshes, making them attractive for rapid integration into clinical platforms. Although these preliminary results are promising, we acknowledge that a modelling limitation was the usage of *global* parameters (i.e., same conductivity or speed within the healthy LV tissue). Better predictions may be obtained if the key parameters in the model will be calibrated locally, using AHA-based 17 segments for LV [18]. Thus, future work will focus on personalizing *local* model parameters per individual heart from EP data using these AHA segments. We envision that this refined approach will improve the model personalization and further reduce the error between simulated and measured activation times. We also aim to include the right ventricle into 3D biventricular heart meshes and simulate the VT inducibility.

Lastly, for rapid scar/GZ imaging we are currently developing and testing a high-resolution 3D MCLE method with navigator, which produces an excellent contrast at blood-tissue interface at a 1.5 mm isotropic spatial resolution [19]. This voxel size can overcome potential partial volume effects in the scar/GZ segmentation obtained at the 1 × 1 × 5 mm resolution. This 3D MCLE method will also help avoiding any potential errors introduced by cardiac and respiratory motion in the segmentation/registration steps. Thus, we anticipate that 3D heart models built from high resolution MCLE images will provide robust identification of the VT substrate and a better guidance of the RF ablation procedure.

5 Conclusion

Our broad goal is to test the predictive power of our 3D T1-based models and predict risk of scar-induced arrhythmias. Here, we proposed a pipeline to build and test predictive T1-mapping image-based computer models using a preclinical pig model of infarction that mimics very well the human pathophysiology of chronic scars. Overall, our novel 3D computer LV models can give superior information compared to the surface EP maps, allowing for visualization of transmural activation times and activation patterns through the myocardial wall, relative to the precise position of the scar in the infarcted hearts.

Acknowledgement. The authors acknowledge funding from CIHR grants (Dr. Pop and Dr. Wright) and Inria *Associate project* (Dr. Sermesant and Dr. Pop). Megyuan Li was supported in part by a summer student UROP – Medical Biophysics award.

References

1. Stevenson, W.G.: Ventricular scars and VT tachycardia. Trans. Am. Clin. Assoc. **120**, 403–412 (2009)
2. Bello, D., Fieno, D.S., Kim, R.J., et al.: Infarct morphology identifies patients with substrate for sustained ventricular tachycardia. J. Am. Coll. Cardiol. **45**(7), 1104–1110 (2005)
3. Codreanu, A., Odille, F., et al.: Electro-anatomic characterization of post-infarct scars comparison with 3D myocardial scar reconstruction based on MRI. J. Am. Coll. Cardiol. **52**, 839–842 (2008)
4. Wijnmaalen, A., et al.: Head-to-head comparison of c-e MRI and electroanatomical voltage mapping to assess post-infarct scar characteristics in patients with VT: real-time image integration and reversed registration. Eur. Heart J. **32**, 104 (2011)
5. Detsky, J.S., Paul, G., Dick, A.J., Wright, G.A.: Reproducible classification of infarct heterogeneity using fuzzy clustering on multi-contrast delayed enhancement MR images. IEEE Trans. Med. Imaging **28**(10), 1606–1614 (2009)
6. Clayton, R.H., Panfilov, A.V.: A guide to modelling cardiac electrical activity in anatomically detailed ventricles. Prog. Biophys. Mol. Biol. Rev. **96**(1–3), 19–43 (2008)

7. Pop, M., et al.: Correspondence between simple 3D MR image-based heart models and in-vivo EP measures in swine with chronic infarction. IEEE Trans. Biomed. Eng. **58**(12), 483–3486 (2011)
8. Pop, M., Ramanan, V., Yang, F., Zhang, L., Newbigging, S., Wright, G.: High resolution 3D T1* mapping and quantitative image analysis of the 'gray zone' in chronic fibrosis. IEEE Trans. Biomed. Eng. **61**(12), 2930–2938 (2014)
9. Guo, F., Li, M., Ng, M., Wright, G., Pop, M.: Cine and multicontrast late enhanced MRI registration for 3D heart model construction. In: Pop, M., et al. (eds.) STACOM 2018. LNCS, vol. 11395, pp. 49–57. Springer, Cham (2019). https://doi.org/10.1007/978-3-030-12029-0_6
10. Ourselin, S., Roche, A., Prima, S., Ayache, N.: Block matching: a general framework to improve robustness of rigid registration of medical images. In: Delp, Scott L., DiGoia, Anthony M., Jaramaz, B. (eds.) MICCAI 2000. LNCS, vol. 1935, pp. 557–566. Springer, Heidelberg (2000). https://doi.org/10.1007/978-3-540-40899-4_57
11. Heinrich, M.P., Jenkinson, M., Papież, B.W., Brady, S.M., Schnabel, J.A.: Towards realtime multimodal fusion for image-guided interventions using self-similarities. In: Mori, K., Sakuma, I., Sato, Y., Barillot, C., Navab, N. (eds.) MICCAI 2013. LNCS, vol. 8149, pp. 187–194. Springer, Heidelberg (2013). https://doi.org/10.1007/978-3-642-40811-3_24
12. Arts, T., Costa, K.D., Covell, J.W., McCulloch, A.D.: Relating myocardial laminar architecture to shear strain and muscle fiber orientation. Am. J. Physiol. Heart Circ. Physiol. **280**(5), H2222–2229 (2001)
13. Aliev, R., Panfilov, A.V.: A simple two variables model of cardiac excitation. Chaos Soliton Fractals **7**(3), 293–301 (1996)
14. Nash, M.P., Panfilov, A.V.: Electromechanical model of excitable tissue to study reentrant cardiac arrhythmias. Prog. Biophys. Mol. Biol. **85**, 501–522 (2004)
15. Sermesant, M., Delingette, H., Ayache, N.: An electromechanical model of the heart for image analysis and simulation. IEEE Trans. Med. Imaging **25**(5), 612–625 (2006)
16. Chinchapatnam, P., Rhode, K.S., Ginks, M., et al.: Model-based imaging of cardiac apparent conductivity and local conduction velocity for planning of therapy. IEEE Trans. Med. Imaging **27**(11), 1631–1642 (2008)
17. Pop, M., Sermesant, M., Liu, G., Relan, J., et al.: Construction of 3D MR image-based computer models of pathologic hearts, augmented with histology and optical imaging to characterize the action potential propagation. Med. Image Anal. **16**(2), 505–523 (2012)
18. Cerqueira, M.D., Weissman, N.J., et al.: Standardized myocardial segmentation and nomenclature for tomographic imaging of the heart: a statement for health-care professionals from the Cardiac Imaging Committee of the Council on Clinical Cardiology of the American Heart Association. Circulation **105**, 539–542 (2002)
19. Zhang, L., Athavale, P., Pop, M., Wright, G.: Multi-contrast reconstruction using compressed sensing with low rank and spatially-varying edge-preserving constraints for high-resolution MR characterization of infarction. Magn. Reson. Med. **78**, 598–610 (2016)

Maximal Conductances Ionic Parameters Estimation in Cardiac Electrophysiology Multiscale Modelling

Yassine Abidi[1], Julien Bouyssier[3], Moncef Mahjoub[1],
and Nejib Zemzemi[2,3]

[1] ENIT-LAMSIN, Tunis El Manar University, Tunis, Tunisia
[2] Inria, Bordeaux Sud-Ouest,
200 Avenue de la vielle Tour 33405, Talence Cedex, France
nejib.zemzemi@inria.fr
[3] IHU-LIRYC, Avenue du Haut Lévêque, 33600 Pessac, France

Abstract. In this work, we present an optimal control formulation for the bidomain model in order to estimate maximal conductances parameters in the physiological ionic model. We consider a general Hodgkin-Huxley formalism to describe the ionic exchanges at the microcopic level. We consider the parameters as control variables to minimize the mismatch between the measured and the computed potentials under the constraint of the bidomain system. The solution of the optimization problem is based on a gradient descent method, where the gradient is obtained by solving an adjoint problem. We show through some numerical examples the capability of this approach to estimate the values of sodium, calcium and potassium ion channels conductances in the Luo Rudy phase I model.

Keywords: Parameters estimation ·
Maximal conductance ionic parameters · Bidomain model ·
Optimal control with pde constraints ·
First order optimality conditions · Physiological ionic model ·
Cardiac electrophysiology

1 Introduction

The bidomain equations are the state-of-the-art model to describe the propagation of the electrical wave in the heart. This model is governed by a system of partial differential equations (PDEs) nonlinearly coupled to a set of non-linear ordinary differential equations (ODEs) describing the dynamics of the cell membrane. These ODEs are usually called the ionic model. The description of these models could be either physiological or phenomenological. In the physiological case, they are in general built using a single cell preparation. Their use in multiscale modeling requires to adjust the parameters. Of particular interest, the

© Springer Nature Switzerland AG 2019
Y. Coudière et al. (Eds.): FIMH 2019, LNCS 11504, pp. 73–84, 2019.
https://doi.org/10.1007/978-3-030-21949-9_9

ion-channels conductances play an important role in the depolarization rate, the conduction velocity, the repolarization times, ... *etc.* They are key parameters in order to proceed to the personalization of a given model. Given the importance of these parameters, theoretical studies were carried out to establish theoretical stability results for the inverse problem of identification of maximum conductances. Brandao et al. are the first who studied the theoretical analysis and the controllability of the optimization using the FitzHugh-Nagumo model [1]. Later, systematic analysis of the optimal control of monodomain and bidomain model is presented in [2–4]. A numerical study for optimal control of the monodomain and the bidomain model allowed to predict optimized shock waveforms in 2D [4] and more recently for the optimal control of bidomain-bath model using Mitchell-Shaeffer model in 3D geometries [5,6]. In those studies the control acts at the boundaries of the bath domain. In an other work [7], authors propose a strategy to optimize a non differentiable cost function obtained from a fit of activation times map. Recently, theoretical studies of the stability of the maximal conductances identification problem in the monodomain [8] and bidomain [9] models have been carried out. Yan and Veneziani proposed a variational procedure for the estimation of cardiac conductivities from measures of the transmembrane and extracellular potentials available at some sites of the tissue [10]. Moreover, the identification from measurements of surface potentials has been tackled in an optimization framework for numerical purposes [11,12]. Recently, drug doses optimization in stem cells preparation has been subject of numerical study following an adjoint procedure [13].

In this study, we propose a variational procedure to the estimation of ionic maximal conductance parameters. The optimal control approach which is based on the minimization of an appropriate cost functional that depends on the maximal conductances and measurments of the transmembrane potentials available at the cardiac tissue. The paper is organized as follows: First, we briefly recall the mathematical equations of the bidomain model describing the electrical wave propagation. In Sect. 3, we present the optimal control formulation approach, a formal derivation of the adjoint system and the first order optimality condition. The numerical approach to solve the optimality system is explained in Sect. 4. Finally, in Sect. 5, we show the numerical results with several test cases and different levels of noise.

2 Mathematical Model

Let $\Omega \subset \mathbb{R}^d$ ($d \geq 1$) be a bounded connected open set whose boundary $\Gamma = \partial\Omega$ is regular enough, ($\Omega \subset \mathbb{R}^3$ being the natural domain of the hearth). Let $T > 0$ be a fixed time horizon. We will use the notation $Q = \Omega \times (0,T)$ and $\Sigma = \Gamma \times (0,T)$.

We introduce a parabolic-elliptic system called *bidomain model*, coupled to a system of ODEs:

$$\begin{cases} A_m\big(C_m\partial_t v + I_{ion}(\bar{\varrho}, v, \mathbf{w}, \mathbf{z})\big) - div(\boldsymbol{\sigma}_i\nabla v) = div(\boldsymbol{\sigma}_i\nabla u_e) + A_m I_{app} & \text{in } Q, \\ -div(\boldsymbol{\sigma}_i\nabla v + (\boldsymbol{\sigma}_i + \boldsymbol{\sigma}_e)\nabla u_e) = 0 & \text{in } Q, \\ \partial_t \mathbf{w} = \mathbf{F}(v, \mathbf{w}) & \text{in } Q, \\ \partial_t \mathbf{z} = \mathbf{G}(\bar{\varrho}, v, \mathbf{w}, \mathbf{z}) & \text{in } Q, \\ \boldsymbol{\sigma}_i\nabla v.\nu + \boldsymbol{\sigma}_i\nabla u_e.\nu = 0 & \text{on } \Sigma, \\ \boldsymbol{\sigma}_i\nabla v.\nu + (\boldsymbol{\sigma}_i + \boldsymbol{\sigma}_e)\nabla u_e.\nu = 0 & \text{on } \Sigma, \\ v(x,0) = v_0(x), \quad \mathbf{w}(x,0) = \mathbf{w}_0(x), \quad \mathbf{z}(x,0) = \mathbf{z}_0(x) & \text{in } \Omega, \end{cases}$$
$$(2.1)$$

where $v : Q \to \mathbb{R}$ is the transmembrane potential, $u_e : Q \to \mathbb{R}$ is the extracellular electric potential, and $\boldsymbol{\sigma}_i, \boldsymbol{\sigma}_e : \Omega \to \mathbb{R}^{d\times d}$ are respectively the intra- and extracellular conductivity tensors. $\mathbf{w} : Q \to \mathbb{R}^k$ represent the gating variables and $\mathbf{z} : Q \to \mathbb{R}^m$ are the ionic intracellular concentration variables. A_m is the surface to volume ratio of the cardiac cells, and $C_m > 0$ is the membrane capacitance per unit area. $I_{app} : Q \to \mathbb{R}$ is the applied current source and $\bar{\varrho} := \{\bar{\varrho}_i\}_{1\leq i\leq N}$ represent a set of maximal conductance parameters. The ionic current I_{ion} and the functions $\mathbf{F}$ and $\mathbf{G}$ depends of the considered ionic model. In isolated heart conditions, no current flows out of the heart, as expressed by the homogeneous Neumann boundary conditions.

2.1   Membrane Models and Ionic Currents

Following the celebrated work by Hodgkin and Huxley [14], many models of Hodgkin-Huxley (HH) type have later been developed for the cardiac action potential. In these models, the ionic current I_{ion} through channels of the membrane, has the following general structure [15]:

$$I_{ion}(\bar{\varrho}, v, \mathbf{w}, \mathbf{z}) = \sum_{i=1}^{N} \bar{\varrho}_i y_i(v) \prod_{j=1}^{k} w_j^{p_{j,i}} (v - E_i(\mathbf{z})), \qquad (2.2)$$

where N is the number of ionic currents, $\bar{\varrho}_i := \bar{\varrho}_i(x)$ is the maximal conductance associated with the ith current, y_i is a gating function depending only on the membrane potentiel v, $p_{j,i}$ are positive integers exponents and E_i is the reversal potential for the ith current, which is the related equilibrium (Nernst) potential and is given by

$$E_i(\mathbf{z}) = \overline{\gamma}_i \log\left(\frac{z_e}{z_i}\right), \quad \mathbf{z} = (z_1, \ldots, z_m), \qquad (2.3)$$

where $\overline{\gamma}_i$ is a constant and z_i, $i = 1, \ldots, m$, are the intracellular concentrations. The constant z_e denotes an extracellular concentration. For each action potential model, the dynamic of the gating variables $\mathbf{w}$ and the intracellular concentrations $\mathbf{z}$ are described by a system of ordinary differential equations (ODEs). In this paper, we consider the Luo-Rudy phase I model (LR1) [16] which extends the Beeler-Reuter model [17] to enhance the representation of depolarization and repolarization phases and their interaction. The time course of the action potential (AP) is governed by $N = 6$ ionic currents:

$$I_{ion} = I_{Na} + I_{si} + I_K + I_{K1} + I_{Kp} + I_b, \tag{2.4}$$

which are fast sodium current (I_{Na}), slow inward calcium current (I_{si}), time dependent potassium current (I_K), time independent potassium current (I_{K1}), plateau potassium current (I_{Kp}) and background current (I_b). The time dependent currents I_{Na}, I_{si} and I_K, depend on six activation and inactivation gates m, h, j, d, f, x, and one intracellular concentration variable of Calcium $[Ca^{2+}]_i$, which are governed by ODEs of the form:

$$\frac{dw}{dt} = \alpha_w(v)(1 - w) - \beta_w(v)w, \quad \text{for } w = m, h, j, d, f, x,$$
$$\frac{d}{dt}[Ca^{2+}]_i = -10^{-4}I_{si} + 0.07(10^{-4} - [Ca^{2+}]_i), \tag{2.5}$$

where α_w and β_w are two positive rational functions of exponentials in v. For details on formulation of those functions and the parameters used in our computations, we refer to the original paper of LR1 model [16]. The existence and uniqueness for the LR1 model and more general of the classical HH model of the couple (v, u_e), with u_e has zero average on Ω, i.e $\int_\Omega u_e dx = 0$, can be found in [18].

3 Optimal Control Problem

In this section, we set the optimal control problem, for which the numerical experiments were carried out. Suppose that v_{meas} is the desired state solution at the cardiac domain, we look for the set/vector of parameters $\bar{\varrho}$ that solves the following minimization problem.

$$(\mathcal{P}) \begin{cases} \min_{\bar{\varrho} \in \mathcal{C}_{ad}} \mathcal{I}(\bar{\varrho}) = \frac{1}{2}\left(\epsilon_1 \int_Q |v(\bar{\varrho}) - v_{meas}|^2 \, dxt + \epsilon_2 \int_\Omega |\bar{\varrho}|^2 \, dx\right), \\ \text{subject to the coupled PDE system (2.1), and } \bar{\varrho} \in \mathcal{C}_{ad}, \end{cases} \tag{3.1}$$

where $\mathcal{I}$ is the quantity of interest, ϵ_1 and ϵ_2 are the regularization parameters, v is the state variable and $\int_\Omega |\bar{\varrho}|^2 \, dx$ denotes a Tikhonov-like regularization term used to weigh the impact of the regularization in the minimize procedure. $\mathcal{C}_{ad}$ is the admissible domain for control given by

$$\mathcal{C}_{ad} = \{\bar{\varrho} \in L^\infty(\Omega)^N : \bar{\varrho}(x) \in [m, M]^N, \forall x \in \Omega\}. \tag{3.2}$$

3.1 Optimal Conditions and Dual Problem

In this paragraph, we formally derive the optimality system associated to (3.1). Let's denote by $\mathcal{J}$ the function

$$\mathcal{J}(\bar{\varrho}, v) = \frac{1}{2}\left(\epsilon_1 \int_Q |v - v_{meas}|^2 \, dx t + \epsilon_2 \int_\Omega |\bar{\varrho}|^2 \, dx\right).$$

If $v(\bar{\varrho})$ is solution of (2.1), then we immediately have $\mathcal{J}(\bar{\varrho}, v(\bar{\varrho})) = \mathcal{I}(\bar{\varrho})$. We follow a Lagrangian approach and introduce the following Lagrange functional:

$$\mathcal{L}(v, u_e, \mathbf{w}, \mathbf{z}, \bar{\varrho}, \lambda^*) = \mathcal{J}(\bar{\varrho}, v) - \int_Q pA_m\big(C_m \partial_t v + I_{ion}(\bar{\varrho}, v, \mathbf{w}, \mathbf{z}) - I_{app}\big) dx dt$$

$$- \int_Q p\big(- div(\boldsymbol{\sigma}_i \nabla v) - div(\boldsymbol{\sigma}_i \nabla u_e)\big) dx dt$$

$$- \int_Q q\big(- div(\boldsymbol{\sigma}_i \nabla v + (\boldsymbol{\sigma}_i + \boldsymbol{\sigma}_e)\nabla u_e)\big) dx dt$$

$$- \int_Q \boldsymbol{r}.\big(\partial_t \mathbf{w} - \boldsymbol{F}(v, \mathbf{w})\big) dx dt - \int_Q \boldsymbol{s}.\big(\partial_t \mathbf{z} - \boldsymbol{G}(\bar{\varrho}, v, \mathbf{w}, \mathbf{z})\big) dx dt, \quad (3.3)$$

where $\lambda^* := (p, q, \boldsymbol{r}, \boldsymbol{s})(x, t)$ denote the Lagrange multipliers. The first order optimality system is given by the Karusch-Kuhn-Tucker (KKT) conditions which result from equating the partial derivatives of $\mathcal{L}$ with respect to the state variables equal to zero. We then obtain the following governing system of the Lagrange multipliers:

$$\begin{cases} -A_m(C_m \partial_t p - p\partial_v I_{ion}) - div(\boldsymbol{\sigma}_i \nabla p) - div(\boldsymbol{\sigma}_i \nabla q) - (\partial_v \boldsymbol{F})^T \boldsymbol{r} - (\partial_v \boldsymbol{G})^T \boldsymbol{s} = \epsilon_1(v - v_{meas}) & \text{in } Q, \\ -div(\boldsymbol{\sigma}_i \nabla p + (\boldsymbol{\sigma}_i + \boldsymbol{\sigma}_e)\nabla q) = 0 & \text{in } Q, \\ -\partial_t \boldsymbol{r} + A_m p\partial_\mathbf{w} I_{ion} - (\partial_\mathbf{w} \boldsymbol{F})^T \boldsymbol{r} - (\partial_\mathbf{w} \boldsymbol{G})^T \boldsymbol{s} = 0 & \text{in } Q, \\ -\partial_t \boldsymbol{s} + A_m p\partial_\mathbf{z} I_{ion} - (\partial_\mathbf{z} \boldsymbol{G})^T \boldsymbol{s} = 0 & \text{in } Q, \end{cases}$$

$$(3.4)$$

with the terminal conditions

$$p(x, T) = 0, \quad \boldsymbol{r}(x, T) = \mathbf{0}, \quad \boldsymbol{s}(x, T) = \mathbf{0} \text{ in } \Omega, \quad (3.5)$$

and the boundary conditions for the adjoint states

$$\begin{cases} -\boldsymbol{\sigma}_i \nabla p.\nu = \boldsymbol{\sigma}_i \nabla q.\nu & \text{on } \Sigma, \\ -\boldsymbol{\sigma}_e \nabla q.\nu = 0 & \text{on } \Sigma. \end{cases} \quad (3.6)$$

In addition, we introduce the compatibility condition for the adjoint variable:
$$\int_\Omega q(t)dx = 0, \text{ for all } t \in (0, T).$$

Based on the adjoint equations, The gradient of the quantity of interest $\mathcal{I}(\bar{\varrho})$ with respect to $\bar{\varrho}$ reads as follows:

$$< \frac{\mathcal{DI}}{\mathcal{D}\bar{\varrho}}, \delta\bar{\varrho} > = < \frac{\partial \mathcal{L}}{\partial \bar{\varrho}}, \delta\bar{\varrho} > = \epsilon_2 \int_\Omega \bar{\varrho}.\delta\bar{\varrho} dx - \int_Q A_m p\frac{\partial}{\partial \bar{\varrho}} I_{ion}.\delta\bar{\varrho} dx dt + \int_Q (\frac{\partial \boldsymbol{G}}{\partial \bar{\varrho}})^T \boldsymbol{s}.\delta\bar{\varrho} dx dt,$$

$$(3.7)$$

where $(\frac{\partial \mathbf{G}}{\partial \bar{\varrho}})^T$ denotes the transpose of the Jacobian matrix of $\mathbf{G} \in \mathbb{R}^m$ in point $\bar{\varrho} \in \mathbb{R}^N$.

4   Numerical Approximation

In this section, we give a brief overview of the space and time discretization techniques to solve the primal (2.1) and adjoint (3.4) equations numerically. We use a finite element method (FEM) for the spatial discretization and a semi-implicit Euler scheme for the temporal discretization. We solve the optimal control problem (3.1) using the gradient descent method.

4.1   Space and Time Discretization

The semi-discretization of the primal equations in space results in the differential algebraic system as follows:

$$A_m C_m M \frac{\partial}{\partial t} \mathbf{V} = -A_i \mathbf{V} - A_i \mathbf{U} + A_m M \big(\mathcal{I}_{app} - \mathcal{I}_{ion}(\bar{\varrho}, \mathbf{V}, \mathbf{W}^{(j)}, \mathbf{Z}^{(j')}) \big), \quad (4.1)$$

$$A_{ie} \mathbf{U} = -A_i \mathbf{V}, \quad (4.2)$$

$$M \frac{\partial}{\partial t} \mathbf{W}^{(j)} = \mathcal{F}^{(j)}(\mathbf{V}, \mathbf{W}^{(j)}), \quad (4.3)$$

$$M \frac{\partial}{\partial t} \mathbf{Z}^{(j')} = \mathcal{G}^{(j')}(\bar{\varrho}, \mathbf{V}, \mathbf{W}^{(j)}, \mathbf{Z}^{(j')}), \quad (4.4)$$

along with initial conditions for $\mathbf{V}, \mathbf{W}^{(j)}$ and $\mathbf{Z}^{(j')}$, where $A_{ie} = \{< (\sigma_i + \sigma_e) \nabla \omega_i, \nabla \omega_{j''} >\}_{i,j''=1}^{M}$ and $A_i = \{< \sigma_i \nabla \omega_i, \nabla \omega_{j''} >\}_{i,j''=1}^{M}$ are the stiffness matrices, $M = \{< \omega_i, \omega_{j''} >\}_{i,j''=1}^{M}$ is the mass matrix, and $\{\omega_i\}_{i=1}^{M}$ denote the basis functions, with M is the number of nodal points at the tissue domain. Analogously, the following semi-discrete form of the dual equations is obtained:

$$-A_m C_m M \frac{\partial}{\partial t} \mathbf{P} + A_m M (\partial_\mathbf{V} \mathcal{I}_{ion})^T \mathbf{P} = -A_i \mathbf{P} - A_i \mathbf{Q} + M (\partial_\mathbf{V} \mathcal{F}^{(j)})^T \mathbf{R}^{(j)} + M (\partial_\mathbf{V} \mathcal{G}^{(j')})^T \mathbf{S}^{(j')}$$
$$+ \epsilon_1 M (\mathbf{V} - \mathbf{V}_{meas}), \quad (4.5)$$

$$A_{ie} \mathbf{P} = -A_i \mathbf{Q}, \quad (4.6)$$

$$-\frac{\partial}{\partial t} \mathbf{R}^{(j)} - (\partial_{\mathbf{W}^{(j)}} \mathcal{F}^{(j)})^T \mathbf{R}^{(j)} = -A_m (\partial_{\mathbf{W}^{(j)}} \mathcal{I}_{ion})^T \mathbf{P} + (\partial_{\mathbf{Z}^{(j')}} \mathcal{G}^{(j')})^T \mathbf{S}^{(j')}, \quad (4.7)$$

$$-\frac{\partial}{\partial t} \mathbf{S}^{(j')} - (\partial_{\mathbf{Z}^{(j')}} \mathcal{G}^{(j')})^T \mathbf{S}^{(j')} = -A_m (\partial_{\mathbf{Z}^{(j')}} \mathcal{I}_{ion})^T \mathbf{P}, \quad (4.8)$$

with terminal conditions $\mathbf{P}(T) = \mathbf{R}^{(j)}(T) = \mathbf{S}^{(j')}(T) = 0$, $\forall j = 1, \ldots, k$, $\forall j' = 1, \ldots, m$.

As concerns the time discretization of the primal problem, we start by computing the ODE system in a semi-implicit way: We use a fourth order Runge-Kutta scheme for the computation $\mathbf{W}$ and $\mathbf{Z}$ while $\mathbf{V}$ is kept constant between t and $t + dt$. Then we solve the PDE system $\mathbf{V}$ and $\mathbf{U}$ sequentially, using a first order semi-implicit scheme where V is taken at time t in the expression of $\mathcal{I}_{ion}$ as in [19]. As concerns the dual equations, although the retrograde problem is fully linear, we use a semi-implicit first order scheme to solve it. The reason is that we separate the ODE system variables R and S from the PDE variables P and Q. We also solve the bidomain problem sequentially, we first compute P and then we compute Q. This follows the same scheme developed for the primal problem in [19].

4.2   Optimization Algorithm

Given an initial guess of maximal conductance parameters $\bar{\varrho}_{guess}$, we solve the optimization problem using the following algorithm based on a gradient descent method.

Algorithm 1. Optimization of the maximal conductance parameters $\bar{\varrho}$

$\bar{\varrho} = \bar{\varrho}_{guess}$,
Solve state problem,
Solve adjoint problem,
while $\mathcal{I}(\bar{\varrho}) > \epsilon_{Func}$ & $\|\frac{\mathcal{DI}}{\mathcal{D}\bar{\varrho}}\| > \epsilon_{Grad}$ & $iter \leq MaxIterNumber$ **do**
$\quad \bar{\varrho} = \bar{\varrho} - \alpha \times \frac{\mathcal{DI}}{\mathcal{D}\bar{\varrho}}$.
$\quad$ Solve state problem,
$\quad$ Solve adjoint problem,
$\quad$ Compute the cost function and its gradient,
end while
$\bar{\varrho}_{opt} = \bar{\varrho}$.

Here, ϵ_{Func} and ϵ_{Grad} are positive constants defining the desired tolerance on the cost function and its gradient respectively. The coefficient α is positive and could be fixed or updated at each iteration and $MaxIterNumber$ stands for the maximal number of iterations in the optimization procedure.

5   Numerical Results

In this section, numerical results on the basis of two different test are presented. In all tests, the computational domain $\Omega = [0,1] \times [0,1] \subset \mathbb{R}^2$ of size $0.1 \times 0.1\,\mathrm{cm}^2$ is fixed and a triangular discretization is used with the mesh parameter $h \approx 25\mu\mathrm{m}$ which consists of 11508 elements and 5835 nodes. The stimulation current is imposed in the right bottom corner of the geometry its magnitude is $I_{app}(t) = 80\,\mu\mathrm{A}/\mathrm{cm}^2$ and its duration is $1\,\mathrm{ms}$. During the simulations, we fix the time step length $\Delta t = 0.1\,\mathrm{ms}$. The termination of the optimization algorithm is based on the following condition:

$$\epsilon_{Func} = 10^{-8} \quad \text{and} \quad \epsilon_{Grad} = 10^{-6}. \tag{5.1}$$

Moreover, if these conditions are not satisfied, the algorithm terminates within a prescribed number of iterations. Here the maximum number of iteration parameter is $MaxIterNumber = 20$. For all the following tests, the desired transmembrane potential v_{meas} are simulated with the physiological Luo Rudy phase I model with its original control parameters. There are six ionic currents in the Luo Rudy phase I model I_{Na}, I_{si}, I_{K1}, I_K, I_{Kp} and I_b. Each of the currents has its corresponding maximal ion-channel conductance $\bar{\varrho}_{Na}$, $\bar{\varrho}_{si}$, $\bar{\varrho}_{K1}$, $\bar{\varrho}_K$, $\bar{\varrho}_{Kp}$ and $\bar{\varrho}_b$. In what follows, we will consider to optimize three of them $\bar{\varrho}_{Na}$, $\bar{\varrho}_{si}$, $\bar{\varrho}_{K1}$ representing three different ion channels: sodium, calcium and potassium, respectively.

5.1   Test 1: Optimize the maximal conductance parameter of the fast inward sodium current $\bar{\varrho}_{Na}$

In this test, we present a numerical results of the estimation of the parameter $\bar{\varrho}_{Na}$. Since this parameter is mainly important in the depolarization phase, we consider the cost function in the time window $[0\,ms, 20\,ms]$ of the simulation. The exact value $\bar{\varrho}_{Na}$ is equal to 23. We generate the measurement v_{meas} by solving the forward problem using the exact value of $\bar{\varrho}_{Na}$ and we start our optimization procedure using a guess value $\bar{\varrho}_{Na,guess} = \dfrac{1}{2}\bar{\varrho}_{Na} = 11.5$. Since the cost function depends on the parameters ϵ_1 and ϵ_2 used to make a balance between the function of interest ($\int_Q |v(\bar{\varrho}_{Na}) - v_{meas}|^2 \, dxt$) and the regularization term ($\int_\Omega |\bar{\varrho}_{Na}|^2 \, dx$), we first run the optimization procedure with $\epsilon_1 = 1$ and we vary ϵ_2 from 0.05 to 0.001. As shown in Fig. 1, for both cases the optimization algorithm converges to the desired control value. But the accuracy is better with

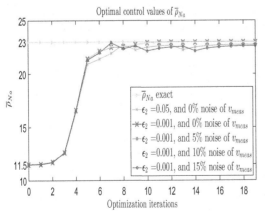

Table 1. Relative error for all cases.

ϵ_2	Noise on v_{meas} (%)	$\dfrac{\|\bar{\varrho}_{Na,exact} - \bar{\varrho}_{Na}\|}{\|\bar{\varrho}_{Na,exact}\|} \times 10^2$
0.05	0 %	1.117 %
0.001	0 %	0.195 %
0.001	5 %	0.196 %
0.001	10 %	0.22 %
0.001	15 %	1.58 %

Fig. 1. The optimal control solution for the optimization of $\bar{\varrho}_{Na}$ for different values of ϵ_2 and different levels of noise.

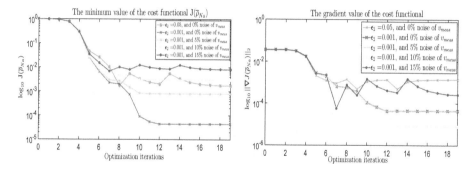

Fig. 2. Left: Log scale plot of the cost function $\mathcal{I}(\bar{\varrho}_{Na})$. Right: Log scale plot of the norm of its gradient during the optimization procedure.

$\epsilon_2 = 0.001$ than $\epsilon_2 = 0.05$ as shown in Table 1. From now on we fix $\epsilon_1 = 1$ and $\epsilon_2 = 0.001$.

In order to test the robustness of the algorithm, we add different levels of gaussian noise to the measured data v_{meas}, and we solve the optimization problem following Algorithm 1 for each value of noise. As shown in Fig. 1, the algorithm converges for all levels of noise. Table 1 shows that the accuracy is altered with the noise. But for 15% of noise, the relative error on the estimated value of $\bar{\varrho}_{Na}$ is under 2%. Figure 2 shows the evolution of the cost function $\mathcal{I}(\bar{\varrho}_{Na})$ and the norm of its gradient with respect to the optimization iterations for different regularization parameter values ϵ_2 and noise levels on the measured potential.

5.2   Test 2: Optimize the Maximal Conductance Parameter of the Slow Inward-Calcium Related Current $\bar{\varrho}_{si}$

In this test, we present a numerical results for the optimization of the parameter $\bar{\varrho}_{si}$. Since this parameter acts on the plateau phase, we performed the optimization on a time window [0 ms, 400 ms]. We consider the initial guess value $\bar{\varrho}_{si,guess} = \dfrac{3}{2}\bar{\varrho}_{si,exact} = 0.135$. Figure 3 (left) shows the evolution of the parameter $\bar{\varrho}_{si}$ during the optimization procedure. The table in Fig. 3 (right) shows the relative error of the obtained solution with respect to the 0%, 5% and 10% noise levels. We can see that it converge from the fourth iteration and the accuracy of the obtained optimal solution of $\bar{\varrho}_{si}$ seems to be less sensitive to noise compared to optimal solution of $\bar{\varrho}_{Na}$.

5.3   Test 3: Optimize the Maximal Conductance Parameter of the Time-Independent Potassium Current $\bar{\varrho}_{K1}$

In this test, we present a numerical results for the optimization of the parameter $\bar{\varrho}_{K1}$. Since this parameter acts on the repolarization phase, we performed the optimization on a time window [0 ms, 400 ms]. The initial guess considered is

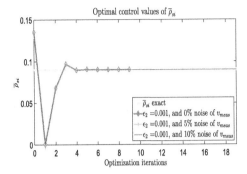

Fig. 3. Left: The evolution of the optimal control solution $\bar{\varrho}_{si}$ during the optimization iteration. Right: Relative errors of the optimal control solution for different noise levels.

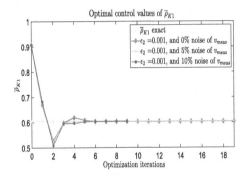

Fig. 4. Left: The evolution of the optimal control solution $\bar{\varrho}_{K1}$ during the optimization iterations. Right: Relative errors of the optimal control solution for different noise levels.

$\bar{\varrho}_{K1,guess} = \dfrac{3}{2}\bar{\varrho}_{K1,exact} = 0.90705$. Figure 4 (left) shows the evolution of the parameter $\bar{\varrho}_{K1}$ during the optimization procedure. The table in Fig. 4 (right) shows the relative error of the obtained solution with respect to the noise level. The results in the table show that the optimal solution of $\bar{\varrho}_{K1}$ is more sensitive to the noise than $\bar{\varrho}_{si}$ and less sensitive to noise than $\bar{\varrho}_{Na}$.

6   Discussion and Conclusions

In this paper, we have presented an approach for the estimation of maximal conductance parameters of the Luo Rudy phase I model. We formulated the problem as an optimization procedure in an optimal control problem where the cost function represents the misfit between the measured signals and the model. Our numerical results shows the capability of this method to estimate the maximal conductance parameter $\bar{\varrho}_{Na}$ (respectively, $\bar{\varrho}_{si}$, $\bar{\varrho}_{K1}$) of the fast sodium current (respectively, slow inward and potassium currents). This study shows also that the optimization procedure is robust with respect of noise. Although, results show also that the optimization of $\bar{\varrho}_{Na}$ is more sensitive to noise than it is for $\bar{\varrho}_{si}$ and $\bar{\varrho}_{K1}$. The challenge is to explore the capability of this method to estimate these physiological parameters when dealing with real life measurement. Finally, we have to say that this study is preliminary and that we didn't explore all of the potential of the optimal control approach. The method here presented allows multiple parameter estimation. It also allows the estimation of space dependent parameters. This would be subject of our future research.

Acknowledgements. This work was supported by the French National Research Agency, grant references ANR-10-IAHU04-LIRYC. This work has also been supported by EPICARD cooperative research program, funded by INRIA international laboratory LIRIMA. The LAMSIN researcher's work is supported on a regular basis by the Tunisian Ministry of Higher Education, Scientific Research and Technology.

References

1. Brandao, A.J., Fernandez-Cara, E., Magalhaes, P., Rojas-Medar, M.A.: Theoretical analysis and control results for the fitzhugh-nagumo equation. Electron. J. Differ. Eqn. (EJDE) [electronic only], 2008:Paper-No 164 (2008)
2. Casas, E., Ryll, C., Tröltzsch, F.: Sparse optimal control of the schlögl and fitzhugh-nagumo systems. Comput. Meth. Appl. Math. **13**(4), 415–442 (2013)
3. Kunisch, K., Wagner, M.: Optimal control of the bidomain system (iii): existence of minimizers and first-order optimality conditions. ESAIM: Math. Model. Numer. Anal. **47**(4), 1077–1106 (2013)
4. Chamakuri, N., Kunisch, K., Plank, G.: Numerical solution for optimal control of the reaction-diffusion equations in cardiac electrophysiology. Comput. Optim. Appl. **49**(1), 149–178 (2011)
5. Chamakuri, N., Kunisch, K.: Primal-dual active set strategy for large scale optimization of cardiac defibrillation. Appl. Math. Comput. **292**, 178–193 (2017)
6. Bendahmane, M., Chamakuri, N., Comte, E., Ainseba, B.: A 3D boundary optimal control for the bidomain-bath system modeling the thoracic shock therapy for cardiac defibrillation. J. Math. Anal. Appl. **437**(2), 972–998 (2016)
7. Ngoma, D., Vianney, P., Bourgault, Y., Nkounkou, H.: Parameter identification for a non-differentiable ionic model used in cardiac electrophysiology. Appl. Math. Sci. **9**(150), 7483–7507 (2015)
8. Abidi, Y., Bellassoued, M., Mahjoub, M., Zemzemi, N.: On the identification of multiple space dependent ionic parameters in cardiac electrophysiology modelling. Inverse Prob. **34**(3), 035005 (2018)
9. Abidi, Y., Bellassoued, M., Mahjoub, M., Zemzemi, N.: Ionic parameters identification of an inverse problem of strongly coupled pdes system in cardiac electrophysiology using carleman estimates. Math. Model. Nat. Phenom. **14**(2), 202 (2019)
10. Yang, H., Veneziani, A.: Estimation of cardiac conductivities in ventricular tissue by a variational approach. Inverse Prob. **31**(11), 115001 (2015)

11. Chávez, C.E., Zemzemi, N., Coudière, Y., Alonso-Atienza, F., Álvarez, D.: Inverse problem of electrocardiography: estimating the location of cardiac ischemia in a 3D realistic geometry. In: van Assen, H., Bovendeerd, P., Delhaas, T. (eds.) FIMH 2015. LNCS, vol. 9126, pp. 393–401. Springer, Cham (2015). https://doi.org/10.1007/978-3-319-20309-6_45

12. Nielsen, B.F., Lysaker, M., Tveito, A.: On the use of the resting potential and level set methods for identifying ischemic heart disease: an inverse problem. J. Comput. Phys. **220**(2), 772–790 (2007)

13. Bouyssier, J., Zemzemi, N.: Parameters estimation approach for the mea/hipsc-cm asaays. In: 2017 Computing in Cardiology (CinC), pp. 1–4. IEEE (2017)

14. Hodgkin, A.L., Huxley, A.F.: A quantitative description of membrane current and its application to conduction and excitation in nerve. J. Physiol. **117**(4), 500–544 (1952)

15. Franzone, P.C., Pavarino, L.F., Scacchi, S.: Mathematical Cardiac Electrophysiology. Springer, Cham (2014). https://doi.org/10.1007/978-3-319-04801-7

16. Luo, C.H., Rudy, Y.: A model of the ventricular cardiac action potential: depolarization, repolarization, and their interaction. Circ. Res. **68**(6), 1501–1526 (1991)

17. Beeler, G.W., Reuter, H.: Reconstruction of the action potential of ventricular myocardial fibres. J. Physiol. **268**(1), 177–210 (1977)

18. Veneroni, M.: Reaction-diffusion systems for the macroscopic bidomain model of the cardiac electric field. Nonlinear Anal. Real World Appl. **10**, 849–868 (2009)

19. Fernández, M.A., Zemzemi, N.: Decoupled time-marching schemes in computational cardiac electrophysiology and ECG numerical simulation. Math. Biosci. **226**(1), 58–75 (2010)

Standard Quasi-Conformal Flattening of the Right and Left Atria

Marta Nuñez-Garcia[1,2(✉)], Gabriel Bernardino[1,3], Ruben Doste[1],
Jichao Zhao[4], Oscar Camara[1], and Constantine Butakoff[1]

[1] Physense, Department of Information and Communication Technologies,
Universitat Pompeu Fabra, Barcelona, Spain
marta.nunez@upf.edu
[2] L'Institut de Rythmologie et de Modélisation Cardiaque
LIRYC/Université de Bordeaux, Bordeaux, France
[3] Philips Research, Medisys, Suresnes, France
[4] Auckland Bioengineering Institute, University of Auckland,
Auckland, New Zealand

Abstract. Two-dimensional standard representations of 3D anatomical
structures are a simple and intuitive way for analysing patient information across populations and image modalities. They also allow convenient
visualizations that can be included in clinical reports for a fast overview
of the whole structure. While cardiac ventricles, especially the left ventricle, have an established standard representation (e.g. bull's eye plot), the
2D depiction of the left (LA) and right atrium (RA) remains challenging
due to their sub-structural complexity. Quasi-conformal flattening techniques, successfully applied to cardiac ventricles, require additional constraints in the case of the atria to correctly place the adjacent structures,
i.e. the pulmonary veins, the vena cava (VC) or the appendages. Some
registration-based methods exist to flatten the LA but they can be time-
consuming and prone to errors if the geometries are very different. We
propose a novel atrial flattening methodology where a quasi-conformal
2D map of both (left and right) atria is obtained quickly and without
errors related to registration. In our approach the RA is mapped to a
standard 2D map where the holes corresponding to superior and inferior
VC are fixed within a disk. Similarly, the LA is divided into 5 regions
which are then mapped to their analogous two-dimensional regions. We
illustrate the application of the method to visualize atrial wall thickness
measurements, and late gadolinium enhanced magnetic resonance data.

Keywords: Conformal flattening · Two-dimensional map ·
Left atrium · Right atrium · Atrial wall thickness · LGE-CMR

1 Introduction

Flattening methods aim to compute an unfolded representation of a 3D surface
mesh by projecting it to a simpler 2D domain, easier to visualize, manage and

© Springer Nature Switzerland AG 2019
Y. Coudière et al. (Eds.): FIMH 2019, LNCS 11504, pp. 85–93, 2019.
https://doi.org/10.1007/978-3-030-21949-9_10

interpret. Additionally, if the 2D map is standardised (e.g. same anatomical regions of different subjects spatially coincide) the 2D unfolded domains can be used as a common reference space to analyse multi-modal data from different patients or from the same patient at different time-steps. The reader is referred to [5] for a thorough review of flattening methods applied to human organs, including the brain, different bones, and the vascular system.

In the case of the heart, the 17 segment AHA bull's eye plot of the left ventricle (LV) has been widely used by clinicians for long time [2,11,15]. The LV's conical shape and the absence of salient sub-structures in 3D anatomies derived from medical images (e.g. ignoring the trabeculations) highly facilitates its flattening. On the contrary, unfolding the atria is challenging due to their more complex morphology. In the case of the LA, its main cavity is connected to several pulmonary veins (PV), the left atrial appendage (LAA) and the LV through the mitral valve (MV). Furthermore, the most common LA morphology (up to 70%) involves 4 PV (with variable size, shape, position and orientation with respect to the main cavity) but in some cases a common left trunk, extra right PV or other oddities may be present [13]. Something similar occurs in the case of the RA, which is connected to the right ventricle through the tricuspid valve (TV), to the superior and inferior vena cava (SVC and IVC), and to the coronary veins (CV).

The first LA flattening technique available in the literature [8] was based on a B-spline with proportional distance between any pair of points in the 3D LA surface and their mapped pairs on the 2D map. Later, the same group proposed to flatten the LA to a square also without constraining the position of the PV and LAA holes [3]. The authors showed how the LA flattened square could be used to display and qualitatively analyse different types of data from the same patient. However, a direct comparison across different patients was not feasible because of the lack of correspondence between the different maps. More recently, Williams et al. [17] proposed a standardised unfold map (SUM), which flattening strategy was based on registering an arbitrary LA to a 3D template and then transferring the data to a 2D LA template using the known 3D-2D point relation between the two templates. The main limitation of this approach is the difficulty of obtaining an accurate and fast registration between different LA surface meshes due to their high shape variability. Together with errors induced by required data projection and interpolation steps, this scheme leads to undesired information loss between the 3D and 2D LA representations. Recently, a universal atrial coordinate system was developed in [14] and used to represent both atria in a two-dimensional domain. The proposed representation is however not completely standardised since the position of the veins and appendages may be distinct for the different cases.

In this paper, we propose a method to represent atrial cavities as 2D disks where the position of the holes corresponding to the different veins and appendages is fixed. Our method is almost real-time and without information loss, i.e. all points in the 3D mesh are represented in the 2D domain.

2   Methodology

The proposed method is based on a quasi-conformal (in general, there is no conformal map compatible with a given map along the boundary) flattening parameterisation [6]. The RA is unfolded with the following boundary constraints:
the TV contour is mapped to the external circumference of a 2D disk, and the
holes corresponding to the SVC and IVC are mapped to predefined circumferences within the disk. Additionally, one constraint is added to fix the position
of the right atrial appendage (RAA). Regarding the LA, the MV contour is also
mapped to the external circumference of a 2D disk, and the PVs and LAA ostia
contours are mapped to predefined circumferences within the disk. Due to the
fact that flattening of surface meshes with holes often results in undesired mesh
self-folding (holes appear covered by adjacent mesh cells), we impose, only for
the LA, additional regional constraints: five anatomical regions are defined in
the 3D LA which are afterwards confined to their 2D counterparts.

It was experimentally found that the use of additional constraints was not
necessary in the case of the RA because mesh self-folding was not found in the
2D maps due to the intrinsically simpler RA shape (as well as the defined 2D
RA template), with, for example, less boundaries compared to the LA. Another
difference is that we decided to include the RAA (while not including the LAA) in
the 2D representation, and constrain only one point since the boundary between
the main body of the RA and RAA is unclear and not as well defined as in the
case of the LAA, where a prominent ostium is typically present.

Given a RA or LA with holes corresponding to the valves (TV, MV) and to
the connected sub-structures (SVC, IVC, PVs, LAA) the pipeline comprises two
main steps: (1) definition of boundary and regional constraints; and (2) regional
flattening (see Fig. 1).

2.1   Definition of Boundary and Regional Constraints

Holes in the main atrial cavities are firstly closed and triangulated as in [7].
Then, seed points are manually placed in specific atrial regions:

1. **Right atrium:** two seed points in the filled SVC and IVC holes, and one
   seed point in the most salient point of the RAA.
2. **Left atrium:** Five seeds at the centre of the filled PV and LAA holes, and 4
   on the MV contour: 2 delimiting the interatrial septal wall and 2 delimiting
   the left lateral wall. Inter-seed geodesic paths (i.e. shortest curve between two
   points on a mesh such that the curve lies on the surface [9], s_1–s_9 in Fig. 1)
   are computed dividing the LA surface into 5 regions.

After that, the final atrial cavities that will be flattened are obtained by
automatically removing the hole covers. The *boundary constrained points* (mesh
vertices of the red curves in Fig. 1) are identified as the SVC, ICV and TV
boundary points in the case of the RA, and as the PV, LAA and MV boundary
points in the case of the LA. Additionally and only for the LA, the *regional*

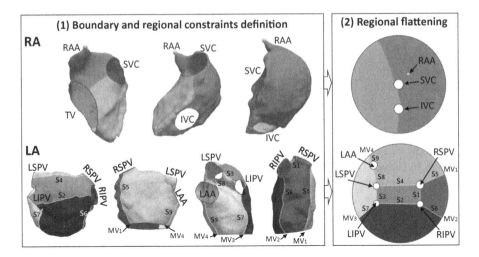

Fig. 1. Pipeline of the proposed RA (top) and LA (bottom) flattening method: (1) definition of boundary and regional constraints: nine boundaries (displayed in red) corresponding to the TV, IVC, SVC, MV, 4 PV, and LAA ostia, and nine segment paths (s_{1-9}, in yellow) are used. (2) Regional flattening. Data mapped into the surface meshes correspond to intensity values from the associated late-gadolinium magnetic resonance image. LAA/RAA = left/right atrial appendage; TV = tricuspid valve; SVC/IVC = superior/inferior vena cava; LSPV = left superior PV; LIPV = left inferior PV; RSPV = right superior PV; RIPV = right inferior PV; MV = mitral valve; $MV_{1,2}$ = seed points in the MV contour delimiting the septal wall, and $MV_{3,4}$ = seed points in the MV contour delimiting the left lateral wall. (Color figure online)

constrained points (points of the yellow curves in Fig. 1) are identified as the projection of the inter-seed paths (s_1–s_9) onto the final LA cavity. Note that for the RA we only consider one *regional constrained point*, the furthest point in the RAA (also shown in yellow in Fig. 1).

2.2   Regional Flattening

With the aim of obtaining a quasi-conformal (i.e. angle-preserving) and standardised flat representation of a 3D atrial surface mesh whose holes are constrained to predefined circumferences within a 2D disk, the corresponding boundary constraints can be added to the quasi-conformal scheme. Unfortunately, the flattened holes will appear often covered by adjacent triangles since the method does not avoid mesh self-folding. To overcome this issue, we include additional regional constraints in the parameterisation. Minor triangle overlapping can still occur near the holes and to further refine the boundary we recompute the point coordinates on the 2D RA or LA with a quasi-conformal parameterisation only constrained with the boundary points and not the regional constraints.

Mathematical Framework. Let M be the atrial cavity surface mesh with N vertices (points) and M_F its corresponding flattened mesh. Let $I_{\partial M}$ be the indices of the *boundary points* of M. Let Δ_M be the $N \times N$ (cotangent [12]) Laplacian of M and let Δ'_M be Δ_M where the off-diagonal elements of the rows defined by the positions of $I_{\partial M}$ set to 0 and corresponding elements on the main diagonal set to 1. Let (b_x, b_y) be the coordinates of the 2D *boundary points* of M_F in the same order as the corresponding indices appear in $I_{\partial M}$ and let b'_x and b'_y be N-dimensional vectors with values b_x and b_y, respectively, in the positions of $I_{\partial M}$ and zeros elsewhere.

Let P be the number of *regional constrained points* (i.e. number of points in the dividing segments (s_1–s_9), yellow lines in Fig. 1) and I_s the corresponding point (vertex) indices. Let (s_x, s_y) be the 2D coordinates of the P vertices in the same order as they appear in I_s and let E_s be a $P \times N$ zero matrix with 1 in each row in the positions corresponding to vertices of I_s (i-th row has 1 in the position given by the i-th element of I_s).

In order to find the coordinates (x^*, y^*) of the vertices of M_F, we propose to solve the following two quadratic programming problems:

$$x^* = \arg\min_x \left(w \left\| \Delta'_M x - b'_x \right\|^2 \right) \quad \text{s.t.} \quad E_s x = s_x \tag{1}$$

$$y^* = \arg\min_y \left(w \left\| \Delta'_M y - b'_y \right\|^2 \right) \quad \text{s.t.} \quad E_s y = s_y \tag{2}$$

with w (set to 1000 in our experiments) penalizing the unfulfillment of the *boundary constraints*. Using Lagrange multipliers these can be rewritten and solved as a system of linear equations:

$$\begin{bmatrix} w^2 \Delta'^T_M \Delta'_M & E^T_s \\ E_s & 0 \end{bmatrix} \begin{bmatrix} x^* \\ \lambda_x \end{bmatrix} = \begin{bmatrix} w^2 \Delta'^T_M b'_x \\ s_x \end{bmatrix} \tag{3}$$

$$\begin{bmatrix} w^2 \Delta'^T_M \Delta'_M & E^T_s \\ E_s & 0 \end{bmatrix} \begin{bmatrix} y^* \\ \lambda_y \end{bmatrix} = \begin{bmatrix} w^2 \Delta'^T_M b'_y \\ s_y \end{bmatrix} \tag{4}$$

To refine the boundary we proceed as follows. Let Δ_F be the $N \times N$ Laplacian of M_F and let Δ'_F be Δ_F where the off-diagonal elements of the rows defined by the positions of $I_{\partial M}$ set to 0 and corresponding elements on the main diagonal set to 1. The refined coordinates $\left(x^{*'}, y^{*'} \right)$ of the vertices of M_F in the final flat representation can be found by solving the following system of linear equations:

$$\Delta'_F x^{*'} = b'_x \tag{5}$$

$$\Delta'_F y^{*'} = b'_y \tag{6}$$

3 Experiments and Results

We used the proposed flattening method to visualise atrial wall thickness (AWT) of both left and right atria. Inspecting heterogeneity of AWT may be useful to

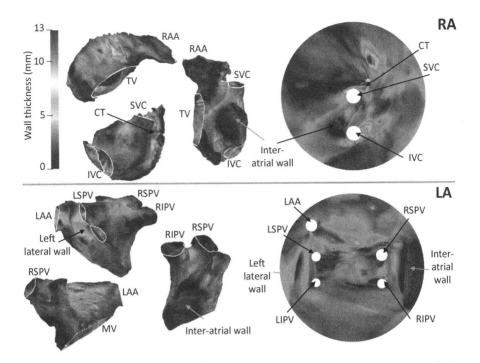

Fig. 2. Right (top) and left (bottom) atrial wall thickness measurements projected onto the endocardial surface meshes and corresponding 2D maps. LAA/RAA = left/right atrial appendage; TV = tricuspid valve; SVC/IVC = superior/inferior vena cava; CT = crista terminalis; LSPV = left superior PV; LIPV = left inferior PV; RSPV = right superior PV; RIPV = right inferior PV; MV = mitral valve.

improve the understanding of the mechanisms underlying atrial fibrillation and to improve radio-frequency ablation treatment planing. Detailed endocardial and epicardial surface meshes were first obtained by segmenting the atrial wall from the MRIs corresponding to an ex vivo human atria and then, AWT was measured as explained in [18]. Wall thickness measurements were projected onto the endocardial atrial meshes which were afterwards flattened with our method. As shown in Fig. 2, atrial unfolding favours fast inspection of AWT measurements: regarding the LA, it can be seen that the inter-atrial septal wall and the posterior wall were thinner than the left lateral and anterior walls; a thin inter-atrial wall was also observed in the RA, where one of the thickest parts was the crista terminalis (CT).

The proposed flattening method was also tested using several manually segmented LA from the 2018 Atrial Segmentation Challenge[1], and corresponding RA segmentations that were automatically obtained applying a multi-atlas whole-heart segmentation technique [19,20]. Triangular meshes were then built

---

[1] http://atriaseg2018.cardiacatlas.org/.

using the marching cubes algorithm. Voxel intensities from LGE-CMR images were mapped onto the obtained RA and LA surface mesh using the maximum intensity projection (MIP) technique: images were sampled along the normals on both sides of the surface mesh, assigning the maximum intensity value to the corresponding vertex on the LA mesh. The depth of the sampling was set to 3 mm. Then, the LA shapes were standardised by only keeping their main cavity after semi-automatically cutting the PVs, the LAA and the MV. This cutting process requires to manually place 5 seeds near the ending points of the PVs and the LAA. The reader is referred to [16] for more details on this method. Regarding the RA, cuts corresponding to the TV, SVC and IVC boundaries were manually applied. Several examples of 3D left and right atrial surfaces with projected LGE-CMR image intensity, and corresponding flattenings can be seen in Fig. 3.

Additional examples and experiments including distortion analysis, sensitivity to seed points selection, and a detailed comparison with the SUM [17] can be found in [10].

4 Discussion and Conclusions

We have presented a method to unfold both the left and right atrium and depict it in an intuitive, standardised, two-dimensional map. Contrary to the state of the art method for LA standardised representation, the SUM [17], our method does not depend on registration techniques being therefore faster and not affected by

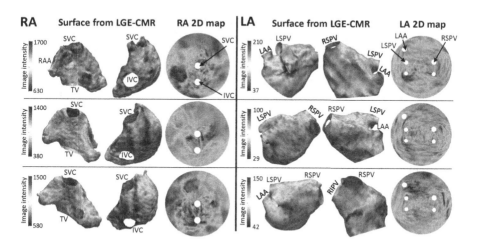

Fig. 3. Three examples of RA (left) and three examples of LA (right) surfaces with projected late gadolinium enhanced magnetic resonance imaging (LGE-CMR) intensities and corresponding 2D maps. LGE-CMR signal intensity is shown in arbitrary units, without any normalization [4]. LAA/RAA = left/right atrial appendage; TV = tricuspid valve; SVC/IVC = superior/inferior vena cava; LSPV = left superior PV; LIPV = left inferior PV; RSPV = right superior PV; RIPV = right inferior PV.

potential registration errors. Additionally, our method does not lose any information from the initial 3D surface, i.e. all points are depicted in the 2D map, while SUM needs to interpolate data when mapping the information from the arbitrary 3D surface to the template. Our algorithm is however highly influenced by the initial manual seed placement step, especially regarding the seeds placed on the MV contour. The seed points and inter-seed paths define the constraints used in the flattening, which is the key part of the method. Automatically computing the constraints, i.e. automatically dividing the LA is challenging due to, for example, bulges in the LA and obliqueness of the MV plane. Nonetheless, since the process of seed placement and unfolding is almost real time it can be repeated several times until a satisfactory result is obtained.

Potential clinical applications of our method include the analysis of the temporal evolution of some parameter (e.g. pre- and post-ablation fibrosis extent) from the same patient; the comparison and correlation of different features from the same patient (e.g. gadolinium enhancement from LGE-CMR data and endocardial voltage from electroanatomical maps); the comparison of the same feature in different patients (e.g. regional or global fibrosis extent [1]), etc. The code (Python) is publicly available at https://github.com/martanunez.

References

1. Benito, E.M., et al.: Preferential regional distribution of artrial fibrosis in posterior wall around left inferior pulmonary vein as identified by late gadolinium enhancement cardiac magnetic resonance in patients with artrial fibrillation. EP Europace (2018)
2. Cerqueira, M.D., et al.: Standardized myocardial segmentation and nomenclature for tomographic imaging of the heart: a statement for healthcare professionals from the cardiac imaging committee of the council on clinical cardiology of the American heart association. Circulation **105**(4), 539–542 (2002)
3. Karim, R., et al.: Surface flattening of the human left atrium and proof-of-concept clinical applications. Comput. Med. Imaging Graph. **38**(4), 251–266 (2014)
4. Khurram, I.M., et al.: Magnetic resonance image intensity ratio, a normalized measure to enable interpatient comparability of left atrial fibrosis. Heart Rhythm **11**(1), 85–92 (2014)
5. Kreiser, J., Meuschke, M., Mistelbauer, G., Preim, B., Ropinski, T.: A survey of flattening-based medical visualization techniques. Comput. Graph. Forum **37**(3), 597–624 (2018)
6. Lévy, B., Petitjean, S., Ray, N., Maillot, J.: Least squares conformal maps for automatic texture atlas generation. ACM Trans. Graph. (TOG) **21**(3), 362–371 (2002)
7. Liepa, P.: Filling holes in meshes. In: Proceedings of the 2003 Eurographics/ACM SIGGRAPH Symposium on Geometry Processing, pp. 200–205. Eurographics Association (2003)
8. Ma, Y.L., et al.: Cardiac unfold: a novel technique for image-guided cardiac catheterization procedures. In: Abolmaesumi, P., Joskowicz, L., Navab, N., Jannin, P. (eds.) IPCAI 2012. LNCS, vol. 7330, pp. 104–114. Springer, Heidelberg (2012). https://doi.org/10.1007/978-3-642-30618-1_11

9. Mitchell, J.S., Mount, D.M., Papadimitriou, C.H.: The discrete geodesic problem. SIAM J. Comput. **16**(4), 647–668 (1987)

10. Núñez García, M.: Left atrial parameterisation and multi-modal data analysis: application to atrial fibrillation. Ph.D. thesis, Universitat Pompeu Fabra (2018)

11. Paun, B., Bijnens, B., Iles, T., Iaizzo, P.A., Butakoff, C.: Patient independent representation of the detailed cardiac ventricular anatomy. Medical Image Anal. **35**, 270–287 (2017)

12. Pinkall, U., Polthier, K.: Computing discrete minimal surfaces and their conjugates. Exp. Math. **2**(1), 15–36 (1993)

13. Prasanna, L., Praveena, R., D'Souza, A.S., Kumar, M.: Variations in the pulmonary venous ostium in the left atrium and its clinical importance. J. Clin. Diagn. Res. **8**(2), 10 (2014)

14. Roney, C.H., et al.: Universal atrial coordinates applied to visualisation, registration and construction of patient specific meshes. arXiv preprint arXiv:1810.06630 (2018)

15. Soto-Iglesias, D., Butakoff, C., Andreu, D., Fernández-Armenta, J., Berruezo, A., Camara, O.: Integration of electro-anatomical and imaging data of the left ventricle: an evaluation framework. Med. Image Anal. **32**, 131–144 (2016)

16. Tobon-Gomez, C., et al.: Benchmark for algorithms segmenting the left atrium from 3D CT and MRI datasets. IEEE Trans. Medical Imaging **34**(7), 1460–1473 (2015)

17. Williams, S.E., et al.: Standardized unfold mapping: a technique to permit left atrial regional data display and analysis. J. Interv. Cardiac Electrophysiol. **50**(1), 125–131 (2017)

18. Zhao, J., et al.: Three-dimensional integrated functional, structural, and computational mapping to define the structural "fingerprints" of heart-specific atrial fibrillation drivers in human heart ex vivo. J. Am. Heart Assoc. **6**(8), e005922 (2017)

19. Zhuang, X., et al.: Multiatlas whole heart segmentation of CT data using conditional entropy for atlas ranking and selection. Med. Phys. **42**(7), 3822–3833 (2015)

20. Zhuang, X., Shen, J.: Multi-scale patch and multi-modality atlases for whole heart segmentation of MRI. Med. Image Anal. **31**, 77–87 (2016)

A Spatial Adaptation of the Time Delay Neural Network for Solving ECGI Inverse Problem

Amel Karoui[1,2,3](✉), Mostafa Bendahmane[1,2,3], and Nejib Zemzemi[1,2,3](✉)

[1] University of Bordeaux, IMB, Bordeaux, France
[2] National Institue of Mathematics and Informatics, Inria Bordeaux, Talence, France
{amel.karoui,nejib.zemzemi}@inria.fr
[3] IHU-Lyric, Bordeaux, France

Abstract. The ECGI inverse problem is still a common area of research. Since the results in the state of the art are not yet satisfactory, exploring new methods for the resolution of the inverse problem of electrocardiography is the main goal of this paper. To this purpose, we suggest to use temporal and spatial constraints to solve the inverse problem using neural networks methods. First, we use a time-delay neural network initialized with the spatial adjacency operator of the heart surface mesh. Then, we suggest a new approach to reconstruct the heart surface potential from the body surface potential using a spatial adaptation of time delay neural network. It consists on taking into account temporal and spatial dependence between potential measures. This allows to exploit the local and dynamic potential propagation properties. We test these approaches on simulated data. Results show that the new approach outperforms the classic time-delay neural network and has considerable improvements with respect to the state-of-the-art methods.

Keywords: Time-delay neural network · Spatial adaptation ·
Adjacency matrix · Inverse problem · Electrocardiography

1 Introduction

The non-invasive electrocardiographic imaging (ECGI) is the procedure carried out nowadays to reconstruct the heart surface potential (HSP) from the body surface potential (BSP) measurements. It provides cardiac information that allows the cardiologists to make better diagnosis of some heart diseases such as atrial and ventricular fibrillations. To date, the state of the art proposes mostly to model the relation between the heart electrical activity and the BSP measurements by a partial differential equation which is solved using a variety of methods based on a transfer matrix. This problem is called inverse problem and known to be ill-posed. In fact, a little perturbation in BSPs can strongly affects the solution. To overcome this problem, several techniques were used especially the Tikhonov regularization. More details about these methods can

© Springer Nature Switzerland AG 2019
Y. Coudière et al. (Eds.): FIMH 2019, LNCS 11504, pp. 94–102, 2019.
https://doi.org/10.1007/978-3-030-21949-9_11

be found in [3]. In recent years, some papers introduced a new vision of the inverse problem by using machine learning algorithms [1,2,5,8,9]. In this paper, we will introduce a new approach to solve the inverse problem using a neural network inspired from the work of Jiaqiu Wang et al. [7] for travel time prediction. Neural network training and assessment are performed using simulated data.

2 Building the Spatial Adaptation of Time-Delay Neural Network

In this section, we first describe the basic architecture of an artificial neural network (ANN) and a time-delay neural network (TDNN). Then we propose the new approach using the spatial adjacency operator.

2.1 The Basic Artificial Neural Network: ANN

A basic ANN has 3 layers: An input layer that reads the input data, a hidden layer, which consists of neurons that learn the relationship between the input and the target and an output layer which produces the output data. A basic neural network can be written as follows:

$$\widehat{O} = \begin{bmatrix} \widehat{O}_1 \\ \vdots \\ \widehat{O}_M \end{bmatrix} = \begin{bmatrix} W_{11} & \cdots & W_{1N} \\ \vdots & \ddots & \vdots \\ W_{M1} & \cdots & W_{MN} \end{bmatrix} \begin{bmatrix} I_1 \\ \vdots \\ I_N \end{bmatrix} + \begin{bmatrix} b_1 \\ \vdots \\ b_M \end{bmatrix} \tag{1}$$

where $\widehat{O}$ is the output of the ANN, I is the input data and W is the weight matrix estimated by the ANN. The vector b is the bias. Figure 1a shows a basic architecture of one hidden layer neural network.

The training process consists of tuning the connection weights W_{ij} between the inputs I_j and the targets O_i using an optimization algorithm to get the optimal prediction $\widehat{O}$.

2.2 Time-Delay Neural Network: TDNN

The main idea is that the body surface potential at a timestep t is highly dependent with its values at previous timesteps $t-1, t-2, \ldots$. Thus, TDNN is a good candidate to get use of this dependence. In fact, each neuron in the TDNN uses the current and its d previous values of the BSP input to estimate the HSP target map at the given timestep t as shown in Fig. 1b where $D^{(d)}$ represents the time delay operation. The value of d is the chosen time-delay window size. Hence, for some timestep t the estimated output $\widehat{HSP}(t)$ is given by:

$$\widehat{HSP}(t) = \sum_{i=0}^{d} W^{(i)} BSP(t-i) + b \tag{2}$$

Here, $W^{(i)}$ is the weight matrix associated with the input BSP at timestep $t-i$, $i = 0 \cdots d$.

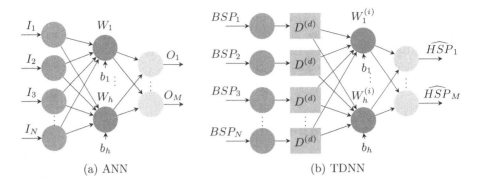

(a) ANN

(b) TDNN

Fig. 1. Architecture diagrams of a ANN and a TDNN models

2.3 Injecting the Spatial Adjacency Operator into the TDNN

Similarly to the temporal correlation, we suppose that the heart surface potential in a given point P is strongly dependent on its recorded values at the adjacent points. Hence, we use the spatial adjacency matrix as a representation of the relation between the target spatial location and its adjacent locations. In our case, this matrix is used in two different ways:

① SATDNN-LL (LL is for Linear Layer): We add a second linear layer to the TDNN model and we initialize its weight matrix by the spatial adjacency matrix.

② SATDNN-AT (AT is for Adjacency Transformation): This model is made with two hidden layers. The first layer is identical to the TDNN expressed by Eq. 2. Then, we perform an element-wise multiplication of the first layer output by the first order adjacency matrix $Adj^{(1)}$. This allows, for each point, to only keep the weights corresponding to its adjacent points and reduces the others to zero. The architecture of this model is provided in Fig. 2. One can see that non adjacent points are not considered (dashed arrows) in the second linear layer. The principle of this method is close to the space-time delay neural network developed for lipreading [4] and traffic forecasting [7]. But to the best of our knowledge, it has not been introduced before for the inverse problem in electrocardiography. The network of Fig. 2 can be written as:

$$\widehat{HSP}(t) = W_2 \left[Adj^{(1)} \left(\sum_{i=0}^{d} W_1^{(i)} BSP(t-i) + b \right) \right] + c \qquad (3)$$

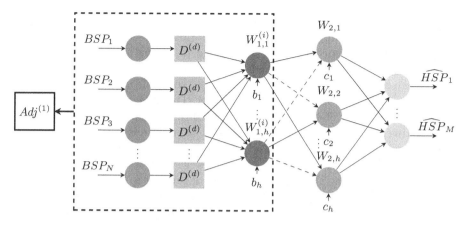

Fig. 2. Architecture diagram of the spatial adaptation of the time-delay neural network (SATDNN-AT).

where b and c are biases. A more detailed matrix form is expressed as:

$$\widehat{HSP}(t) = \begin{bmatrix} \widehat{HSP}_1(t) \\ \widehat{HSP}_2(t) \\ \vdots \\ \vdots \\ \widehat{HSP}_M(t) \end{bmatrix} = \begin{bmatrix} \left[Adj^{(1)} \left(\sum_{i=0}^{d} W_{1,1}^{(i)} BSP(t-i) + b_1 \right) \right] W_{2,1} + c_1 \\ \left[Adj^{(1)} \left(\sum_{i=0}^{d} W_{1,2}^{(i)} BSP(t-i) + b_2 \right) \right] W_{2,2} + c_2 \\ \vdots \\ \vdots \\ \left[Adj^{(1)} \left(\sum_{i=0}^{d} W_{1,h}^{(i)} BSP(t-i) + b_h \right) \right] W_{2,h} + c_h \end{bmatrix} \quad (4)$$

3   Data

3.1   Simulated Data

Simulated data is obtained by considering a realistic 3D heart-torso geometry segmented from CT-Scan images (see [3] for more details). The propagation of the electrical wave was computed using the monodomain reaction-diffusion model. The transmembrane currents used to compute the extracellular potential distribution throughout the torso were computed by solving a static bidomain problem in an homogeneous, isotropic torso model. Synchronized electrical potential on the epicardium and on the body surface were extracted on coarse meshes in order to test the inverse methods. The torso mesh contains 2873 nodes and the heart mesh 519 nodes.

3.2   Model Implementation

The different models are implemented using Python and the machine learning library PyTorch [6]. The spatial adjacency operator is implemented using VTK library. To train and test the neural network, the dataset is split into 3 subsets: a training dataset for training the model, a validation dataset to avoid overfitting during the training process and a testing one to evaluate the trained model performance. They correspond respectively to 75%, 20% and 5% of the whole dataset. In each epoch of the training process, the trained model is assessed using the validation dataset to avoid overfitting. The training phase stops when the maximum number of epochs is reached. Finally, the trained model is applied to the testing dataset and evaluated using error and correlation metrics.

The three models are implemented using the mean squared error as an optimization criterion and the stochastic gradient descent with momentum as an optimization algorithm. The hyperparameters of this latter are fixed empirically by executing the training process with different values and choosing the best one. Table 1 contains the chosen values of the learning rate and the momentum for training the different models. It is important to mention that we use the same partition of data randomly generated for the assessment and comparison of the three different models.

Table 1. Hyperparameters' values used to train the different models: TDNN, SATDNN-LL and SATDNN-AT

	TDNN	SATDNN-LL	SATDNN-AT
Learning rate	0.01	0.001	0.01
Momentum	0.8	0.8	0.8

4   Results

In this section, we present the results obtained after training and then testing the three models using the simulated dataset.

4.1   Heart Surface Potential Reconstruction

Results of the training and testing processes in terms of relative error (RE) and correlation coefficient (CC) are reported in Figs. 3 and 4. Figure 3 shows HSP maps reconstructed using the three models at the timestep $t = 418$ chosen randomly of the total simulation time. We notice that the SATDNN-AT gives the minimum $RE = 0.23$ and $CC = 0.97$ while the SATDNN-LL comes second with $RE = 0.25$ and $CC = 0.96$ compared to TDNN ($RE = 0.33$, $CC = 0.94$). This approves that our approach improves the performance of the classic TDNN. Figure 4a shows that the SATDNN-AT outperforms the two other models in the testing phase with a mean relative error equal to 20% compared to 24% for the SATDNN-LL which yields the second best result and 27% for the TDNN. Concerning the correlation coefficient, the Fig. 4b shows a little distinction between the three models.

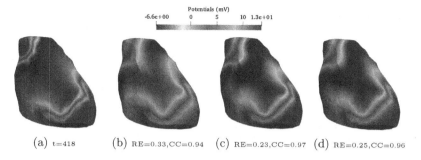

(a) t=418   (b) RE=0.33,CC=0.94   (c) RE=0.23,CC=0.97   (d) RE=0.25,CC=0.96

Fig. 3. Simulated (a) and estimated heart surface potential maps with: (b) TDNN, (c) SATDNN-AT, (d) SATDNN-LL at the timestep t = 418 of the simulation

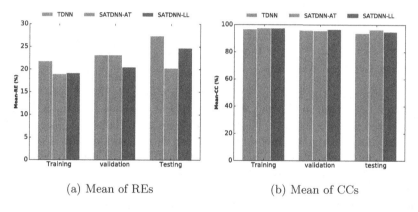

(a) Mean of REs   (b) Mean of CCs

Fig. 4. Means of RE and CC of the reconstructed BSPs during the training, validation and testing phases.

4.2 Robustness Analysis

The sensitivity issue is of crucial importance in this study due to the ill-posedness of the ECGI inverse problem caused by the measurement errors. To assess the robustness of the three models, we tested them using noisy dataset with different SNR values going from 2 dB to 50 dB. Figure 5a shows the means of the relative errors obtained by using the SNRs 2 dB, 5 dB, 10 dB, 20 dB and 50 dB. We observe that the behavior of the TDNN and SATDNN-AT with respect to the SNR is consistent, except for the lowest value 2 dB. However, the SATDNN-LL shows a high degradation for low SNRs. For example, it goes from 38% to 70% for 5 dB and 2 dB respectively.

In terms of correlation coefficient, we observe in Fig. 5b that all the models presented a similar response to the variation of SNR.

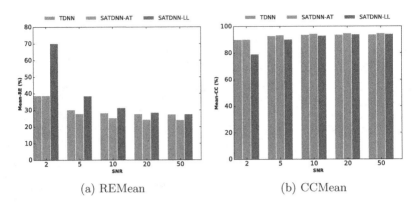

(a) REMean   (b) CCMean

Fig. 5. Means of RE and CC of the reconstructed HSPs with respect to the SNR between the original BSPs and the noisy ones.

4.3   Convergence Analysis

Figure 6 represents the training and validation loss in terms of mean squared error as a function of the training number of epochs. It shows that the SATDNN-LL initialized with the adjacency matrix converges after 200 iterations compared to 350 iterations for the SATDNN-LL initialized with a random matrix generated by a Gaussian distribution (SATDNN-LL-RI) and 400 or more for the TDNN. First, this indicates that adding a hidden linear layer improves the performance of the neural network in terms of convergence. It also proves that the SATDNN-LL initialized with the adjacency matrix reaches stable training and validation errors faster than the others. The observed spike in the SATDNN-LL curves is due to the use of stochastic gradient descent with momentum as an optimization algorithm.

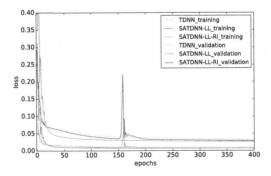

Fig. 6. Evolution of training and validation losses with respect to the number of epochs during the training process using TDNN, SATDNN-LL and SATDNN-LL-RI (random initialization)

5   Discussion and Conclusion

In this paper, we presented a proof of concept showing the ability of machine learning techniques to solve the ECGI inverse problem. We presented three different approaches. The first is based on a pure time delayed neural network. The second and the third methods are variants of a spatial adaptation taking into account the influence of adjacent points. Numerical results show that SATDNN-AT is more accurate in terms of RE and that the three methods are almost equivalent in terms of CC. We also tested the robustness of the methods by adding noise on the data. We have seen that the third model SATDNN-LL is more sensitive to the noise and seems to be unstable for low values of SNR. Compared to the state-of-the-art methods, neural network based methods show an improvement in terms of RE and CC evaluated in [3]. This confirms the usefulness of machine learning approach for solving the ECGI inverse problem for few reasons. In one hand, it provides a better generalization of the problem since it doesn't depend on a transfer matrix and its induced errors. In the other hand, machine learning methods are less time-consuming than traditional methods since they are trained once and then can be used multiple times. However, we have to say that introducing this method into real life applications would be challenging. First we will need a training dataset collected on patients containing the BSPs and their correspondent HSPs. We also need to standardize the geometries between patients and map the electrical information on a template geometry. This is one of the challenges of using machine learning in this ECGI application. Furthermore, in order to be efficient the database should be also sufficiently rich, which means that it should contain data for different heart conditions. The last limitation is related to the high computational cost of the training, because of the number of degrees of freedom in the heart geometry and the number of electrodes collecting the potential data on the body surface. The last one could be easily reduced by selecting a subset of the electrodes. However, reducing the number of the nodes describing the heart geometry will have an impact on the resolution of the reconstructed electrical information. This would be subject of future works.

Acknowledgements. This work was supported by the French National Research Agency, grant references ANR-10-IAHU04- LIRYC and ANR-11-EQPX-0030.

References

1. Alawad, M., Wang, L.: Learning domain shift in simulated and clinical data: localizing the origin of ventricular activation from 12-leadelectrocardiograms. IEEE Trans. Med. Imaging **38**, 1172–1184 (2018)
2. Giffard-Roisin, S., et al.: Transfer learning from simulations on a reference anatomy for ECGI in personalized cardiac resynchronization therapy. IEEE Trans. Biomed. Eng. **66**(2), 343–353 (2019)
3. Karoui, A., Bear, L., Migerditichan, P., Zemzemi, N.: Evaluation of fifteen algorithms for the resolution of the electrocardiography imaging inverseproblem using ex-vivo and in-silico data. Front. Physiol. **9**, 1708 (2018)

4. Lin, C.T., Nein, H.W., Lin, W.C.: A space-time delay neural network for motion recognition and its application to lipreading. Int. J. Neural Syst. **9**(04), 311–334 (1999)
5. Malik, A., Peng, T., Trew, M.L.: A machine learning approach to reconstruction of heart surface potentials from body surface potentials. In: 2018 40th Annual International Conference of the IEEE Engineering in Medicine and Biology Society (EMBC), pp. 4828–4831. IEEE (2018)
6. Paszke, A., et al.: Automatic differentiation in Pytorch. In: NIPS-W (2017)
7. Wang, J., Tsapakis, I., Zhong, C.: A space-time delay neural network model for travel time prediction. Eng. Appl. Artif. Intell. **52**, 145–160 (2016)
8. Zemzemi, N., Dubois, R., Coudiere, Y., Bernus, O., Haissaguerre, M.: A machine learning regularization of the inverse problem in electrocardiography imaging. In: Computing in Cardiology 2013, pp. 1135–1138. IEEE (2013)
9. Zemzemi, N., Labarthe, S., Dubois, R.D., Coudière, Y.: From body surface potential to activation maps on the atria: a machine learning technique. In: 2012 Computing in Cardiology, pp. 125–128. IEEE (2012)

On Sampling Spatially-Correlated Random Fields for Complex Geometries

Simone Pezzuto[1(✉)] ⓘ, Alessio Quaglino[1,2], and Mark Potse[3,4,5] ⓘ

[1] Center for Computational Medicine in Cardiology,
Institute of Computational Science, Università della Svizzera italiana,
Lugano, Switzerland
simone.pezzuto@usi.ch
[2] NNAISENSE SA, Lugano, Switzerland
alessio@nnaisense.com
[3] Univ. Bordeaux, IMB, UMR 5251, Talence, France
mark@potse.nl
[4] CARMEN Research Team, Inria Bordeaux Sud-Ouest, Talence, France
[5] IHU Liryc, fondation Bordeaux Université, Pessac, France

Abstract. Extracting spatial heterogeneities from patient-specific data is challenging. In most cases, it is unfeasible to achieve an arbitrary level of detail and accuracy. This lack of perfect knowledge can be treated as an uncertainty associated with the estimated parameters and thus be modeled as a spatially-correlated random field superimposed to them. In order to quantify the effect of this uncertainty on the simulation outputs, it is necessary to generate several realizations of these random fields. This task is far from trivial, particularly in the case of complex geometries. Here, we present two different approaches to achieve this. In the first method, we use a stochastic partial differential equation, yielding a method which is general and fast, but whose underlying correlation function is not readily available. In the second method, we propose a geodesic-based modification of correlation kernels used in the truncated Karhunen-Loève expansion with pivoted Cholesky factorization, which renders the method efficient even for complex geometries, provided that the correlation length is not too small. Both methods are tested on a few examples and cardiac applications.

Keywords: Random fields · Geodesic distance · Stochastic PDE · Fibrosis · Heterogeneity

1 Introduction

The ubiquitous presence of tissue heterogeneities affects the electrophysiological and mechanical function of the heart. An example of the severity of these effects is seen in the atria, whose tissue is a characterized by a complex structure of several fibre bundles [15]. At the micro-structural level, atria are often affected by fibrosis [1].

ⓒ Springer Nature Switzerland AG 2019
Y. Coudière et al. (Eds.): FIMH 2019, LNCS 11504, pp. 103–111, 2019.
https://doi.org/10.1007/978-3-030-21949-9_12

Despite their relevance in cardiac modeling, heterogeneities in the cardiac tissue are often neglected in patient-specific studies because they cannot be extracted from clinical imaging. An alternative is to include such heterogeneities as *random* variables, reflecting the inability to accurately describe the spatial distribution of tissue properties. It would be desirable to explicitly quantify this lack of knowledge in the output of cardiac simulations, providing confidence intervals of the quantities of interest [14].

Spatial heterogeneities in the parameters of the model typically exhibit a certain degree of spatial correlation. Generating correlated, stationary, and isotropic random fields for simple geometries, e.g. a box domain, is relatively straightforward. Efficient methods based on circulant embedding can be exploited to quickly sample random fields for a given correlation function [6]. Problems arise, however, in non-convex domains such as the heart muscle. If the shape of the domain were just ignored and the field were sampled in the bounding box of the original domain, anatomically close but functionally distant regions may be more correlated than expected.

A solution is offered by stochastic partial differential equations (SPDEs) [13]. Each sample is the solution of a PDE. The correlation function is encoded implicitly in the equation, as the solution of the second-moment equation.

Alternatively, random fields are commonly sampled using the truncated Karhunen-Loève (KL) expansion, a linear combination with random coefficients of eigenfunctions of the Hilbert-Schmidt operator associated with the correlation function [11]. Obviously, we rely on the assumption that the correlation kernel can be evaluated easily, e.g. when the domain is geometrically simple and the kernel is expressed in terms of the Euclidean distance. However, for more complicated domains this is not possible.

Here we analyze these two approaches to generate random, spatially-correlated heterogeneities on complex geometries. In Sect. 2, we recall the SPDE approach, including non-stationary and anisotropic random fields. In Sect. 3 we propose generalized correlation kernels, obtained by replacing the Euclidean distance with the *geodesic* one. We conclude with a comparison between the two methods (in Sect. 4) and an application to atrial fibrosis (Sect. 5).

2   Random Fields via SPDE

A simple, yet powerful method to sample random fields on arbitrarily complex geometries is based on SPDEs. We consider the following one:

$$\begin{cases} (\kappa^2 - \nabla \cdot \mathbf{D}\nabla)^{\frac{\alpha}{2}} u = \mathcal{W}, & x \in \Omega \subset \mathbb{R}^d, \\ \mathbf{D}\nabla(\kappa^2 - \nabla \cdot \mathbf{D}\nabla)^j \cdot \mathbf{n} = 0, & x \in \partial\Omega, \, j = 0, \dots, \lfloor \frac{\alpha-1}{2} \rfloor. \end{cases} \tag{1}$$

where $\kappa > 0$, $\alpha = \nu + d/2$ with $\nu > 0$, and $\mathbf{D}(x)$ is a uniformly elliptic tensor field. On the right hand side we have a Gaussian white noise $\mathcal{W}$ [13, Def. 6, sec. B.2]. The linear fractional SPDE (1) was analyzed in detail by Lindgren *et al.* [13]. In summary, the correlation function of u is related to the Matérn kernel.

The parameter ν, therefore, is a measure of the smoothness of the random field, while κ is inversely proportional to the correlation length.

The solution for general α is discussed by Bolin *et al.* [2]. A simpler setting occurs when $\alpha/2 = K \in \mathbb{N}$, which corresponds to the choice $\nu = 2K - d/2$. Denoting by u_K the solution of (1) with this assumption, the following iterative scheme applies:

$$u_0 = \mathcal{W}, \quad \begin{cases} (\kappa^2 - \nabla \cdot \mathbf{D}\nabla)u_k = u_{k-1}, & x \in \Omega,\ k = 1, \ldots, K, \\ \mathbf{D}\nabla u_k \cdot \mathbf{n} = 0, & x \in \partial\Omega,\ k = 1, \ldots, K. \end{cases} \tag{2}$$

The Galerkin discretization of (2) follows by selecting a finite-dimensional V_h subspace of $\mathrm{H}^1(\Omega)$, for instance a finite-element basis. Denoting by $\{\phi_i\}_{i=1}^N$ a basis for V_h, of dimension N, we define the matrices:

$$[\mathbf{K}]_{ij} = \kappa^2\langle\phi_j, \phi_i\rangle + \langle\nabla\phi_j, \nabla\phi_i\rangle, \quad [\mathbf{M}]_{ij} = \langle\phi_j, \phi_i\rangle.$$

Then the algorithm (2) reduces to:

$$\begin{cases} \mathbf{K}\mathbf{u}_1 = \mathbf{w}, \\ \mathbf{K}\mathbf{u}_k = \mathbf{M}\mathbf{u}_{k-1}, & k = 2, \ldots, K, \end{cases}$$

where $\mathbf{w}$ is the discrete Gaussian white noise, that is an N-dimensional Gaussian sample with zero mean and covariance $\mathbf{M}$. Efficient methods to sample the discretized white noise $\mathbf{w}$ have been proposed. Mass lumping diagonalizes the mass matrix using reduced quadrature, rendering the sampling procedure trivial. Mass lumping is easy for linear finite elements, but not for higher-order polynomials. Croci *et al.* computed the Cholesky decomposition of the mass matrix element-wise, making the evaluation of white noise very efficient and parallelizable [5]. Alternatively, one can select a Galerkin space V_h with an orthonormal basis, as for instance in spectral methods [3].

Figure 1 provides an example of a random field sampled with the SPDE approach for increasingly values of ν. The simulation also shows the effect of non-constant anisotropic $\mathbf{D}(x)$ for sampling non-stationary random fields.

3   Random Fields via KL with Geodesic Distance

An alternative method to sample random fields is based on the KL expansion, for which we have to choose an appropriate correlation function. Given a square-integrable correlation function $r(x, y)$, i.e. r is in $\mathrm{L}^2(\Omega \times \Omega)$, the random field $u(x, \omega)$ with mean $\bar{u}(x)$ and covariance $r(x, y)$ reads as follows:

$$u(x, \omega) = \bar{u}(x) + \sum_{i=1}^{\infty} \sqrt{\lambda_i}\psi_i(x)Z_i(\omega),$$

where Z_i are jointly Gaussian random variables with zero mean and unit variance and $\{\lambda_i\}$ and $\{\psi_i\}$ are respectively eigenvalues and eigenvectors of the

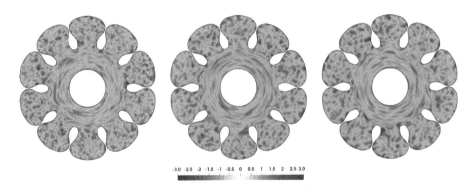

Fig. 1. Example of a random field sampled with the SPDE approach on acomplex domain. From left to right we increased the value of ν, using $K = 1$, $K = 5$ and $K = 20$. The corresponding values for ν, given that $d = 2$, are respectively 1, 9 and 39. In the central area, we imposed an anisotropy factor 100 in the angular direction. The anisotropy ratio is reduced to isotropy towards the "leaves." The correlation length is 5 % of the domain size.

Hilbert-Schmidt operator associated with $r(x, y)$. In the discrete setting, as above, the eigenvalue problem reduces to:

$$\mathbf{A}\mathbf{v} = \lambda \mathbf{M}\mathbf{v}, \qquad [\mathbf{A}]_{ij} = \int_\Omega \int_\Omega r(x, y)\phi_j(y)\phi_i(x) \, \mathrm{d}x\mathrm{d}y.$$

To compute the eigendecomposition efficiently we apply the low-rank pivoted Cholesky decomposition to replace the (large) matrix $\mathbf{A} \in \mathbb{R}^{N \times N}$ by the low-rank matrix $\mathbf{A}_m := \mathbf{L}_m \mathbf{L}_m^\mathrm{T}$, with $\mathbf{L}_m \in \mathbb{R}^{N \times m}$ and such that $\|\mathbf{A} - \mathbf{A}_m\| < \varepsilon$ [10]. In general, the performance of the method strongly depends on the decay rate of the spectrum of $\mathbf{A}$. Otherwise, methods based on $\mathcal{H}$-matrices for directly computing the square root of the covariance matrix show better performance [7].

A major limitation of this approach is that, for kernels depending on the distance function, that is $r(x, y) = h(\|x - y\|)$, the Euclidean distance does not always account for the geometry of the domain. It would be preferable to adopt the *geodesic distance*. In this way, geometrically-close but topologically distinct regions correctly show small correlation. There are two major problems to be addressed: (1) how to efficiently evaluate the distance function, and (2) how to ensure positive-definiteness of the corresponding Hilbert-Schmidt operator.

When relying on low-rank pivoted Cholesky decomposition, it is not necessary to assemble the full matrix $\mathbf{A}$: the algorithm only needs the diagonal entries and the function returning the i-th row of the matrix. Therefore, we approximate such functions with

$$[\mathbf{A}]_{i,j} \approx \int_\Omega \int_\Omega h(\delta(x_i, y))\phi_j(y)\phi_i(x) \, \mathrm{d}x\mathrm{d}y.$$

where $\delta(x_i, y)$ is the solution of the eikonal equation:

$$\begin{cases} \|\nabla_y \delta(x,y)\| = 1, & x \in \Omega \setminus \{x_i\}, \\ \delta(x_i, x_i) = 0. \end{cases} \tag{3}$$

The number of required eikonal evaluations matches the rank of the low-rank approximation of $\mathbf{A}$. In Fig. 2, we compare random fields generated with geodesic distance (A) and with Euclidean distance (B).

The second issue, regarding the positive-definiteness of the Hilbert-Schmidt operator, is more subtle. From our experience, the pivoted Cholesky procedure often ends prematurely because of a negative pivot. It also occurs on simple domains with Euclidean distance approximated by the eikonal solution, suggesting that the numerical error may play a role. Nonetheless, the resulting eigenfunctions are sufficiently accurate to sample random fields via KL expansion. There are examples of correlation functions, e.g. the square-exponential, which are not positive definite even with exact geodesic distance: for instance, the sphere with great circle distance [8].

The geodesic distance can also be generalized to include varying velocity and anisotropy, substituting the norm in (3) with $\sqrt{\mathbf{G}(x)\mathbf{p}\cdot\mathbf{p}}$, with $\mathbf{G}$ symmetric positive definite. An example is provided in Fig. 2C, where $\mathbf{G} = \sigma_f \mathbf{f}\otimes\mathbf{f} + \sigma_t(\mathbf{I} - \mathbf{f}\otimes\mathbf{f})$, $\sigma_f = 1 = 10\sigma_t$, $\mathbf{f} = \cos\alpha(z)\mathbf{e}_1 + \sin\alpha(z)\mathbf{e}_2$ and $\alpha(z) = \frac{\pi}{3}(2z-1)$.

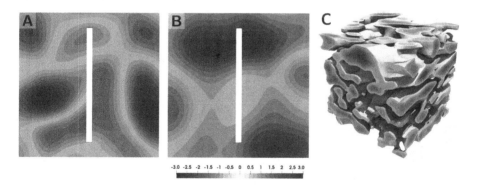

$$-3.0 \;\; -2.5 \;\; -2 \;\; -1.5 \;\; -1 \;\; -0.5 \;\; 0 \;\; 0.5 \;\; 1 \;\; 1.5 \;\; 2 \;\; 2.5 \;\; 3.0$$

Fig. 2. Random fields generated with geodesic distance. A: a domain with a hole in the central area. Geometrically close regions around the cut are uncorrelated, as expected. B: when adopting the Euclidean distance, a correlation around the cut is present. C: a more complex example with non-constant and anisotropic metric for the distance function. The correlation length is 20% of the domain size in all the examples.

4   Comparison of the Two Methods

In this section we compare the SPDE and the geodesic-based KL (geoKL) approaches in terms of quality of the samples and computational cost. In the test, we drew 10 000 samples from a random field defined on a hollow square domain, using both methods (see Fig. 3). The domain was discretized with a uniform mesh of size $h = 1/100$. A squared-exponential kernel $h(d) = e^{-d^2/\rho^2}$ with fixed correlation length of $\rho = 0.2$ was used in geoKL. Correspondingly, we set the number of iterations K for the SPDE method to $K = 5$ and $\kappa = 2\sqrt{\nu}/\rho$.

Estimated mean and variance for each method are reported in Fig. 3. The variance with the geoKL method was uniform across the domain, with some numerical artifacts (Fig. 3A). Such oscillations also present in the trace of $\mathbf{A}_m$ and are likely due to numerical error in the geodesic distance. For problem (3), the singularity at the origin $y = x_i$ is indeed responsible for severe degradation of the convergence rate. The variance with the SPDE method (Fig. 3B) showed instead a strong boundary effect, being significantly larger close to the boundary. This is a known effect of Neumann boundary conditions [12]. Finally, we computed the correlation with respect to the geodesic distance from point $(0.25, 0.25)$ towards point $(0.75, 0.25)$: see Fig. 3C, reporting excellent agreement.

Method geoKL took 26 s to approximate 87 eigenfunctions (tolerance was 10^{-8}). Sampling via KL expansion took a fraction of a second. At shorter correlation lengths, the geoKL method took significantly longer time: respectively 87 s with $\rho = 0.1$ and 270 s with $\rho = 0.05$. In contrast, the computational cost of drawing one sample with the SPDE method was constant, regardless of the value of ρ. In this case the total time was 50 min (3 s per sample). As usual, timings are purely indicative, as they are greatly affected by implementation. Moreover, the SPDE-based samples were not computed in parallel.[1]

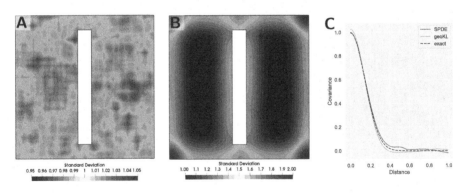

Fig. 3. Comparison between geoKL and SPDE methods for sampling of random fields. A: estimated standard deviation for geoKL method, B: estimated standard deviation for SPDE method, C: estimated correlation as function of distance along geodesic connecting points $(0.25, 0.25)$ and $(0.75, 0.25)$. A total of 10 000 samples were drawn.

---

[1] Python code is available at the address https://github.com/pezzus/fimh2019.

5   Application to Atrial Fibrosis

Initiation and perpetuation of atrial fibrillation (AF) is highly associated to heterogeneities in the substrate [9]. Hence, computational AF models for clinical applications should account for uncertainty in the anatomy and heterogeneity in the parameters. In this section we provide two examples of spatially-correlated heterogeneities for the atria.

Cardiac myocytes form a structure of branching and merging fibers. Structural damage and fibrosis have a tendency to follow the main fibre orientation. Most commonly, these fibrosis patterns have dimensions in the order of the size of a few myocytes, i.e. ten to a few hundred micrometer. Such small structures cannot be assessed by clinical imaging methods; at best, techniques such as late gadolinium-enhanced magnetic resonance imaging provide the *density* of fibrosis averaged over a volume of several cubic millimetres [16]. Several modelling studies of cardiac electrophysiology have therefore used rule-based methods to generate anisotropic correlated random patterns to implement fibrosis in models [4, 16].

Figure 4A shows a simulation of a random field with strong and nonstationary anisotropy in the correlation length. The interstitial space in the figure may represent fibrosis at microscale, for instance. We used the SPDE approach with $N = 10$, $\rho = 0.05$ (unit-length), and diffusion coefficient in the fibre direction 100 times larger than in the cross-fibre direction.

Figure 4B involves a more complex geometry. We simulated a fibrosis pattern on a patient-specific atrial anatomy using two random fields, one with short correlation length (4 mm), representing tissue-scale fibrosis, and another with longer correlation (2 cm), for organ-scale "patchiness." The two random fields were eventually averaged. The synthetic fibrosis pattern in Fig. 4 resembles patterns observed with late gadolinium enhanced MRI [1].

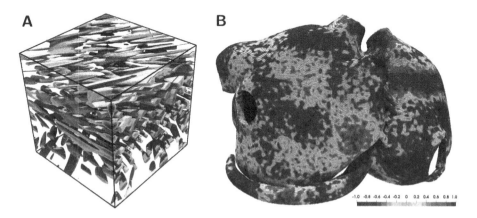

Fig. 4. Examples of randomly-generated fibrosis patterns on 3D tissues. A: the same configuration described in the last example of Fig. 2, but with shorter correlation and stronger anisotropy. B: a fibrosis pattern on a realistic atrial geometry.

6   Final Remarks

We presented two alternative approaches to simulate spatially-correlated random fields for complex geometries. The SPDE approach is fast and sufficiently general, is easy to implement within any finite-element code, and handles general geometries. However, it does not provide much flexibility in the choice of the correlation function. In contrast, sampling the random fields with KL expansion provides the freedom to select the kernel, including the geodesic distance, but it could be computationally demanding in some cases.

A potential application of such techniques is the automatic generation of realistic fibrosis patterns for the atria. This is important, for instance, for quantifying the uncertainty in cardiac electrophysiology, an emerging trend in the community and fundamental in our opinion in view of clinical applications.

Acknowledgement. The authors acknowledge financial support by the Theo Rossi di Montelera Foundation,the Metis Foundation Sergio Mantegazza, the Fidinam Foundation, and the Horten Foundation to the Center for Computational Medicine in Cardiology. This work was also supported by grants from the Swiss National Supercomputing Centre (CSCS) under project ID s778. Finally, we would like to thank Prof. Michael Multerer for the fruitful discussion.

References

1. Benito, E.M., et al.: Left atrial fibrosis quantification by late gadolinium-enhanced magnetic resonance: a new method to standardize the thresholds for reproducibility. Europace **19**(8), 1272–1279 (2017). https://doi.org/10.1093/europace/euw219
2. Bolin, D., Kirchner, K., Kovács, M.: Numerical solution of fractional elliptic stochastic PDEs with spatial white noise. IMA J. Numer. Anal. 1–23 (2018). https://doi.org/10.1093/imanum/dry091
3. Canuto, C., Hussaini, M.Y., Quarteroni, A., Zang, T.A.: Spectral Methods. Springer, Heidelberg (2006). https://doi.org/10.1007/978-3-540-30726-6
4. Clayton, R.H.: Dispersion of recovery and vulnerability to re-entry in a model of human atrial tissue with simulated diffuse and focal patterns of fibrosis. Front. Physiol. **9**(AUG), 1–16 (2018). https://doi.org/10.3389/fphys.2018.01052
5. Croci, M., Giles, M.B., Rognes, M.E., Farrell, P.E.: Efficient white noise sampling and coupling for multilevel monte carlo with nonnested meshes. SIAM/ASA J. Uncertainty Quant. **6**(4), 1630–1655 (2018). https://doi.org/10.1137/18M1175239
6. Dietrich, C.R., Newsam, G.N.: Fast and exact simulation of stationary gaussian processes through circulant embedding of the covariance matrix. SIAM J. Sci. Comput. **18**(4), 1088–1107 (1997). https://doi.org/10.1137/S1064827592240555
7. Feischl, M., Kuo, F.Y., Sloan, I.H.: Fast random field generation with H-matrices. Numer. Math. **140**(3), 639–676 (2018). https://doi.org/10.1007/s00211-018-0974-2
8. Gneiting, T.: Strictly and non-strictly positive definite functions on spheres. Bernoulli **19**(4), 1327–1349 (2013). https://doi.org/10.3150/12-BEJSP06
9. Haissaguerre, M., et al.: Intermittent drivers anchoring to structural heterogeneities as a major pathophysiological mechanism of human persistent atrial fibrillation. J. Physiol. **594**(9), 2387–2398 (2016). https://doi.org/10.1113/JP270617

10. Harbrecht, H., Peters, M., Schneider, R.: On the low-rank approximation by the pivoted Cholesky decomposition. Appl. Numer. Math. **62**(4), 428–440 (2012). https://doi.org/10.1016/j.apnum.2011.10.001
11. Harbrecht, H., Peters, M., Siebenmorgen, M.: Efficient approximation of random fields for numerical applications. Numer. Linear Algebra Appl. **22**(4), 596–617 (2015). https://doi.org/10.1002/nla.1976
12. Khristenko, U., Scarabosio, L., Swierczynski, P., Ullmann, E., Wohlmuth, B.: Analysis of boundary effects on PDE-based sampling of Whittle-Matérn random fields. arXiv e-prints arXiv:1809.07570 (2018)
13. Lindgren, F., Rue, H., Lindström, J.: An explicit link between Gaussian fields and Gaussian Markov random fields: the stochastic partial differential equation approach. J. Roy. Stat. Soc. B (Stat. Methodol.) **73**(4), 423–498 (2011). https://doi.org/10.1111/j.1467-9868.2011.00777.x
14. Pathmanathan, P., Gray, R.A.: Validation and trustworthiness of multiscale models of cardiac electrophysiology. Front. Physiol. **9**, 1–19 (2018). https://doi.org/10.3389/fphys.2018.00106
15. Schotten, U., Verheule, S., Kirchhof, P., Goette, A.: Pathophysiological mechanisms of atrial fibrillation: a translational appraisal. Physiol. Rev. **91**(1), 265–325 (2011). https://doi.org/10.1152/physrev.00031.2009
16. Vigmond, E., Pashaei, A., Amraoui, S., Cochet, H., Haïssaguerre, M.: Percolation as a mechanism to explain atrial fractionated electrograms and reentry in a fibrosis model based on imaging data. Heart Rhythm **13**, 1536–1543 (2016)

Interpolating Low Amplitude ECG Signals Combined with Filtering According to International Standards Improves Inverse Reconstruction of Cardiac Electrical Activity

Ali Rababah[1]([⊠]), Dewar Finlay[1], Laura Bear[2,3,4], Raymond Bond[1], Khaled Rjoob[1], and James Mclaughlin[1]

[1] Faculty of Computing, Engineering and the Built Environment,
Ulster University, Shore Road, Newtownabbey, UK
Rababah-A@ulster.ac.uk
[2] IHU Liryc, Electrophysiology and Heart Modeling Institute,
Fondation Bordeaux Université, 33600 Pessac, Bordeaux, France
[3] Univ. Bordeaux, Centre de recherche Cardio-Thoracique de Bordeaux, U1045,
33000 Bordeaux, France
[4] INSERM, Centre de recherche Cardio-Thoracique de Bordeaux, U1045,
33000 Bordeaux, France

Abstract. In this paper, the effect of reducing noise from ECG signals is investigated by applying filters in compliance with the standards for ECG devices, removing and interpolating low amplitude signals. Torso-tank experiment data was used with electrical activity recorded simultaneously from 128 tank electrodes and 108 epicardial sock electrodes. Subsequently, 10 representative beats were selected for analysis. Tikhonov zero-order regularization method was used to solve the inverse problem for the following groups; raw fullset, filtered fullset, raw low amplitude removed, filtered low amplitude removed, raw low amplitude interpolated, filtered low amplitude interpolated torso signals. Pearson's correlation was used for comparison between measured and computed electrograms and between activation maps derived from them. Filtering the signal according to the standards improved the reconstructed electrograms. In addition, removal of low amplitude signals and replacing them with interpolated signals combined with filtering according to the standard significantly improved the reconstructed electrograms and derived activation maps.

1 Introduction

Electrocardiographic imaging (ECGI) is a reconstruction of cardiac electrical activity from high number of body surface electrodes and a patient specific heart-torso geometry. It is noticeable that ECGI has attracted the attention of both industry and academic researchers in the recent decade. It is a promising technique for guiding ablation therapy for atrial and ventricular arrhythmias [1]. The error in non-invasive electrocardiographic imaging (ECGI) reconstruction can be partially attributed to two

© Springer Nature Switzerland AG 2019
Y. Coudière et al. (Eds.): FIMH 2019, LNCS 11504, pp. 112–120, 2019.
https://doi.org/10.1007/978-3-030-21949-9_13

main sources of signal noise. Firstly, low amplitude ECG signal recorded from the tank electrodes located away from the heart. Secondly, noises that contaminate ECG signal such as baseline drift, muscle artifact, and power line interference. In this paper, we will investigate the effect of removal or interpolating these low amplitude ECG signals on the inverse reconstruction of cardiac electrical activity which has not been investigated in the literature yet. In addition, we investigate the impact of applying a designed digital filter that complies with International and American standards for diagnostic ECG devices on the inverse reconstruction of epicardial potentials. We evaluate both filtering and interpolation based on a data recorded from both heart and tank surfaces in torso-tank experiment that contain an explanted pig's heart [2].

2   Methods

2.1   Torso-Tank Experimental Setup

All experimental data was obtained in accordance with the guidelines from Directive 2010/63/EU of the European Parliament on the protection of animals used for scientific purposes and approved by the local ethical committee and has previously been described in [2]. An explanted pig's heart was suspended in an instrumented, human shaped, electrolytic torso tank. Electrical potentials were recorded simultaneously from 128 tank electrodes and 108 sock electrodes at 2 kHz (BioSemi, the Netherlands) and referenced to a Wilson's central terminal. Following that, 3D rotational fluoroscopy (Artis, Siemens) was used to acquire heart and tank geometries and locations of electrodes.

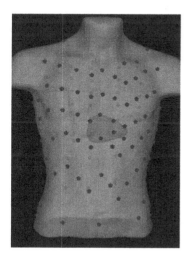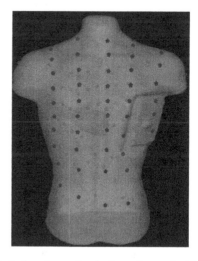

Fig. 1. The geometry of the heart and torso tank with the locations of tank electrodes marked in blue (a) anterior view. (b) posterior view [3]. (Color figure online)

2.2 Removal of Low Amplitude ECG Signals

After manual investigating of ECGs recorded from all tank electrodes, it appeared that the signals recorded from the inferior anterior and inferior posterior region (Fig. 1) of the tank have low amplitude (<0.2 mV peak to peak). To investigate the impact of these low amplitude signals (21 tank signals) on ECGI reconstruction, the inverse problem was solved before and after discarding these signals and solutions compared.

2.3 Interpolating Low Amplitude ECG Signals

Laplacian interpolation [4, 5] has been reported in the literature for interpolating missing information in body surface potential maps. In this paper, 21 low amplitude ECG signals (<0.2 mV peak to peak) were removed, and the remaining 107 tank signals were used to interpolate signals at these locations. Epicardial potentials were calculated by solving the inverse problem before and after the interpolation and solution compared.

2.4 ECG Signal Filtering

High-Pass Filter for Baseline Wander Removal. The change of electrical impedance at the skin-electrode interface which is resulted from subject movement, respiration, and perspiration appears in the recorded ECG signal as a low frequency noise called baseline wander (Fig. 2a and b). This noise may hinder important ECG features such as ST-segment which is an important indicator for the diagnosis of myocardial infarction and ischemia. Hence, the suppression of baseline wander while preserving the ST-segment and other ECG components is the desired goal.

Current international IEC 60601-2-51, and American Standard AAMI EC11 for diagnostic electrocardiographic devices adopt the following recommendation [6]:

– cut-off frequency (−3 dB) of the high pass filter shall not exceed 0.5 Hz.
– applying a 0.3 mV-s test signal to the high pass filter shall not result in a displacement in the baseline of more than 0.1 mV.
– passband ripple within 0.67–40 Hz range shall not exceed 0.9 dB [6].

There are plenty of available digital filters to pick from. Infinite Impulse Response (IIR) Butterworth Filter is the right selection when it comes to the flatness in the passband. A high pass Butterworth filter with a cut-off frequency (−3 dB) of 0.5 Hz and order 3 was designed and implemented in the forward and backward direction to ensure the compliance with the previously mentioned standards.

Power Line Frequency Removal. The contamination of ECG signal by 50/60 Hz power line interference is a major problem that prevents accurate interpretation and may lead to a misdiagnosis (Fig. 2d). Digital filters are used to eliminate this noise or reduce its impact. However, inaccurate implementation of these filters can cause an unacceptable ringing effect in the ECG signal after the QRS complex. This is caused when using a filter with high order. AAMI standards does not have an item for testing the ringing effect of a notch filter [7]. However, the British and European standard BS EN 60601-2-51:2003 states that the ringing peak to peak noise should not exceed

50 μV when a test ECG signal ANE20000 is used to test the filter. Since this signal is a closed source and it is difficult to access, we have used a simulated triangular signal of 3 mV amplitude and 100 ms duration to test the ringing effect of our notch filter. The resulting peak to peak ringing noise is 24 μV which is below 50 μV requested by the European and British standard [7].

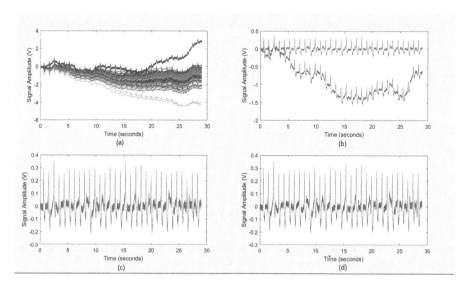

Fig. 2. (a) Raw signal recorded from 128 torso electrodes. (b) ECG signal at electrode number 1 before (blue) and after (red) baseline wander (0.5 Hz, order 3, 0-phase). (c) ECG signal at electrode number 1 before (blue) and after (red) removing High frequency noise (150 Hz, order 3, 0-phase). (d) ECG signal at electrode number 1 before (blue) and after (red) removing 50 Hz line interference. (Color figure online)

Low-Pass Filter for the Suppression of High Frequency Interference. High frequency noises such as muscle artifacts corrupt ECG signals and may hide some important features needed for ECG interpretations (Fig. 2c). The recommendation of American Heart Association (AHA) for diagnostic electrocardiographic devices is to use low-pass filters with cut-off frequency at 150 Hz to prevent the distortion of the QRS complex and maintain the small high frequency notches needed for diagnosis [8]. We have used a Butterworth low-pass filter with cut-off frequency at 150 Hz and order 3 to eliminate the high frequency noise from the signal.

2.5   Inverse Reconstruction of Epicardial Potentials

To solve the inverse problem, a forward transformation matrix that relates the epicardial potentials (894 nodes) to torso tank surface potentials (107 or 128 nodes) was calculated using the boundary element method (BEM) assuming the conductivity of the volume between heart surface and tank surface is homogenous [9]. We solved the inverse problem using Tikhonov zero-order regularization method from a toolkit for forward/inverse problems in electrocardiography within the SCIRun environment [3].

Ten beats were selected for the analysis (Fig. 3). These signals were processed in different ways; filtering, removal of low amplitude signals, and interpolating low amplitude signals. The resulting signals; raw fullset signals, filtered fullset signals, raw (low amplitude removed) signals, filtered (low amplitude removed) signals, raw (low amplitude interpolated) signals, filtered (low amplitude interpolated) signals, were used for solving the inverse problem leading to 6 different solutions for each beat (Fig. 4). Comparison was performed between the computed electrograms and the measured electrograms (Ground Truth) using Pearson's correlation coefficient. Activation times were defined by fitting a global activation field to activation delays between electrograms [10]. Activation was also compared using Pearson's correlation. Student's t-test was used for statistical analysis with statistical significance defined for $p < 0.05$.

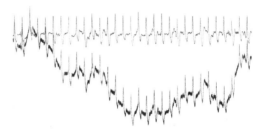

Fig. 3. Raw vs filtered ECG signal recorded at one of the electrodes. Ten representative beats marked in red were selected for analysis. (Color figure online)

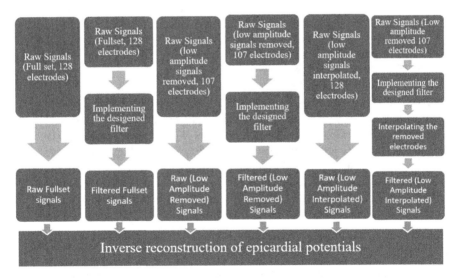

Fig. 4. Flow diagram which shows the protocol followed to prepare tank signals for the inverse reconstruction.

3   Results

The median correlation between computed and measured EGMs of the 10 beats are as shown in Fig. 5. For the three categories of signals (fullset, low amplitude removed, low amplitude interpolated), filtering the signal according to the standards significantly improved the inverse reconstruction of epicardial potentials (p < 0.01). Comparing between filtered signals showed that epicardial reconstruction better for filtered LA-interpolated signals than for filtered fullset signals which was better than that for filtered LA-removed signals (p < 0.05).

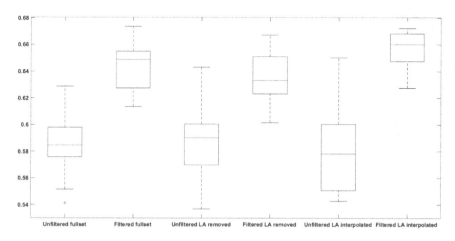

Fig. 5. Comparison of median correlation between different groups. Filtered fullset > unfiltered fullset (p < 0.01), Filtered LA removed > unfiltered LA removed (p < 0.01), Filtered LA interpolated > unfiltered LA interpolated (p < 0.01), Filtered LA interpolated > Filtered fullset > Filtered LA removed (p < 0.05).

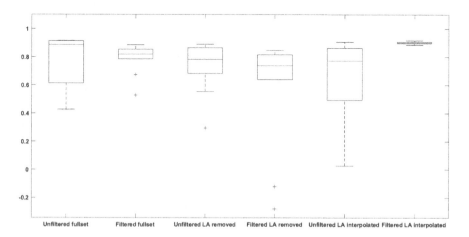

Fig. 6. Correlation between computed activation times and measured activation times for different groups. Filtered LA interpolated > all other groups (p < 0.05). Difference between any other groups is insignificant (p > 0.1)

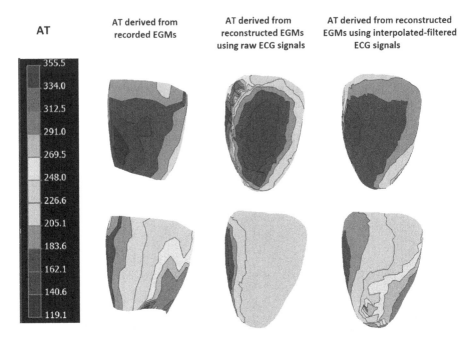

Fig. 7. Activation maps derived from recorded (left) and reconstructed electrograms using raw fullset (middle) and filtered LA-interpolated (right) torso signals for the fifth selected beat [12].

Figure 6 shows that when comparing between computed and measured activation times, the filtered LA-interpolated signals resulted in the best solution as compared to any other filtered or unfiltered signals ($p < 0.05$). There is no significant different between any other groups shown in Fig. 6 ($p > 0.1$). Figure 7 shows the high correlation between Activation maps derived from recorded and reconstructed electrograms using filtered LA-interpolated torso signals for the seventh selected beat.

4   Discussion

Filtering tank ECG signals is an important processing method for a better inverse reconstruction of epicardial potentials. However, careful filtering (zero-phase or linear phase response, 0.5–150 Hz notched at 50 Hz, lower than 50 uV peak to peak ringing artifact) that is according with international and American standard for ECG diagnostic devices [6] should be implemented to guarantee this improvement in the reconstruction. Low amplitude ECG signals (<0.2 mV) is another source of error that impacts the inverse solution. Removal of these signals and replacing them with interpolated signals significantly improved the reconstructed EGMs and activation times as shown in Figs. 5, 6 and 7.

Bear et al. have shown in their work about the impact of filtering on inverse reconstruction of epicardial potentials that the removal of high frequency noise by applying signal averaging or low pass filtering improved the computation of activation times [9]. This suggests investigating the effect of signal averaging and low pass filtering at 40 Hz combined with interpolating low amplitude signals on the reconstructed activation maps. In addition, other methods of interpolation such as inverse forward interpolation method [12] should be investigated in the future work to compare that with Laplacian method used in this study.

5   Conclusions

Filtering ECG signals according to international and American standards improves electrogram reconstruction. In addition, removal of low amplitude signals and replacing them with interpolated signals combined with filtering according to the standard significantly improves reconstructed electrograms and derived activation maps. This suggests the need for adopting these two pre-processing steps for consistently and accurately reconstructing of epicardial potentials and derived activation maps.

Acknowledgments.   This project is supported by the European Union's INTERREG VA Programme, managed by the Special EU Programmes Body (SEUPB). This work is also supported by the French National Research Agency (ANR-10-IAHU04-LIRYC).

References

1. Cluitmans, M., et al.: Validation and opportunities of electrocardiographic imaging: from technical achievements to clinical applications. Front. Physiol. **9**, 1305 (2018). https://doi.org/10.3389/fphys.2018.01305
2. Bear, L.R., Huntjens, P.R., Walton, R.D., Bernus, O., Coronel, R., Dubois, R.: Cardiac electrical dyssynchrony is accurately detected by noninvasive electrocardiographic imaging. Heart Rhythm **15**, 1058–1069 (2018). https://doi.org/10.1016/j.hrthm.2018.02.024
3. Burton, B.M., et al.: A toolkit for forward/inverse problems in electrocardiography within the SCIRun problem solving environment. In: Proceedings of 2011 Annual International Conference of the IEEE Engineering in Medicine and Biology Society EMBS, pp. 267–270 (2011). https://doi.org/10.1109/iembs.2011.6090052
4. Oostendorp, T.F., van Oosterom, A., Huiskamp, G.: Interpolation on a triangulated 3D surface. J. Comput. Phys. **80**, 331–343 (1989). https://doi.org/10.1016/0021-9991(89)90103-4
5. Rababah, A.S., Finlay, D.D., Guldenring, D., Bond, R., Mclaughlin, J.: An Adaptive Laplacian Based Interpolation Algorithm for Noise Reduction in Body Surface Potential Maps
6. Briller, S., Mortara, D.: Diagnostic electrocardiographic devices (2000)
7. Luo, S., Johnston, P.: A review of electrocardiogram filtering. J. Electrocardiol. **43**, 486–496 (2010). https://doi.org/10.1016/j.jelectrocard.2010.07.007
8. Kligfield, P., Gettes, L.: Recommendations for the standardization and interpretation of the electrocardiogram. J. Am. Coll. Cardiol. **49** (2007). https://doi.org/10.1016/j.jacc.2007.01.024
9. Bear, L.R., et al.: Effects of ECG signal processing on the inverse problem of electrocardiography. In: Computing in Cardiology Conference, vol. 45 (2018)

10. Duchateau, J., Potse, M., Dubois, R.: Spatially coherent activation maps for electrocardiographic imaging. IEEE Trans. Biomed. Eng. **64**, 1149–1156 (2017). https://doi.org/10.1109/TBME.2016.2593003
11. MacLeod, R.S., Johnson, C.R.: Map3d: interactive scientific visualization for bioengineering data
12. Burnes, J.E., Kaelber, D.C., Taccardi, B., Lux, R.L., Ershler, P.R., Rudy, Y.: A field-compatible method for interpolating biopotentials. Ann. Biomed. Eng. **26**, 37–47 (1998). https://doi.org/10.1114/1.49

Model Assessment Through Data Assimilation of Realistic Data in Cardiac Electrophysiology

Antoine Gérard[1,3,4]($\boxtimes$), Annabelle Collin[2,3,4], Gautier Bureau[5,6], Philippe Moireau[5,6], and Yves Coudière[1,2,4]

[1] IHU Liryc, Electrophysiology and Heart Modeling Institute, F-33000, Pessac-Bordeaux, France
[2] Université Bordeaux, IMB UMR 5251, Talence, France
[3] Bordeaux INP, IMB UMR 5251, Talence, France
[4] Inria, Inria Bordeaux - Sud-Ouest, Talence, France
antoine.gerard@inria.fr
[5] Inria, Université Paris-Saclay, Palaiseau, France
[6] LMS, Ecole Polytechnique, CNRS UMR 7649, Institut Polytechnique de Paris, Palaiseau, France

Abstract. We consider a model-based estimation procedure – namely a data assimilation algorithm – of the atrial depolarization state of a subject using data corresponding to electro-anatomical maps. Our objective is to evaluate the sensitivity of such a model-based reconstruction with respect to model choices. The followed data assimilation approach is capable of using electrical activation times to adapt a monodomain model simulation, thanks to an ingenious model-data fitting term inspired from image processing. The resulting simulation smoothes and completes the activation maps when they are spatially incomplete. Moreover, conductivity parameters can also be inferred. The model sensitivity assessment is performed based on synthetic data generated with a validated realistic atria model and then inverted using simpler modeling ingredients. In particular, the impact of the muscle fibers definition and corresponding anisotropic conductivity parameters is studied. Finally, an application of the method to real data is presented, showing promising results.

Keywords: Cardiac modeling · Electrophysiology · Data assimilation

1 Introduction

Data assimilation consists in coupling a dynamical model with available measurements in order to register the model on the data and identify model parameters of interest for a specific diagnosis. This approach is now recognized as a potential key ingredient to personalize computational models on clinical measurements allowing to produce predictive patient specific simulations, in particular in electrophysiology [2,3,6,11,13]. The purpose of this work is to evaluate

© Springer Nature Switzerland AG 2019
Y. Coudière et al. (Eds.): FIMH 2019, LNCS 11504, pp. 121–130, 2019.
https://doi.org/10.1007/978-3-030-21949-9_14

the robustness of assimilating electrical catheter data based on such a model-based estimation procedure when relying on different modeling assumptions. To this end, we need a state-of-the-art data assimilation procedure, some controlled or real measurements, and the model options considered. For the data assimilation strategy, we choose the sequential approach introduced in [2]. For the available measurements, we generate realistic synthetic datasets, produced by a controlled refined model. For the modeling options, we propose to use two variants of a simpler model: one which does not take into account the fiber direction and one which does. The result of the data assimilation will be in the first case the estimation of an isotropic conductivity tensor without building the fiber distribution of the left atrium. In the second case, two conductivity parameters are identified and the conductivity tensor is reconstructed using the fiber distribution. Eventually, these estimated parameters allow to compute activation times which are compared to the available measurements. Finally, we propose a first attempt of the data assimilation procedure on a real patient's dataset from RHYTHMIA HDx^{TM} recording. Here, five anatomical regions are defined on the electro-anatomical recording mesh and an isotropic conductivity tensor is estimated in each region. This example allows to understand the remaining limitations of the procedure in a clinical context.

2   Problem Setting and Estimation Methodology

2.1   Models Formulation

Model Used to Generate Realistic Synthetic Data. Realistic synthetic data are computed as solutions to the bilayer atrial model defined in [7]: transmembrane voltages are defined on two layers of the atrial surface coupled in a resistive manner, but with independent ionic currents and fiber directions. They may be seen as the endo- and epicardial layers, and the two transmembrane voltages $u^{(k)}$ ($k = 1, 2$) solve the following monodomain equations with the coupling coefficient $\gamma > 0$:

$$\partial_t u^{(k)} + \frac{I_{\text{ion},k}}{C_m}(u^{(k)}, \cdot) = \text{div}\left(\frac{\sigma^{(k)}}{\xi_m C_m}\nabla u^{(k)}\right) + (-1)^k \gamma(u^{(1)} - u^{(2)}), \quad (1)$$

for all $t > 0, x \in \Omega$ where Ω is the atrial surface. The ratio ξ_m of surface of membrane per unit volume of tissue and the membrane capacitance C_m are fixed. The conductivity tensors $\sigma^{(k)}(x)$ can be different on each layer. To obtain realistic simulations, we consider – like in [7] – two different conductivity coefficients defined from endo- and epicardial fiber directions as $\sigma^{(k)}(x) = \sum_{i=1}^{2} d_i \nu_i^{(k)}(x)\nu_i^{(k)}(x)^T$, where $(\nu_1^{(k)}, \nu_2^{(k)})$ define the local directions parallel and perpendicular to the fiber on the surface Ω. The values of the conductivity parameters d_i are piecewise constant on a few subdomains of the atria allowing to considerer specific atrial structures as sinus node, Bachmann bundle and pectinate muscles. Concerning the ionic model given by the functions $I_{\text{ion},k}$, the CRN [4] ionic model (16 gating variables and 5 ion concentrations) is considered on each of the two layers.

Model Used for the Estimation. To study the robustness of the data assimilation strategy faced with modeling errors, the model used for the estimation is simplified in several ways. Indeed, instead of having a two layers model with local electrophysiological parameters, as the one used to create data, we will consider only one layer

$$\partial_t u + \frac{I_{\text{ion}}}{C_m}(u, \cdot) = \text{div}\,(d\nabla u), \quad t > 0, x \in \Omega. \tag{2}$$

The equation is a simplified version of the bilayer model (1) in which only the (endocardial) transmembrane voltage u is considered. It is coupled with the *much simpler Mitchell-Schaeffer ionic model* (MS) [8] (compared to the CRN model). Furthermore, the conductivity tensor d is assumed to be homogeneous (*i.e.* specific atrial structures are not considered). Moreover, in order to analyze the importance of knowing the fiber distribution, we will compare results of two data assimilation models: one in which fiber distribution is unknown (isotropic), *i.e.* $d := d_0 \,\text{Id}$ and the other where it is known (anisotropic), *i.e.* $d := \sum_{i=1}^{2} d_i \nu_i^{(1)}(x)\nu_i^{(1)}(x)^T$ (defined by the endocardial fiber direction from the realistic model described above).

2.2   Presentation of the Estimation Method

Our objective is to complete and regularize the activation maps recorded by clinical catheter systems, and to adjust the conductivity tensor d in Eq. (2) based only on the observed activation maps information. Indeed, our data are the values of the activation map $0 \le t_a(x) \le T$, where $t_a(x)$ is the time of first arrival of the electrical activation at $x \in \Omega$. It also defines our working time interval $[0, T]$. From a given data t_a, we introduce the following activation function

$$z(x, t) = \begin{cases} 1 & \text{if } t > t_a(x) \text{ (activated region)} \\ -1 & \text{if } t < t_a(x) \text{ (region at rest)}. \end{cases}$$

This new quantity can then be compared to the solution u of the model (Eq. (2)) by defining the activated region $\Omega_u^+(t) := \{x;\ u(x, t) > u_{\text{th}}\}$, the region at rest $\Omega_u^-(t) := \{x;\ u(x, t) < u_{\text{th}}\}$, and the activation front $\Gamma_u(t) = \{x;\ u(x, t) = u_{\text{th}}\}$ (here $u_{\text{th}} >$ is the given activation threshold). The discrepancy D between the solution u and the observation z (or t_a), a key ingredient of the estimation strategy, is then given by

$$D(z, u) = (1 - H(u - u_{\text{th}}))\left(z - c^-(z, u)\right) + H(u - u_{\text{th}})\left(z - c^+(z, u)\right), \tag{3}$$

where H is the Heaviside function and $c^{\pm}(z, u) = \frac{\int_{\Omega_u^{\pm}} z(x,t)dx}{|\Omega_u^{\pm}|}$. The objective is that (1) $D = 0$ when the model is registered on the data and (2) the sensitivity of D to u and z allows to find a direction of model correction when model and data do not coincide.

State Observer. In order to take into account the modeling errors and the initial condition uncertainties, we first introduce a corrected dynamics

$$\partial_t u + \frac{I_{\text{ion}}}{C_m}(u, w) = \text{div}\,(d\nabla u) + \mathcal{L}(z, u), \quad t \in [0, T], \quad x \in \Omega, \tag{4}$$

with $u(0, x) = u_0(x)$ any arbitrary initial data. The correction filter $\mathcal{L}(z, u)$ – initially introduced in [2] – is directly inferred from the discrepancy D sensitivity with respect to the front evolution:

$$\mathcal{L}(z, u) = \lambda \delta_{u-u_{\text{th}}} \left(\left(z - c^-(z, u) \right)^2 - \left(z - c^+(z, u) \right)^2 \right), \tag{5}$$

where $\lambda > 0$ is the gain parameter (and can be related to the data confidence) and $\delta_{u-u_{\text{th}}}$, the Dirac distribution on the surface $\Gamma_u(t)$. As wished, the filter effect vanishes when simulated observations and recorded observations coincide.

Reduced Order Joint State and Parameter Filtering. As we also want to estimate the conductivity tensor d, we follow the strategy introduced in [9]: the state u is corrected with the state observer (5) whereas the conductivity tensor d is estimated with a reduced Kalman filter. We actually use a Reduced-order Unscented Kalman Filter (RoUKF) [10], implemented in the Verdandi library [1]. Moreover, the parameter estimation is realized following an iterative Kalman filter strategy [5] which consists in several consecutive estimations using the previous estimated parameters as an *a priori* for the next one.

3   Synthetic and Clinical Data

3.1   Computation of Realistic Synthetic Data

We compute realistic activation times based on solutions of the bilayer model recalled in Sect. 2.1, and detailed in [7]. The equations are discretized on two coupled meshes of the endo- and epicardial layers of the right and left atria (total of 348657 vertices and 690117 triangles). Each layer has its own fiber distribution, and regional electrophysiology with 13 different regions, e.g. the CRN ionic model is tuned to have a short APD in the pulmonary veins, inactive regions around the sinus node, fast propagation in the Bachmann bundle, specific properties in the pectinate muscles.

Here, we simulate numerically three different catheter pacing scenarii. We pace the sinus node at 1.33 Hz in order to mimic the sinus rhythm, and, after 1.6s of free running sinus rhythm:

LIPV: pace from a location in the left inferior pulmonary vein at 3.33 Hz;
LAR: pace from a location in the roof of the left atrium at 3.33 Hz;
LAA: pace from a location in the appendage of the left atrium at 3.33 Hz.

Each simulation runs for a total time of 5s, which results in successive focal activations of the left atria. In each case, a unique activation time map $t_a(x)$ is computed from an average of these activations. At last, the activation maps are projected on a coarse mesh of 20773 nodes and 41129 triangles.

3.2   Clinical Data

The clinical dataset (Fig. 1) consists in a left atrium electro-anatomical map acquired on a patient suffering from atrial tachycardia, using the RHYTHMIA HDx^{TM} system. The mesh contains 10140 nodes and 20032 triangles.

4   Results and Discussion

4.1   Synthetic Data

Computation of Local Activation Times, and Total Activation Duration. The estimation of the conductivity d_0 (resp. (d_1, d_2)) is realized using the iterative strategy with five consecutive simulations from an initial guess $d_0 = 5\mathrm{S\,cm}^{-1}$ (resp. $(d_1, d_2) = (5, 5)$). Each of the five estimated parameters (resp. 10) is used to reconstruct an activation map using the classical monodomain model presented in Sect. 2.1. Modeled activation times will be denoted by $\overline{t_a}(x)$ and computed as $\overline{t_a}(x) = \min\{t | u(x, t) \geq u_{th}\}$. Then, these computed activation times are compared with the data activation times $t_a(x)$. In Table 1, we display the evolution of the estimated parameters along iteration of the iterative Kalman filter as well as the relative error – in percent – between total activation duration for both data and reconstructed times. This total activation duration is nothing else than the range of $t_a(x)$ (resp. $\overline{t_a}(x)$). If we denote by TAD the total activation duration of the dataset and $\overline{TAD}$ the modeled one, the "error" entry in Table 1 is computed as $\frac{|TAD - \overline{TAD}|}{TAD} \times 100$. In Table 2, we expose the minimum and maximum of the pointwise difference $t_a(x) - \overline{t_a}(x)$ as well as 25th, 50th and 75th percentiles denoted as Q1, Q2, and Q3.

Convergence of the Conductivity Coefficients. By looking at the parameter evolutions, we can notice that – contrary to the anisotropic case, – the isotropic case seems to depend on the pacing site. Indeed, at the fifth step d_0 is enclosed between 2.27 and 4.28S cm^{-1}. Furthermore, it appears that in the isotropic case, d_0 converges for both LIPV and LAR cases whereas it oscillates in LAA case.

Total Activation Duration. If we now look at the relative error between total activation duration, LAA is still the worst case with 31.9% of error at the fifth iteration. For both LIPV and LAR, we decrease the relative error to respectively 4.8% and 12.7% after five iterations. The parameter identification process seems to be more efficient when estimating an anisotropic tensor. At the fifth iteration, each of the three couples (d_1, d_2) are similar, and we recover an anisotropy ratio which is consistent with the one used to create data. Again, LAA relative error of total activation duration increases in the anisotropic case but stays lower than 6% over all iterations.

Table 1. Conductivity coefficients, and relative errors on the activation duration along the iterates.

Without fiber (isotropic) case						With fiber (anisotropic) case					
Iter.	1	2	3	4	5	Iter.	1	2	3	4	5
LIPV											
d_0	2.4	4.3	2.9	3.8	3.5	d_1	5.6	9.8	6.9	8.5	7.7
–	–	–	–	–	–	d_2	1.3	2.4	1.9	2.1	2.0
Error (%)	14	14	3.9	9.7	4.8	Error (%)	20	8.5	3.0	2.9	0.10
LAR											
d_0	5.1	4.9	4.6	4.3	4.3	d_1	7.3	8.7	7.5	8.3	7.7
–	–	–	–	–	–	d_2	1.5	2.3	1.9	2.1	2.0
Error (%)	19	18	16	13	13	Error (%)	5.2	9.4	2.3	6.6	3.9
LAA											
d_0	3.0	5.6	2.3	5.6	2.3	d_1	7.3	7.7	7.6	7.6	7.7
–	–	–	–	–	–	d_2	2.8	2.4	2.5	2.5	2.5
Error (%)	16	14	30	14	32	Error (%)	2.8	6.3	5.6	5.8	5.8

Distribution of Local Activation Times. If we now look at point-wise differences in Table 2, the same conclusion prevails: knowing the fiber distribution is essential to build a model faithful to the observations. For example, point-wise differences for LIPV and LAR are respectively enclosed in intervals $[-10, 11]$ and $[-11.5, 8.8]$ for anisotropic cases, whereas it was in $[-22.8, 16]$ and $[-15.4, 20.4]$ intervals for isotropic cases. Again, LAA case is less convincing than the others but we can notice an improvement of the three quartiles Q1, Q2 and Q3 for this case when we estimate an anisotropic tensor.

Importance of Fiber Structure. These results show us how essential is the fiber distribution in cardiac modeling if we want models to be consistent with observed activation times. Even if we reduce error on total activation duration for two of the three cases in the isotropic context, the range of point-wise differences between activation times stays quite large for each of the three cases. Moreover, the isotropic case gives us three different conductivity parameters d_0. This is surely due to the difference of anisotropy of paced region but also because data were created using two layers and regional electrophysiological parameters while we are trying to estimate an isotropic conductivity tensor on a one layer model with global electrophysiological parameters. Incidentally, when we take into account the fiber distribution, we are able to reduce the range of the point-wise differences for the three cases. In the anisotropic case, the relative error on total activation duration is enclosed between 0.1 and 5.8% which is promising. Nevertheless, even though prescribing a fiber architecture allows us to reduce the point-wise differences between activation times, there is still some contrast surely due to model simplification. Indeed, as we already said before, data were

Table 2. Statistics for $t_a(x) - \overline{t_a}(x)$ – Q1, Q2 and Q3 respectively represent first, second and third quartiles.

Without fiber (isotropic) case					With fiber (anisotropic) case				
Iter.	1	2	3	4	5	Iter.	1	2	3	4	5

LIPV

	1	2	3	4	5		1	2	3	4	5
Min.	−42	−14	−32	−19	−23	Min.	−33	−7.3	−12	−8.9	−10.0
Q1	−20	−1.5	−13	−4.3	−6.9	Q1	−18	1.7	−5.7	−1.8	−3.8
Q2	−12	2.4	−6.3	0.17	−1.7	Q2	−14	3.7	−3.9	0.13	−1.8
Q3	−5.2	8.0	−0.26	5.2	3.4	Q3	−8.7	5.8	−1.3	2.4	0.53
Max.	5.0	22	10	18	16	Max.	4.0	15	9.1	13	11

LAR

	1	2	3	4	5		1	2	3	4	5
Min.	−9.7	−11	−13	−15	−15	Min.	−16	−8.1	−12	−9.8	−12
Q1	−2.3	−3.0	−4.2	−6.0	−6.2	Q1	−4.9	2.2	−1.3	1.2	−0.34
Q2	3.5	2.9	1.8	0.38	0.24	Q2	−1.8	3.8	0.44	2.5	1.1
Q3	11	10	9.3	8.0	7.8	Q3	0.34	5.7	2.4	4.3	3.0
Max.	23	23	22	21	20	Max.	4.1	12	7.8	11	8.8

LAA

	1	2	3	4	5		1	2	3	4	5
Min.	−39	−15	−52	−14	−53	Min.	−21	−22	−22	−22	−22
Q1	−17	2.0	−28	2.1	−29	Q1	−2.5	−3.1	−2.9	−3.0	−3.0
Q2	−9.8	5.0	−18	5.1	−18	Q2	1.8	1.1	1.3	1.2	1.2
Q3	−3.4	8.6	−8.8	8.7	−9.4	Q3	3.7	2.5	2.8	2.6	2.6
Max.	13	25	11	25	11	Max.	19	19	19	19	19

created using local parameters whereas we estimate global parameters in order to see if we could build a simpler model which fits at best observed activation times. In this way, estimating an anisotropic tensor seems to be the best strategy and gives us encouraging results.

4.2 Clinical Data

Computation on Clinical Dataset. The same work is carried out on the clinical dataset presented in Sect. 3.2. For this dataset, fiber distribution is not known, but we have seen in previous section how important it is. We will try to overcome this by defining five different regions in the atria and estimate an isotropic tensor on each one. Those five regions were manually created using the formalism of [12] and can be resumed as: anterior wall, lateral wall, septum, inferior wall and roof.

Distribution of Local Activation Times, and Total Activation Duration. As for the synthetic cases, we display in Table 3 the point-wise differences between the dataset and the estimated activation map. An additional column gives the

Table 3. Statistics of $t_a(x) - \overline{t_a}(x)$, and relative error on the activation duration along the iterates – clinical dataset.

Clinical dataset			
Iter.	1	2	3
Min	−26.70	−26.70	−43.32
Q1	0.65	2.79	−3.38
Q2	8.35	10.90	3.37
Q3	19.40	21.27	16.02
Max	66.26	70.46	63.55
Error (%)	19.60	7.60	1.70

relative error of total activation duration. In this real case, only the relative error on activation duration is improved. The point-wise difference between activation times stays large and at the third iteration, we are between −43.32 and 63.55 ms with 50% of the data between −3.38 and 16.02 ms.

Anisotropy and Measure Artefact Effect. These differences could be explained in several ways. One of them lies in the apparent strong anisotropy in the data. We emphasized this anisotropy on Fig. 1 by drawing in black the isolines $\{t_a(x) = t_i, \ t_i = 10i, \ i = 0\ldots12\}$. Therefore, even if we split the atria in several regions, we will not be able to model the anisotropy without prescribing a fiber distribution. Moreover, some measure artefacts like high front slowdowns or accelerations showed on Fig. 1 are not reproduced by the model which leads to more error between target and reconstructed times. These slowdowns and acceleration are probably measurement artefacts, due to the electrical signals processing methods that detect the activation.

Fig. 1. High front slowdowns (red rectangles) and high accelerations (blue rectangles) on real data isolines. (Color figure online)

5   Conclusion

In this paper, we assess the robustness of a data assimilation method when facing with modeling simplifications. To do so, we firstly estimate conductivity parameters in two different ways, using realistic synthetic data. The first strategy consists in considering a unique conductivity parameter without taking into account the fiber distribution of the left atrium. This method allows us to decrease the relative error of total activation duration in two cases but was not reliable in the third case. The second one – consisting in estimating two conductivity parameters and using the fiber direction – is promising and allows to recover a large anisotropy ratio in the three cases. This illustrates how much the fiber architecture and the anisotropy of the conductivity are essential.

The application of our strategy on real, but rather noisy, data – which does not contain the fiber architecture – illustrates the sensitivity of the procedure to modeling choices. Moreover, in this case, due to the inherent acquisition and signal processing errors, there is no guarantee that the clinical activation map is consistent with the solution of a monodomain model, and hence the data quality is also in question. To conclude, this first attempt paves the way of evaluating modeling choices and data acquisition quality through real data assimilation model-data coupling.

References

1. Chapelle, D., Fragu, M., Mallet, V., Moireau, P.: Fundamental principles of data assimilation underlying the Verdandi library: applications to biophysical model personalization within euHeart. Med. Biol. Eng. Comput. **51**, 1–13 (2012)
2. Collin, A., Chapelle, D., Moireau, P.: A Luenberger observer for reaction-diffusion models with front position data. J. Comput. Phys. **300**, 288–307 (2015)
3. Corrado, C., et al.: Personalized models of human atrial electrophysiology derived from endocardial electrograms. IEEE Trans. Biomed. Eng. **64**(4), 735–742 (2017)
4. Courtemanche, M., Ramirez, R., Nattel, S.: Ionic mechanisms underlying human atrial action potential properties: insights from a mathematical model. Am. J. Physiol. **275**, H301–H321 (1998)
5. Jazwinski, A.H.: Stochastic Processes and Filtering Theory. Academic Press, Cambridge (1970)
6. Konukoglu, E., et al.: Efficient probabilistic model personalization integrating uncertainty on data and parameters: application to Eikonal-Diffusion models in cardiac electrophysiology. Prog. Biophys. Mol. Bio. **107**(1), 134–146 (2011)
7. Labarthe, S., et al.: A bilayer model of human atria: mathematical background, construction, and assessment. Europace **16**(Suppl. 4), iv21–iv29 (2014)
8. Mitchell, C., Schaeffer, D.: A two-current model for the dynamics of cardiac membrane. Bull. Math. Bio. **65**, 767–793 (2003)
9. Moireau, P., Chapelle, D., Le Tallec, P.: Joint state and parameter estimation for distributed mechanical systems. Comput. Methods Appl. Mech. Eng. **197**, 659–677 (2008)
10. Moireau, P., Chapelle, D.: Reduced-order unscented kalman filtering with application to parameter identification in large-dimensional systems. ESAIM Control Optimisation Calc. Var. **17**(2), 380–405 (2011)
11. Moreau-Villeger, V., Delingette, H., Sermesant, M., Ashikaga, H., McVeigh, E., Ayache, N.: Building maps of local apparent conductivity of the epicardium with a 2-D electrophysiological model of the heart. IEEE Trans. Biomed. Eng. **53**(8), 1457–1466 (2006)

12. Prabhu, S., et al.: Biatrial electrical and structural atrial changes in heart failure: electroanatomic mapping in persistent atrial fibrillation in humans. JACC Clin. Electrophysiol. **4**(1), 87–96 (2018)
13. Talbot, H., Cotin, S., Razavi, R., Rinaldi, C., Delingette, H.: Personalization of cardiac electrophysiology model using the unscented kalman filtering. In: Computer Assisted Radiology and Surgery (CARS 2015) (2015)

Fibrillation Patterns Creep and Jump in a Detailed Three-Dimensional Model of the Human Atria

Mark Potse[1,2,3]($\boxtimes$) (ID), Alain Vinet[4] (ID), Ali Gharaviri[5], and Simone Pezzuto[5] (ID)

[1] Univ. Bordeaux, IMB, UMR 5251, Talence, France
[2] CARMEN Research Team, Inria Bordeaux Sud-Ouest, Talence, France
mark@potse.nl
[3] IHU Liryc, fondation Bordeaux Université, Pessac, France
[4] Institut de génie biomédical, Université de Montréal, Montreal, Canada
alain.vinet@umontreal.ca
[5] Center for Computational Medicine in Cardiology,
Institute for Computational Science, Università della Svizzera italiana,
Lugano, Switzerland
{ali.gharaviri,simone.pezzuto}@usi.ch

Abstract. Activation mapping in animal models of atrial fibrillation (AF) has shown that activation patterns can repeat for several cycles and then be followed by different changing or repetitive patterns. Subsequent clinical studies have suggested a similar type of activity in human AF. Our purpose was to investigate whether a computer model of human AF can reproduce this behavior.

We used a three-dimensional model of the human atria consisting of 0.2-mm volumetric elements with a detailed representation of bundle structures and fiber orientations. Propagating activation was simulated over 9 s with a monodomain reaction-diffusion model using human atrial membrane dynamics. AF was induced by rapid pacing. Pattern recurrence was quantified using a similarity measure based on transmembrane voltage at 1000 uniformly distributed points in the model.

Recurrence plots demonstrated a continuous evolution of patterns, but the speed of evolution varied considerably. Groups of upto 15 similar cycles could be recognized, separated by rapid changes. In a few cases truly periodic patterns developed, which were related to macroscopic anatomical reentry. The drivers of non-periodic patterns could not be clearly identified. For example, a spiral wave could appear or disappear without an obvious impact on the similarity between cycles.

We conclude that our model can indeed reproduce strong variations in similarity between subsequent cycles, but true periodicity only in case of anatomical reentry.

Keywords: Atrial fibrillation · Reentry · Recurrence plot · Computer model

© Springer Nature Switzerland AG 2019
Y. Coudière et al. (Eds.): FIMH 2019, LNCS 11504, pp. 131–138, 2019.
https://doi.org/10.1007/978-3-030-21949-9_15

1   Introduction

Optical mapping of atrial fibrillation (AF) in sheep atria has shown that activation patterns can repeat for several cycles and then suddenly give way to a sequence of dissimilar cycles or another repetitive pattern [12]. Clinical studies have suggested a similar type of activity in human AF [8]. In the sheep atria these repetitive patterns were related to temporarily stable rotor activity. In human AF such a link is harder to establish, because AF patterns in patients can only be observed with coarse resolution by noninvasive mapping [8] or in a small portion of the atria by invasive mapping [1].

A computer model can help to gain more insight in this pseudo-stable behavior of AF, in its determinants, and in possible methods to recognize it with limited data, such as an ordinary electrocardiogram. The purpose of this study was therefore to investigate whether a computer model can reproduce the AF dynamics observed in animal studies, and if such behavior can be linked to meandering spiral waves.

2   Methods

A three-dimensional model of the human atria was previously constructed [5,10]. The model included the crista terminalis, pectinate muscles, Bachmann's bundle, posterior interatrial bundles, bundle structures in the left atrial appendage, and up to three layers of fiber orientation (Fig. 1).

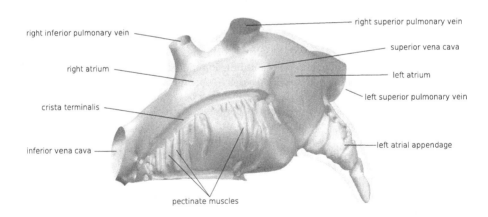

Fig. 1. Anatomy of the atrial model. The epicardium is rendered semi-transpararently to show the variations in wall thickness.

Propagating activation was simulated with a monodomain reaction-diffusion model using human atrial membrane dynamics [2], with minor adaptation to avoid numerical instabilities due to discontinuous coefficients, on a hexahedral mesh with 0.2-mm spacing. The model was integrated with the Rush-Larsen method for gating variables and an explicit Euler method for the other variables, using a fixed time step of 0.01 ms.

To simulate electrically remodeled atria we modified the maximum conductivities of the calcium current, g_{CaL}, from 0.1238 to 0.037, g_{TO} from 0.1652 to 0.026, g_{K1} from 0.09 to 0.180, and g_{Ks} from 0.129 to 0.1 nS/pF. [4].

Simulations were performed in a "fibrotic" model and in a "normal" model. In the fibrotic model 50% of the model elements had zero transverse conductivity, to mimic the uncoupling of lateral connections in ageing myocardium [13]. The affected elements were chosen randomly, with a uniform probability density.

With each model 10 simulations, each with 9 s duration, were performed. AF was induced by rapid pacing, from a different location in each simulation. Pattern recurrence and similarity between simulations were quantified at 1-ms intervals using a state variable consisting of transmembrane voltage sampled from 1000 distributed points in the model. The points were chosen randomly, regardless of whether they were endocardial, epicardial, or intramural. For each pair of time samples i, j, a similarity measure S_{ij} was computed as

$$S_{ij} = (\Phi_i \cdot \Psi_j)/(|\Phi_i| \cdot |\Psi_j|) \tag{1}$$

where Φ_i is the state at time i in one simulation and Ψ_j the state at time j in another; $\Phi \equiv \Psi$ for self-similarity or "recurrence." This similarity was visualized using recurrence plots [3,9], which show S_{ij} as a function of i and j.

Phase singularity filaments were identified using a phase measure computed with a time embedding of the transmembrane potential [6] with a delay of 10 ms. On each face of the computational mesh, the topological charge was computed as the line integral of the phase over the corners. If the integral was $\pm 2\pi$ a singularity filament was assumed to cross the face. Crossings that occurred on different faces of the same element were assumed to be part of the same filament.

Simulations were performed with the Propag-5 software [7] on 96 compute nodes each containing two 14-core Intel Xeon E5 processors. A full 9-s simulation took 55 min.

3   Results

3.1   Normal Model

In the normal model, 3 out of 10 simulations yielded AF until the end of the simulation (9 s). In the other 7 activity stopped between 536 and 1398 ms after the last pacing pulse (approximately 4 to 9 cycles), either before any reentrant pattern had established or after all wavefronts encountered refractory tissue.

In the 3 simulations that yielded 9 s of AF, periods of higher and lower similarity between subsequent cycles could be discerned. An example is shown in Fig. 2. Panel A shows S_{ij} together with all phase singularity filaments that lasted more than 150 ms. The first such filaments appear shortly before the end of the pacing interval. Subsequently, several intervals with simultaneous filaments are seen, separated (around 3.6 s) by a short interval without any long-living

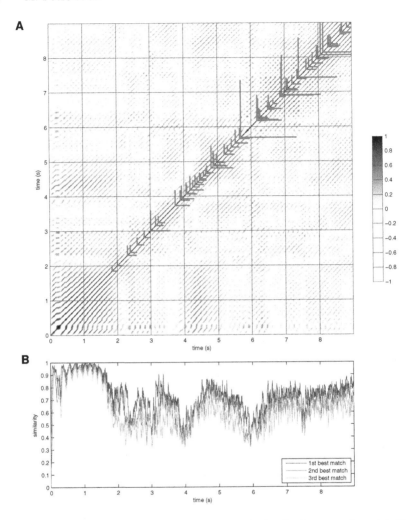

Fig. 2. A: Recurrence plot, showing S_{ij} for a simulation in the normal model. Darker shades of gray represent higher similarity between the potential patterns at each pair of time instants. A dark diagonal line outside the main diagonal indicates a repeating activation pattern. Here, after 2 s of rapid pacing, periods of higher and lower similarity follow. Red lines indicate the intervals during which long-living phase singularity filaments were present. B: Graph showing the similarity of the three best-matching time instants for each time instant in the simulation. (Color figure online)

filaments. Panel B shows, for each time instant in the simulation, the similarity of the three best-matching time instants in the simulation separated by at least 50 ms; i.e. $S_{ij_1} > S_{ij_2} > S_{ij_3} > \ldots$ where $|j_n - i| > 50$ and $|j_n - j_m| > 50$ for any j_n, j_m. Maximum similarity is seen to oscillate roughly between 0.4 and 0.8. There is no obvious relation with periods of rotor activity.

In all 10 simulations with the normal model combined, 7940 separate singularity filaments were detected. 2324 of these (29%) were detected for only a single time instant, 138 filaments lived longer than 150 ms (approximately one cycle length), and 68 lived longer than 300 ms. The longest-living filament lasted for 5601 ms.

3.2 Fibrotic Model

In the fibrotic model, 5 out of 10 simulations yielded AF until the end of the simulation. In the other 5, activity stopped between 655 and 1268 ms after the last pacing pulse.

As in the normal model, recurrence plots demonstrated a continuous evolution of patterns, with a varying speed of evolution. A "creeping" progression during groups of upto 15 similar cycles could be recognized, separated by rapid "jumps." An example is shown in Fig. 3.

In the fibrotic model, three of the 5 simulations that showed activity until the end of the simulation arrived at the same periodic regime, which corresponded to an anatomical reentry that passed from the coronary sinus to the left atrium and returned through the interatrial septum. A comparison of two of these simulations is shown in Fig. 4.

In all 10 simulations with the fibrotic model combined, 20873 separate filaments were found. 7898 of these (38%) were detected for only a single time instant, 31 filaments lived longer than 150 ms, and 11 lived longer than 300 ms. The longest-living filament lasted for 2901 ms.

4   Discussion

This study shows that a computer model can reproduce variations in similarity between subsequent cycles of AF. The results also highlight that it is difficult to define when a pattern is "periodic," as it was called by Skanes et al. [12], because the similarity between subsequent cycles was never perfect, and ranged between 0.4 and 0.98 in our results. Since published studies have not quantitatively defined when patterns were considered periodic or nonperiodic, a quantitative comparison to their results was not attempted here.

In addition to the similarity of patterns between different time instants we also studied the occurrence of phase singularity or "vortex" filaments. Here we found surprising results. The total number of detected filaments was 2.5 times larger in the fibrotic model than in the normal model, while the total time spent in AF was only 1.6 times larger. In striking contrast, the number of filaments that lived longer than 150 ms was 5 times smaller in the fibrotic model, despite the longer time in AF. Clearly the presence of fibrosis in this model promotes the generation of phase singularities but also reduces their lifetime. Thus, if one were to use the number of phase singularities to demonstrate that fibrosis changes the complexity of AF, the result would greatly depend on the definition of a separate phase singularity.

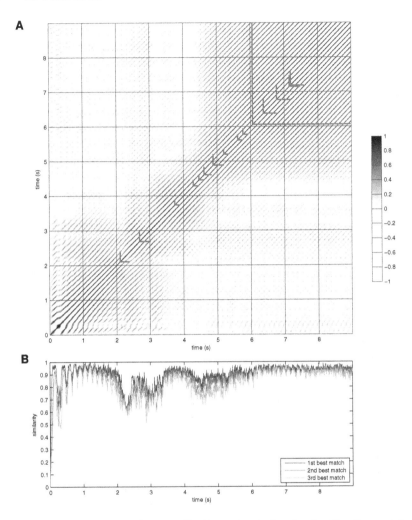

Fig. 3. A: Recurrence plot, showing S_{ij} for a simulation in the fibrotic model. The format is as in Fig. 2. After 2 s of rapid pacing, the development of 3 groups of similar cycles can be recognized. Long-living phase singularities (red lines) are present but do not coincide with the groups of similar cycles. B: Similarity of the three best-matching time instants for each time instant in the simulation. (Color figure online)

A clear relation between nearly-periodic patterns and the existence of long-lived phase singularities could not be established. This may be partly explained by methodological restrictions. For example, when a boundary between excitable and refractory tissue crosses an anatomical obstacle such as a pulmonary vein, a vortex filament traveling along this boundary may enter the obstacle and reappear on the other side. In such cases, our methods will see this as two different filaments. It is in general very difficult to determine the type of reentry

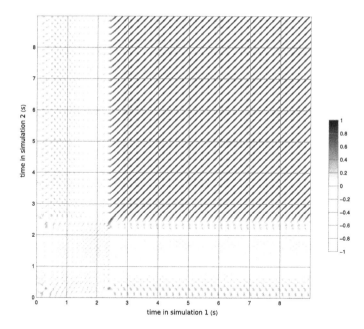

Fig. 4. Similarity plot for two simulations in the fibrotic model, which ended up in the same periodic regime. The format is the same as in Fig. 3. The simulations were stimulated at different locations, leading to different patterns during the first 2.3 s. Subsequently the two simulations produced exactly the same patterns, albeit with a phase delay (none of the lines of maximum intensity coincides exactly with the main diagonal).

that takes place in this complex anatomy. Similarly, the precise mechanism of termination of a reentry could not always be understood.

Our results concerning vortex filaments likely depend on the methods chosen to detect the intersections of filaments with mesh elements [14] and the decisions whether different intersections belong to the same filament and whether filaments in subsequent time steps are related. A comparison of these methods, however, is outside the scope of this study. The type of fibrosis and the method to model it [11] will also affect the results. We chose to model a diffuse type of fibrosis in which there are no gaps in the myocardium on the scale of a model element: all our elements represented myocardium. Fibrotic elements only had zero transverse conductivity, which was to mimic fine-grained fibrosis that interrupts side-to-side coupling between myofibers [13].

In summary, we have shown that pseudo-periodic AF patterns can be simulated and that this stability can be quantified. A similar quantification of measured results is needed to establish whether the model truly agrees with reality.

Acknowledgments. We thank Sarah Peris for the implementation of the singularity detection algorithm. This work was supported by the French National Research Agency, grant reference ANR-10-IAHU04-LIRYC. This work was granted access to HPC resources of CINES under GENCI allocation 2019-A0050307379.

References

1. Allessie, M.A., et al.: Electropathological substrate of long-standing persistent atrial fibrillation in patients with structural heart disease: longitudinal dissociation. Circ. Arrhythm. Electrophysiol. **3**, 606–615 (2010)
2. Courtemanche, M., Ramirez, R., Nattel, S.: Ionic mechanisms underlying human atrial action potential properties: insights from a mathematical model. Am. J. Physiol. Heart Circ. Physiol. **275**, H301–H321 (1998)
3. Eckmann, J.P., Oliffson Kamphorst, S., Ruelle, D.: Recurrence plots of dynamical systems. Europhys. Lett. **4**, 973–977 (1987)
4. Gharaviri, A., et al.: How disruption of endo-epicardial electrical connections enhances endo-epicardial conduction during atrial fibrillation. Europace **19**, 308–316 (2017)
5. Gharaviri, A., et al.: Acute changes in P-wave morphology by pulmonary vein isolation in atrial fibrillation patients. In: Computing in Cardiology. Maastricht, The Netherlands (2018)
6. Gray, R.A., Pertsov, A.M., Jalife, J.: Spatial and temporal organization during cardiac fibrillation. Nature **392**, 75–78 (1998)
7. Krause, D., Potse, M., Dickopf, T., Krause, R., Auricchio, A., Prinzen, F.: Hybrid parallelization of a large-scale heart model. In: Keller, R., Kramer, D., Weiss, J.-P. (eds.) Facing the Multicore - Challenge II. LNCS, vol. 7174, pp. 120–132. Springer, Heidelberg (2012). https://doi.org/10.1007/978-3-642-30397-5_11
8. Lim, H.S., et al.: Complexity and distribution of drivers in relation to duration of persistent atrial fibrillation. J. Am. Coll. Cardiol. **69**, 1257–1269 (2017)
9. Marwan, N., Kurths, J., Foerster, S.: Analysing spatially extended high-dimensional dynamics by recurrence plots. Phys. Lett. A **379**, 894–900 (2015)
10. Potse, M., et al.: P-wave complexity in normal subjects and computer models. J. Electrocardiol. **49**, 545–553 (2016)
11. Roney, C.H., et al.: Modelling methodology of atrial fibrosis affects rotor dynamics and electrograms. Europace **18**, iv146–155 (2016)
12. Skanes, A.C., Mandapati, R., Berenfeld, O., Davidenko, J., Jalife, J.: Spatiotemporal periodicity during atrial fibrillation in the isolated sheep heart. Circulation **98**, 1236–1248 (1998)
13. Spach, M.S., Dolber, P.C.: Relating extracellular potentials and their derivatives to anisotropic propagation at a microscopic level in human cardiac muscle; evidence for electrical uncoupling of side-to-side fiber connections with increasing age. Circ. Res. **58**, 356–371 (1986)
14. Zhou, R., Kneller, J., Leon, L.J., Nattel, S.: Development of a computer algorithm for the detection of phase singularities and initial application to analyze simulations of atrial fibrillation. Chaos **12**, 764–778 (2002)

Tissue Drives Lesion: Computational Evidence of Interspecies Variability in Cardiac Radiofrequency Ablation

Argyrios Petras[1(✉)] ⓘ, Massimiliano Leoni[1,2] ⓘ, Jose M. Guerra[3] ⓘ,
Johan Jansson[1,2] ⓘ, and Luca Gerardo-Giorda[1] ⓘ

[1] BCAM - Basque Center for Applied Mathematics, 48009 Bilbao, Spain
apetras@bcamath.org
[2] KTH Royal Institute of Technology, 11428 Stockholm, Sweden
[3] Hospital de la Santa Creu i San Pau, CIBERCV, 08041 Barcelona, Spain

Abstract. Radiofrequency catheter ablation (RFCA) is widely used for the treatment of various types of cardiac arrhythmias. Typically, the efficacy and the safety of the ablation protocols used in the clinics are derived from tests carried out on animal specimens, including swines. However, these experimental findings cannot be immediately translated to clinical practice on human patients, due to the difference in the physical properties of the types of tissue. Computational models can assist in the quantification of this variability and can provide insights in the results of the RFCA for different species. In this work, we consider a standard ablation protocol of 10 g force, 30 W power for 30 s. We simulate its application on a porcine cardiac tissue, a human ventricle and a human atrium. Using a recently developed computational model that accounts for the mechanical properties of the tissue, we explore the onset and the growth of the lesion along time by tracking its depth and width, and we compare the lesion size and dimensions at the end of the ablation.

Keywords: Radiofrequency catheter ablation · Mathematical model · Tissue properties · Interspecies variability

1 Introduction

Radiofrequency ablation (RFA) is a common treatment for cardiac arrhythmias. Through a catheter advanced into the patient's heart, radiofrequency current is delivered to the tissue which produces resistive heating in the neighborhood of the electrode, while conduction propagates the heat to the immediate surrounding tissue. At a temperature of 50 °C the tissue is irreversibly damaged and a permanent lesion develops [3]. RFA is generally safe, however life-threatening

Supported by the Basque Government through BERC 2019–2021 and by the Spanish Ministry of Economy and Competitiveness MINECO through BCAM Severo Ochoa excellence accreditation SEV-2017-0718.

© Springer Nature Switzerland AG 2019
Y. Coudière et al. (Eds.): FIMH 2019, LNCS 11504, pp. 139–146, 2019.
https://doi.org/10.1007/978-3-030-21949-9_16

complications can occur, including the formation of thrombi due to blood over-heating at $80\,°C$ and steam pops at temperatures of $100\,°C$ within the tissue, which are among the most severe complications [3].

Typically, ablation protocols are designed and tested using *in-vitro*, *ex-vivo* and *in-vivo* experiments on animals. Porcine cardiac tissue is commonly used in the experiments to assess the efficacy and the safety of the RFA treatment [1,5]. However, the biophysical, mechanical and physiological properties of the porcine cardiac tissue differ from the corresponding human ones [6], thus a direct translation of the experimental results to clinical practice can lead to insufficient treatment or potential life-threatening complications. Computational models can be a valuable asset in the assessment of RFA protocols. The efficacy and safety of the procedure can be directly assessed on simulated human tissue using the reported physical properties in the literature, thus avoiding the translation of data from experiments that use porcine tissue.

In this study, we investigate the interspecies variability in the RFA treatment using a recently developed computational model [5] which includes the mechanical properties of the tissue. We simulate a commonly used ablation protocol on cardiac tissue of two species: porcine and human. Two different ablation sites are considered in the case of the simulated human tissue: an atrium and a ventricle.

2   Computational Model

2.1   Geometry

Based on an experimental setup similar to [1], an $80\,mm \times 80\,mm \times 80\,mm$ box is constructed that includes blood (top $40\,mm$), cardiac tissue of thickness H mm and an external factors board (bottom $40 - H$ mm). Within the blood compartment, the catheter is placed perpendicularly to the tissue at the middle of the box. We consider a 6-hole electrode with a hemispherical tip and a thermistor inspired by RFA catheters commonly used in clinics, as described in [5]. Figure 1 shows a sample computational geometry for a tissue thickness of $H = 20\,mm$.

Due to the contact with the catheter, the tissue undergoes a mechanical deformation. The vertical deformation is described in [5] and is formulated as

$$\omega(r) = \begin{cases} \omega_{\max} - (R - \sqrt{R^2 - r^2}), & r \leq a, \\ \frac{a^2}{\pi} \int_0^1 \frac{2\omega_{\max} - ax \log\left(\frac{R+ax}{R-ax}\right)}{r^2 - a^2 x^2} \, dx, & r > a, \end{cases}$$

where R is the radius of the electrode, $\omega_{\max}$ is the maximum indentation depth, a is the contact radius of the electrode with the tissue and r is the planar distance from the center of the electrode. The indentation depth and contact radius can be computed using the contact force F, the Young's modulus E and the Poisson's ratio ν of the tissue, as follows:

$$\omega_{\max} = \frac{a}{2} \log\left(\frac{R + a}{R - a}\right),$$

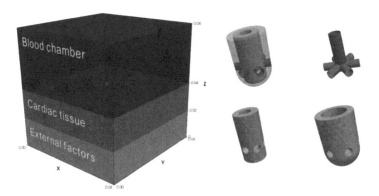

Fig. 1. Left: The full computational geometry. Right: The computational tip of the catheter (top left), the saline tubes (top right), the thermistor (bottom left) and the electrode (bottom right).

$$F = \frac{E}{2(1-\nu^2)}\left((a^2+R^2)\log\left(\frac{R+a}{R-a}\right)-2aR\right).$$

The deformation of the tissue is directly embedded in the construction of the computational geometry.

2.2 Governing Equations

The blood flow and its interaction with the irrigated saline in the blood compartment Ω_{blood} of the geometry are modelled using the incompressible Navier-Stokes equations

$$\frac{\partial \mathbf{u}}{\partial t} + \mathbf{u} \cdot \nabla \mathbf{u} - \text{div}\,\sigma(\mathbf{u},p) = \mathbf{0},$$

$$\text{div}\,\mathbf{u} = 0,$$

where $\mathbf{u}$ is the velocity, $\sigma(\cdot,\cdot)$ is the stress tensor and p is the scaled pressure. A constant inflow is considered in one of the sides of Ω_{blood} with its corresponding outflow conditions at the opposite side. A constant inflow is also considered from the saline irrigation holes on the electrode flowing radially within Ω_{blood}. The remaining interfaces of Ω_{blood} are equipped with no slip conditions.

A modified Penne's bioheat equation tracks the changes in the temperature of the geometry Ω over time

$$\rho c(T)\left(\frac{\partial T}{\partial t} + \mathbf{u} \cdot \nabla T\right) - \text{div}(k(T)\,\nabla T) = \sigma(T)|\nabla\Phi|^2,$$

where T is the temperature, t is the time, ρ is the density, $c(T)$ is the specific heat, $\mathbf{u}$ is the velocity, $k(T)$ is the thermal conductivity, $\sigma(T)$ is the electrical conductivity and Φ is the electrical potential. A temperature of $22\,°\text{C}$ is assumed on the interface between the irrigation holes and the blood compartment Ω_{blood}

to model the cooling effect of the saline, while insulation boundary conditions are applied on the irrigation tubes and the catheter body. A constant body temperature of $37\,^\circ\mathrm{C}$ is imposed on all outer boundaries of the computational domain Ω.

The spatial distribution of the potential follows a quasi-static equation, augmented with a power constraint for constant power ablations

$$\begin{cases} \mathrm{div}(\sigma(T)\,\nabla\Phi) = 0, \\ \int_\Omega \sigma(T)|\nabla\Phi|^2\,dx = P, \end{cases}$$

where P is the power dissipated within the computational domain Ω. A potential V_0 is applied on the interface of the catheter and the electrode, tuned to match the power P dissipated in the computational domain Ω. The dispersive electrode is placed at the bottom of Ω, where zero potential is imposed. All the remaining surfaces are considered electrically insulated. More details on the computational model can be found in [5].

2.3 Parameters

The parameters considered in the computational model are summarized in Table 1. The specific heat c, thermal and electrical conductivities k and σ are considered temperature dependent within the tissue:

$$c(T) = c_0(1 + c_1(T - 37))$$
$$k(T) = k_0(1 + k_1(T - 37))$$
$$\sigma(T) = \sigma_0(1 + \sigma_1(T - 37))$$

The electrical conductivity of the external effects board σ_b is tuned to match the initial resistance of the system, and the power delivered to the tissue, which is calculated as follows:

$$P_{tissue} = \frac{A_{tissue}\sigma_0^{(tissue)}}{A_{blood}\sigma_0^{(blood)} + A_{tissue}\sigma_0^{(tissue)}} P_{abl} =: \alpha P_{abl},$$

where A_{tissue} and A_{blood} are the surface areas of the electrode in contact with the tissue and the blood respectively, $\sigma_0^{(tissue)}$ and $\sigma_0^{(blood)}$ are the electrical conductivities of the tissue and the blood at body temperature and P_{abl} is the total power set by the ablation protocol. The initial resistance of the system is set as $120\ \Omega$. More details on the calculation of σ_b can be found in [5].

In our extensive literature review, no work has been found that addresses differences in biophysical properties on different regions of the human heart. Thus, we assume that the human atrium and ventricle have the same biophysical properties, which are drawn from [2] at body temperature. The temperature dependence of the specific heat and thermal conductivity can be calculated from [10]. No data were found for the dependence of the electrical conductivity of the

Table 1. The summary of the biophysical and mechanical parameters that appear in the computational model.

Parameters	Blood	Electrode	Thermistor	Board	Tissue	
					Porcine	Human
ρ (kg m^3)	1050	21500	32	1076	1076	1081
c_0 (J kg^{-1} K^{-1})	3617	132	835	3017	3017	3686
c_1 ($^\circ$C^{-1})	-	-	-	-	-0.0042	-0.0011
k_0 (W m^{-1} K^{-1})	0.52	71	0.038	0.518	0.518	0.56
k_1 ($^\circ$C^{-1})	-	-	-	-	-0.0005	0.0022
σ_0 (S m^{-1})	0.748	4.6×10^6	10^{-5}	σ_b	0.54	0.381
σ_1 ($^\circ$C^{-1})	-	-	-	-	0.015	0.015
ν (-)	-	-	-	-	0.499	0.499
E (kPa)	-	-	-	-	75	40

myocardium on the temperature, thus values for the human liver are considered, since a similar behavior is reported for other species [6]. The mechanical properties for human cardiac tissue are summarized in [7,9].

The thickness of the porcine cardiac wall is considered as $H = 20$ mm [4]. Two different substrates are modelled for the simulated human cardiac tissue: an atrium of $H = 6$ mm thickness and a ventricle of $H = 12.5$ mm [8,11].

3   Results

We simulate a standard constant power ablation protocol of 30 W, with a contact force of 10 g for a total of 30 s of ablation. The blood flow protocol is set to 0.1 m s^{-1} and the saline irrigation rate to 17 ml min^{-1}. This ablation protocol is typically used in RFA experiments [1].

We identify the computational lesion by the 50 °C isotherm contour: the quantities measured are the depth (D), the width (W), the depth at which the maximum width occurs (DW), the surface area of the lesion (SA) and the volume (V). In addition to the lesion size dimensions mentioned above, we track the maximum temperature in the tissue and the blood. Details on the measurement of these quantities can be found in [5].

The computational results are shown in Table 2. The lesion on the human atrium appears to be the biggest in volume, while the smallest is on the human ventricle. The depth in the porcine cardiac tissue and the human atrium are comparable, while the lesion is more shallow in the human ventricle. A bigger difference appears in the measured width. In particular, a difference of nearly 1 mm is observed between the cases of human atrium and ventricle, while the porcine lesion width is comparable with the human ventricle one. The morphology of all three lesions appears to be different, with significant changes in the depth at which the maximum width occurs. Though the maximum width of the

Table 2. The summary of the temperature and lesion size dimensions after the completion of 30 s of ablation for the simulated tissue considered.

	Porcine	Human ventricle	Human atrium
D (mm)	3.76	3.35	3.75
W (mm)	6.73	6.56	7.42
DW (mm)	1.38	0.86	0.94
SA (mm^2)	0.11	0.0	0.0
V (mm^3)	103.4	92.3	124.9
T_{max} tissue (°C)	77.7	71.9	77.3
T_{max} blood (°C)	51.5	56.1	62.0

lesions on the human specimen appears to be more shallow, there is no surface burning. A small surface burn is present in the case of the porcine cardiac tissue.

The temperature of the tissue is similar in the cases of the porcine cardiac tissue and the human atrium, while it is about 5 °C lower in the human ventricle. On the contrary, the blood temperature is much lower in the porcine cardiac tissue. The temperature difference with respect to the human tissue is bigger than 10 °C for the atrium and around 5 °C for the ventricle.

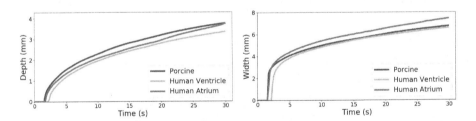

Fig. 2. The lesion depth (left) and width (right) over time for the simulated porcine cardiac tissue, human ventricle and human atrium.

To further compare the results on the three types of tissue considered, we explore in Fig. 2 the evolution of the width and the depth of the lesion throughout the duration of the ablation. We observe that the lesion is created almost at the same time on the porcine cardiac tissue and the human atrium, while the one in the human ventricle is delayed by 0.8 s. The depth of the lesion in the porcine case is consistently larger than the one in the human tissue. The depth of the lesion in the human atrium case is initially similar to the human ventricle case, but after 20 s it increases to a depth comparable to the porcine case. On the contrary, the width of the lesion in the human atrium is larger than the other cases throughout the duration of the ablation, while comparable values appear between the human ventricle and the porcine tissue.

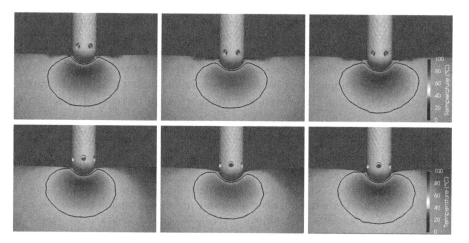

Fig. 3. The medial (top) and lateral (bottom) view of the lesion for porcine cardiac tissue (left), human ventricle (middle) and human atrium (right) at the final time of the ablation.

Finally, to explore morphological differences of the lesions, we show in Fig. 3 the medial and lateral view of the 50 °C isotherm contour within the tissue at the final time of the ablation. All three lesions appear tilted towards the direction of the blood in the lateral view, while they are symmetric in the medial view, as previously observed in [5]. The lesions in the human tissue appear elongated and less spherical than the ones in the porcine tissue, which is reflected in the difference in the depth of the maximum width in Table 2.

4   Conclusions

We explored interspecies variability during RFA treatment, comparing the resulting lesion size dimensions of the ablation process on a porcine cardiac tissue, a human ventricle and a human atrium. Our computational results indicate that the lesion characteristics are different in all three cases. The human atrium provides the largest lesion overall. The depth of the lesion in the porcine cardiac tissue case is larger than the one on the human tissue throughout the ablation process. Thus, a direct application to a human patient of a protocol based on the achievement of a specific lesion depth in porcine tissue would result in too small lesions and insufficient treatment in cases that transmural lesions are targeted.

The width of the lesion in the human atrium is at least 0.7 mm larger than the other two cases. In addition, the lesions in the simulated human tissue appear less spherical than the porcine one. A direct translation from porcine data to the human atrium would affect the efficiency of the RFA process, indicating that a larger number of lesions are required when targeting isolation lines in the atrium.

The maximum tissue temperature in the atrium and the porcine tissue are comparable, however there is a difference of more than 10 °C the maximum blood temperature. Regarding the human ventricle, the tissue temperature is lower than the porcine one, however the maximum blood temperature is about 5 °C higher. This indicates that the results on porcine tissue can lead to blood overheating and thrombus formation if directly translated to human tissue. On the contrary, steam pops would be avoided as the maximum tissue temperature is overestimated, though the efficiency of the process can be reduced by inducing smaller lesions in the human ventricle.

References

1. Guerra, J.M., et al.: Effects of open-irrigated radiofrequency ablation catheter design on lesion formation and complications: in vitro comparison of 6 different devices. J. Cardiovasc. Electrophysiol. **24**(10), 1157–1162 (2013)
2. Hasgall, P., Neufeld, E., Gosselin, M., Klingenböck, A., Kuster, N.: ITIS database for thermal and electromagnetic parameters of biological tissues, Version 3.0 (2015)
3. Huang, S.K.S., Wood, M.A.: Catheter Ablation of Cardiac Arrhythmias E-book. Elsevier Health Sciences, Philadelphia (2014)
4. Liu, S.K., et al.: Hypertrophic cardiomyopathy in pigs: quantitative pathologic features in 55 cases. Cardiovasc. Pathol. **3**(4), 261–268 (1994)
5. Petras, A., Leoni, M., Guerra, J.M., Jansson, J., Gerardo-Giorda, L.: A computational model of open-irrigated radiofrequency catheter ablation accounting for mechanical properties of the cardiac tissue. arXiv preprint arXiv:1810.09157 (2018)
6. Rossmann, C., Haemmerich, D.: Review of temperature dependence of thermal properties, dielectric properties, and perfusion of biological tissues at hyperthermic and ablation temperatures. Crit. Rev. Biomed. Eng. **42**(6), 467 (2014)
7. Rump, J., Klatt, D., Braun, J., Warmuth, C., Sack, I.: Fractional encoding of harmonic motions in MR elastography. Magn. Reson. Med. **57**(2), 388–395 (2007)
8. Sjögren, A.L.: Left ventricular wall thickness determined by ultrasound in 100 subjects without heart disease. Chest **60**(4), 341–346 (1971)
9. Strachinaru, M., et al.: Cardiac shear wave elastography using a clinical ultrasound system. Ultrasound Med. Biol. **43**(8), 1596–1606 (2017)
10. Valvano, J.W., Cochran, J.R., Diller, K.R.: Thermal conductivity and diffusivity of biomaterials measured with self-heated thermistors. Int. J. Thermophys. **6**(3), 301–311 (1985)
11. Whitaker, J., et al.: The role of myocardial wall thickness in atrial arrhythmogenesis. Ep Eur. **18**(12), 1758–1772 (2016)

Correcting Undersampled Cardiac Sources in Equivalent Double Layer Forward Simulations

Jess D. Tate[1](✉) ⓘ, Steffen Schuler[2]ⓘ, Olaf Dössel[2]ⓘ, Robert S. MacLeod[1]ⓘ, and Thom F. Oostendorp[3]ⓘ

[1] Scientific Computing and Imaging Institute, University of Utah,
Salt Lake City, USA
jess@sci.utah.edu
[2] Institute of Biomedical Engineering, KIT, Karlsruhe, Germany
[3] Donders Centre for Neuroscience, Radboud University, Nijmegen, Netherlands

Abstract. Electrocardiographic Imaging (ECGI) requires robust ECG forward simulations to accurately calculate cardiac activity. However, many questions remain regarding ECG forward simulations, for instance: there are not common guidelines for the required cardiac source sampling. In this study we test equivalent double layer (EDL) forward simulations with differing cardiac source resolutions and different spatial interpolation techniques. The goal is to reduce error caused by undersampling of cardiac sources and provide guidelines to reduce said source undersampling in ECG forward simulations. Using a simulated dataset sampled at 5 spatial resolutions, we computed body surface potentials using an EDL forward simulation pipeline. We tested two spatial interpolation methods to reduce error due to undersampling triangle weighting and triangle splitting. This forward modeling pipeline showed high frequency artifacts in the predicted ECG time signals when the cardiac source resolution was too low. These low resolutions could also cause shifts in extrema location on the body surface maps. However, these errors in predicted potentials can be mitigated by using a spatial interpolation method. Using spatial interpolation can reduce the number of nodes required for accurate body surface potentials from 9,218 to 2,306. Spatial interpolation in this forward model could also help improve accuracy and reduce computational cost in subsequent ECGI applications.

Keywords: ECG forward simulation · Activation times ·
Body surface potentials · Spatial interpolation ·
Boundary element method · Equivalent double layer

1 Introduction

Electrocardiographic Imaging (ECGI) non-invasively describes the electrical activity on the surface of the heart and is increasingly being used to diagnosis

Supported by KIT, Radboud University, and the National Institute of General Medical Sciences of the National Institutes of Health under grant number P41 GM103545-18.

Y. Coudière et al. (Eds.): FIMH 2019, LNCS 11504, pp. 147–155, 2019.
https://doi.org/10.1007/978-3-030-21949-9_17

and treatment of cardiac arrhythmias. ECGI relies heavily on ECG forward models, yet recent research has shown these models to produce more error than commonly thought [1] and that properly sampled cardiac sources can play a role in reducing error [7]. Despite the risk of increased error, ECG forward models are often subsampled to reduce computational cost. Guidelines for cardiac source resolution and interpolation tools are needed to improve the accuracy of ECG forward and subsequent ECGI calculations.

Mitigating error due to insufficient cardiac source resolution can be dependent on the type ECG forward model being used. The challenge originates from the nature of the cardiac activation wave because it is characterized by high gradients of transmembrane and extracellular potentials in a small region of space. Undersampling the activation wavefront can produce temporal noise from large regions of the tissue becoming active simultaneously. A previous study was able to reduce error due to spatial undersampling of the transmembrane currents in a finite element (FEM) model by spatially averaging the values [5]. For other types of source models, such as using activation times with the equivalent double layer (EDL) [3], spatial averaging may not be effective in reducing error due to spatial resolution. However, using EDL with a boundary element method (BEM) formulation provides an other methods to spatially interpolate cardiac surface data from undersampled sources for use in the ECG forward simulation.

In this paper, we propose new spatial interpolation schemes for use in an EDL forward simulation and tested them with varying source resolution. The goal of the study was to use these interpolation schemes to reduce or eliminate errors due to undersampling the cardiac sources, including temporal irregularities. We also sought to provide some guidelines for appropriate sampling resolutions for the cardiac surface in order to effectively use the EDL simulation with different spatial interpolations. Our results show that these techniques can be an effective solution to reducing error due to cardiac sampling.

2   Methods

We analyzed the effect of different cardiac source resolutions and different spatial interpolation methods on body surface potentials generated with the ECG forward simulation. We used five mesh resolutions and computed the EDL model with two spatial interpolation methods, triangle weighting and triangle splitting, in addition to testing without added interpolation. We evaluated computed body surface potentials against a set of ground truth potentials generated using a BEM transmembrane potential forward model.

2.1   ECG Forward Simulation

The ECG forward simulation used in this study is the EDL forward model [3]. EDL transmembrane potential (TMP) sources are parameterized to allow reformulation of the forward simulation in term of the local activation times of the cardiac tissue. For this study, the other phases of the action potential were ignored. Therefore, only the ventricular activation was simulated on the body surface.

Spatial Interpolation. In addition to the constant and linear interpolation of the TMP over each facet of the cardiac surface, we tested two methods for performing spatial interpolation of the activation times to reduce error due to low spatial resolution: triangle weighting and triangle splitting.

Triangle Weighting incorporates the activation times for the three vertices for each triangular face then scales that face's contribution to the source geometry based on how much of it is activated. This implementation requires solving the EDL for the triangular faces, instead of the vertices, yet activation times for each vertex.

Triangle Splitting is effectively adaptively refining the cardiac mesh along the activation wave front. For each time step, partially activated faces are subdivided into activated and inactivated regions. Using this adapted ventricular tessellation at each sample moment, the transfer function from ventricular surface to body surface is computed anew to find the ECGs at that moment.

2.2   Comparison Studies

The ground truth dataset used in this study consisted of simulated TMPs forward computed to the torso. The TMP were calculated at the maximum cardiac mesh resolution (147,458 nodes, mean edge length of 0.78 mm) using a bidomain simulation with a Ten Tusscher ionic model, with rule-based tissue anisotropy and conduction velocity of 1 m/s. The body surface potentials were then computed from the TMP using BEM. The torso mesh used contained 1520 nodes and had a mean edge length of 23 mm. The torso was modeled as homogeneous with a conductivity of 1 S/m.

To test various cardiac sampling resolutions, the cardiac mesh and TMPs were subsampled at 4 resolutions with edge lengths of 12.4 mm (578 nodes, resolution 0), 6.21 mm (2,306 nodes, resolution 1), 3.1 mm (9,218, resolution 2), and 1.6 mm (36,866 nodes, resolution 3). These subsampled cardiac sources combined with the full resolution mesh (resolution 4) comprised the 5 source resolutions tested in this study. For each resolution, we computed body surface potentials using the EDL model without spatial interpolation, with triangle weighting, and with triangle splitting. Simulated body surface potentials were compared to the ground truth using root mean squared voltage ($\bar{E}$), relative root mean square error ($rRMSE$), and correlation (ρ), as defined as follows:

$$\bar{E} = \frac{||\Phi_{\{gt,s\}}||}{\sqrt{n}} \tag{1}$$

$$rRMSE = \frac{||\Phi_{gt} - \Phi_s||}{||\Phi_{gt}||} \tag{2}$$

$$\rho = \frac{\Phi_{gt}^T \Phi_s}{||\Phi_{gt}|| ||\Phi_s||}, \tag{3}$$

where $\Phi_{\{gt,s\}}$ is a vector of either ground truth or simulated body surface potentials, n is the number of body surface electrodes, Φ_{gt} is a vector of the ground truth body surface potentials, and Φ_s is a vector of the associated simulated body surface potentials.

3   Results

Using different cardiac source sampling altered the computed body surface potentials in that the error in computed body surface potentials was reduced as the spatial resolution increased. Using spatial interpolation techniques also affected body surface potentials, yet the results were less dependent on the technique.

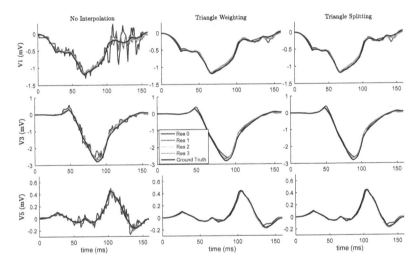

Fig. 1. Effect of spatial resolution and interpolation on ECG recordings. Ground truth and three cardiac surface resolutions are shown in each plot. The three rows represent precordial leads (v1, v3, and v5). Columns represent interpolation method. The ECGs simulated with Res 4 were nearly indistinguishable from those of Res 3.

The resolution of the cardiac sources and spatial interpolation affected the ECG signals computed with the forward simulation (Fig. 1). When the lowest resolution cardiac surface was used, the ECG signals contained high frequency artifacts that are reduced in amplitude as resolution increases, and are missing altogether from the highest resolution and the ground truth. These artifacts are greatly reduced or eliminated using either of the spatial interpolation methods. There is no observable difference between the two methods' effects on the ECG signals.

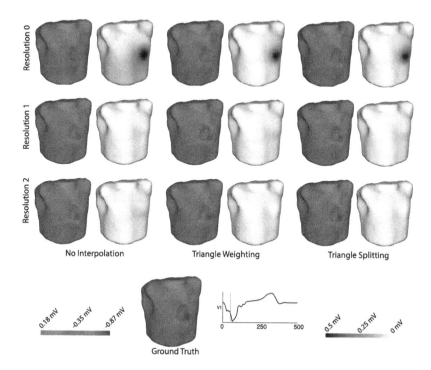

Fig. 2. Effect of spatial resolution and interpolation on body surface potential maps. Ground truth and three cardiac surface resolutions are shown. Difference maps are also included. The three rows represent cardiac surface resolution. Columns represent interpolation method. There were minimal qualitative differences between potential maps generated with Res 2 and those generated with higher source sampling (Res 3 & 4) maps generating higher

The predicted body surface potential maps were also affected by cardiac mesh resolution and interpolation method (Fig. 2). Although the predicted potential maps are overall qualitatively similar to the ground truth, the greatest difference occurs near the extrema. As the resolution increases, the potential maps more closely align with those of the ground truth. Using either spatial method also reduces the difference between the predicted potential map and the ground truth.

Figure 3 illustrates the $\bar{E}$, $rRMSE$, and the ρ for the predicted body surface potentials over the ventricle activation. Similar to the ECG signal tracings, there is high frequency oscillations in the metrics over time using the low resolution cardiac surfaces (resolution 0 and 1) in the ECG forward simulation. These oscillations are reduced with mesh resolutions 2 and higher for the $\bar{E}$ and ρ,

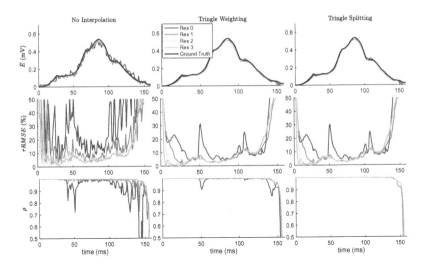

Fig. 3. Effect of spatial resolution and interpolation on all body surface potentials. Three cardiac surface resolutions are shown in each plot, and the ground truth is shown in the $\bar{E}$ plots. The three rows represent a metric ($\bar{E}$, $rRMSE$ and ρ). Columns represent interpolation method. The plot of the temporal metrics of Res 4 were nearly indistiguishable from Res 3.

yet $rRMSE$ still shows oscillations with resolution 2. These oscillations are reduce or eliminated with when spatial interpolation methods are used, even with the lowest cardiac source resolution. The two methods produced similar levels of improvement, with the exception of two remaining transient reductions in correlation with resolution 0 and triangle weighting, which are not present with triangle splitting. Interestingly, the $rRMSE$ error remains high at the beginning and end of ventricular activation, regardless of resolution or spatial interpolation. Likewise, the ρ is very low at the end of ventricular activation for all resolutions and spatial interpolation methods.

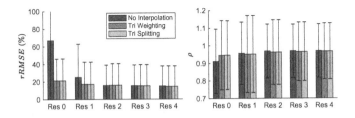

Fig. 4. Mean effect of spatial resolution and interpolation on all body surface potentials. The mean metric value over ventricular activation is shown, with the error bars representing standard deviation.

Figure 4 illustrates the mean $rRMSE$ and ρ in simulated body surface potentials for all cardiac surface resolutions and spatial interpolation methods. Similar to the trend seen in Fig. 3, the $rRMSE$ is highest and the ρ the lowest with resolution 0 and no spatial interpolation. The $rRMSE$ reduces and ρ increases when mesh resolution increases or spatial interpolation is used. There was no observable difference between spatial interpolation methods based on the mean error metrics of the predicted body surface potentials. The $rRMSE$ remained somewhat high despite the increased resolution and interpolation (minimum $rRMSE$ of 14.4 %). When we excluded the last 10 ms of the body surface potentials from the metric calculation, the $rRMSE$ reduced by ~ 5 % for all cases except with resolution 0 and no interpolation.

4 Discussion

The results presented in this study demonstrate that the sampling resolution of the cardiac sources can effect the ventricular activation body surface potentials computed with an EDL forward model. Additionally, the results show that spatial interpolation methods can mitigate or eliminate errors due to sampling resolution. The presented data also provide some insight into the resolution needed for EDL simulations.

The high frequency oscillation observed when predicting ECG signals with lower resolution cardiac sources (Fig. 1) are expected when using EDL and BEM theory. The relatively low resolution of the source mesh increases the relative contribution of each source point or face within the simulation and the time difference between neighboring regions becoming activated. Each of these effect contribute to increased discontinuity in the electric field that is manifest as high frequency oscillations. These effects can be smoothed through the forward calculation, yet with lower resolutions, this smoothing does not adequately eliminate the high frequency artifacts. The error in the forward simulation was also manifested through changes in extrema location in the body surface potential maps (Fig. 2). Similar behavior was demonstrated in previous studies analyzing the effect of source sampling error on forward simulations [5,7]. Interestingly, the error generated from reduced cardiac sampling is easier to observe by analyzing temporal signals than body surface potential maps, despite the error originating from spatial processes. While this discrepancy could be attributed to the complex nature of visualizing multidimensional data, it highlights the need for careful analysis of the sampling requirements for ECG forward simulations.

The spatial interpolation methods, triangle weighting and triangle splitting, both improved the accuracy of the predicted body surface potentials (Figs. 1, 2, 3, and 4) to a similar level. Either method improved simulation accuracy from a low resolution (resolution 0) cardiac source to match or exceed the accuracy of a simulation from a higher resolution (resolution 1, Fig. 4). Furthermore, using spatial interpolation, ECG signals calculated from the coarsest cardiac sources could be smoother and more closely match the ground truth than those generated with 20 times the number of nodes (resolution 2, Fig. 1). Both interpolation methods performed comparably, despite triangle splitting requiring more computational resources to compute.

The results from this dataset also indicate some possible guidelines for the spatial resolution cardiac source data for EDL simulations. Without spatial interpolation, the forward simulation required at least 9,218 nodes (resolution 2) to generate ECG signals with minimal artifact and reduced noise. However, with spatial interpolation, similar smoothness in ECG signals can be achieved with as few 578 nodes (resolution 0) and a similar level of error can be achieved with 2,306 nodes (resolution 1).

The guidelines and the interpolation tools presented have only been tested on a limited dataset. To expand the scope of these guidelines and tools, more geometries and activation profiles need to be tested. However, we anticipate a similar outcome based on previous published [5,7] results, as well as observations obtained from other, unpublished, simulations. This study also only evaluated the effect of cardiac sampling on body surface potentials during ventricular activation, excluding the plateau and repolarization stages of cardiac activity. It is possible to apply the interpolation methods to the plateau and repolarization phases, yet we expect the effect to be negligible because the spatial gradients of the cardiac sources is much lower than during activation. However, the ground truth dataset did include these stages, which provides an explain for the large error near the end of ventricular activation (Fig. 3). Excluding this time points reduced the overall error (Fig. 4, Sect. 3), suggesting that overall accuracy metrics could be improved by including the plateau phase and non-zero ST potentials to better represent late ventricular activation.

Although this study focused on the effect of cardiac source sampling on the ECG forward simulation, the result also have implications for ECGI applications. First, accurate forward simulations are needed for ECGI because many studies use simulated data to test new techniques [4,9]. Second, the errors associated with cardiac source sampling highlight a growing need to provide quantified uncertainty to both ECG forward models and ECGI [2,6,8], especially those designed for clinical use. Finally, similar strategies of generating accurate results with lower mesh resolution can improve ECGI by making complex solutions less computationally expensive, e.g., when using non-linear optimization with forward computed solutions to guess the next iterations [3].

We conclude that cardiac source resolution is an important consideration in EDL forward simulations. This study also suggested a possible minimum resolution for EDL and we introduced some spatial interpolation tools that may help mitigate errors from reduced cardiac sampling. These results will help improve ECG forward models which will lead to more accurate and new ECGI techniques to diagnose and treat cardiac arrhythmias.

References

1. Bear, L.R., et al.: The forward problem of electrocardiography: is it solved? Circ. Arrhythm. Electrophysiol. **8**(3), 677–684 (2015). https://doi.org/10.1161/CIRCEP.114.001573, https://www.ahajournals.org/doi/10.1161/CIRCEP.114.001573
2. Coll-Font, J., Roig-Solvas, B., van Dam, P., MacLeod, R.S., Brooks, D.H.: Can we track respiratory movement of the heart from the ECG itself - and improve inverse solutions too? J. Electrocardiol. **49**(6), 927 (2016)
3. van Dam, P., Oostendorp, T., Linnenbank, A., van Oosterom, A.: Non-invasive imaging of cardiac activation and recovery. Ann. Biomed. Eng. **37**(9), 1739–1756 (2009)
4. Erem, B., Ghodrati, A., Tadmor, G., MacLeod, R., Brooks, D.: Combining initialization and solution inverse methods for inverse electrocardiography. J. Electrocardiol. **44**(2), e21 (2011)
5. Potse, M., Kuijpers, N.H.L.: Simulation of fractionated electrograms at low spatial resolution in large-scale heart models. In: 2010 Computing in Cardiology, pp. 849–852 (2010)
6. Swenson, D., Geneser, S., Stinstra, J., Kirby, R., MacLeod, R.: Cardiac position sensitivity study in the electrocardiographic forward problem using stochastic collocation and BEM. Ann. Biomed. Eng. **30**(12), 2900–2910 (2011)
7. Tate, J., et al.: Reducing error in ECG forward simulations with improved source sampling. Front. Physiol. **9**, 1304 (2018). https://doi.org/10.3389/fphys.2018.01304
8. Tate, J.D., Zemzemi, N., Good, W.W., van Dam, P., Brooks, D.H., MacLeod, R.S.: Effect of segmentation variation on ECG imaging. In: Computing in Cardiology, vol. 45, September 2018. https://doi.org/10.22489/CinC.2018.374
9. Wang, Y., et al.: Noninvasive electroanatomic mapping of human ventricular arrhythmias with electrocardiographic imaging. Sci. Transl. Med. **3**(98), 98ra84 (2011)

Novel Imaging Tools and Analysis Methods for Myocardial Tissue Characterization and Remodelling

Left Ventricular Shape and Motion Reconstruction Through a Healthy Model for Characterizing Remodeling After Infarction

Mathieu De Craene[1](✉), Paolo Piro[1], Nicolas Duchateau[2], Pascal Allain[1], and Eric Saloux[3]

[1] Philips Research Paris, Suresnes, France
mathieu.de_craene@philips.com
[2] CREATIS, UMR5220, INSERM U 1206 Univ. Lyon 1, Lyon, France
[3] Caen University Hospital, Caen, France

Abstract. We introduce a framework for the statistical characterization of heart remodeling from both shape and dynamics of the left ventricle. Shape was characterized by thickness and radius maps, unfolded in a two-dimensional dense Bull's eye. Motion was represented as a mixture of affine transformations in an anatomical space of coordinates. Using this representation, a population can be projected (after defining spatiotemporal correspondences) to an atlas space built for a given reference population - here, healthy subjects using a classic PCA approach - yielding a joint model of healthy shape and motion statistics. The reconstruction error on shape and motion can then be exploited to quantify remodeling abnormalities. We demonstrate these concepts on 48 healthy subjects and 62 patients with infarct (29 with one year follow-up) imaged with 3D echocardiography, analyzing a total of 139 sequences.

Keywords: Cardiac · Shape · Motion · Abnormality

1 Introduction

Cardiac remodeling can be defined as the capacity of the cardiac muscle to adapt its structure, shape and dynamics to adversarial stresses. Statistical atlases can describe the variability of normal and pathological cardiac shape and motion. Therefore, they appear as a potential way of quantifying cardiac remodeling in a population. If statistical cardiac shape models are commonly used for segmenting images in clinical practice [2], motion atlases have been mostly used as a research tool for comparing or stratifying populations [1,4,5]. The following two issues actually condition their translation to clinical use.

How to compute a distance to normality? Computing a statistical distance between one subject and a control group made of healthy individuals, previously aligned to the same reference, can highlight suspicious regions. Such a concept

© Springer Nature Switzerland AG 2019
Y. Coudière et al. (Eds.): FIMH 2019, LNCS 11504, pp. 159–167, 2019.
https://doi.org/10.1007/978-3-030-21949-9_18

was often applied in neurologic images for revealing subtle volume changes. For cardiac imaging, it was applied to represent abnormalities of myocardial velocities [1] and shapes [10] as p-values.

How to compute a distance between subjects based on several features? Depending on the pathology and its severity, shape and motion will be affected differently. A complete analysis therefore requires combining these features in a relevant manner. A straightforward way is to concatenate features in a single vector per subject. For example, in a recent shape modelling challenge [9] aiming at classifying infarcted subjects, some authors combined end-systolic and end-diastolic shapes, thicknesses or radial displacements using PCA. An alternative when corroborating similar information for various modalities is to look for a common lower dimensional space where both information sources are projected. This unified lower-dimensional space can be used to reconstruct either inputs (for example in case of missing information) [6]. Finally, an alternative way of integrating heterogeneous features is to weigh their relative importance, as recently demonstrated with MKL, a non-linear dimensionality reduction technique that has been exploited to reveal relevant groups of subjects [7].

In this paper, we propose to exploit the reconstruction of cardiac shape and motion from low-dimensional coordinates to express both shape and motion abnormalities against a reference population of healthy subjects. We explicitly generate abnormality maps that quantify cardiac remodeling in a practical and intuitive way.

2   Methods

Patient Population and Data Acquisition. This paper combines data from two research protocols, approved by the ethical committee of the hospital (DIRECT and Reve II). Both protocols were conducted at the same institution and involved the same imaging team. Enrolled patients all gave informed consent. In total, 48 control subjects and 62 patients with infarcts were included. Table 1 summarizes characteristics and inclusion criteria of all populations.

Table 1. Patient characteristics for control and infarct populations.

	Controls	Infarcts	
		REVE II	DIRECT
n	**48**	**42**	**20**
Type of infarct		Anterior	Anterior / Inferior
Time points	$t = 0$	$t = 0$	$t = 0, 12$ months
Inclusion criteria	No known cardiovascular disease	Previous heart attack assessed by either scintigraphy or cardiac MRI	Acute infarct assessed by STEMI on ECG. Echo at most 5 days after infarct. $EF < 50\%$. > 3 akinetic ant. AHA segments on echo

Shape Representation. Left Ventricular (LV) meshes were obtained from the Philips Qlab (3DQA) software. All meshes were resampled as described in [11]. The endocardium and the epicardium were represented by two distance maps storing the distance to the long axis. Different samplings were used at different mesh locations: spherical for the top and bottom and cylindrical for the middle parts. Averaging the distance maps of the endocardial and epicardial surfaces gave a representation of the mid-LV surface. We defined this map as the LV *radius map*. The difference of the two distance maps was defined as the *thickness* map. Both maps can be visualized as Bull's eye plots (see Fig. 1).

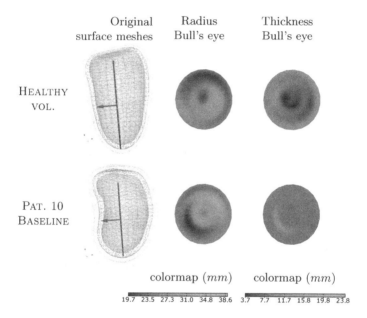

Fig. 1. Bull's eye representation of radius (center) and thickness maps (right) for one healthy volunteer and one infarcted patient (baseline).

Motion Model. We chose to represent cardiac motion by a poly-affine transformation model [4]. For this, we projected a dense non-rigid displacement field as obtained by a fast implementation of the Sparse Demons algorithm [8] on the regularization model from [11]. This regularization consists of three steps. First, a volumetric hexahedral mesh is built by connecting the endocardial, epicardial and mid-layers. Second, an anatomical system of coordinates is computed by orthogonalizing mesh edges at each vertex to define the Radial, Circumferential and Longitudinal (R, C, L) directions. Third, by geodesic integration of these vectors on the mesh, we obtain three scalar fields (r, c, and l) giving the local anatomical coordinates. A smooth partition of the mesh is then obtained through the definition of window functions.

$$\varphi^w(\mathbf{x}) = \frac{g^w(\mathbf{x})}{\sum_n g^n(\mathbf{x})} \quad . \tag{1}$$

where $g^w(\mathbf{x}) = \frac{1}{2\pi\sigma_w}\exp\left(-\frac{\xi(\mathbf{x},\mathbf{x}_w)^2}{2\sigma_w{}^2}\right)$ and ξ represents the geodesic distance on the mesh to the center of the wth window. We used $N_w = 17$ windows corresponding to the AHA segments. In each window, we estimated deformation and translation parameters to approximate the frame to frame displacement field $\mathbf{v}$ [8] by a linear model expressed in anatomical coordinates

$$\boldsymbol{\ell}^w(\mathbf{x}) = \mathbf{A}^w\left(r(\mathbf{x})\,c(\mathbf{x})\,l(\mathbf{x})\right)^T + \mathbf{b}^w \quad \text{for } w \in [0, N_w - 1]. \tag{2}$$

Minimizing the following error for each window

$$E_w = \sum_{\mathbf{x}\in\Omega} \varphi^w(\mathbf{x})\,\|\boldsymbol{\ell}^w(\mathbf{x}) - \mathbf{v}(\mathbf{x})\|^2 \quad , \tag{3}$$

is equivalent to solving a linear system in each direction $d \in \{r, c, l\}$:

$$\begin{pmatrix} \sum_{\mathbf{x}} \varphi^w(\mathbf{x})r^2(\mathbf{x}) & \sum_{\mathbf{x}} \varphi^w(\mathbf{x})r(\mathbf{x})l(\mathbf{x}) & \sum_{\mathbf{x}} \varphi^w(\mathbf{x})r(\mathbf{x})c(\mathbf{x}) & \sum_{\mathbf{x}} \varphi^w(\mathbf{x})r(\mathbf{x}) \\ \sum_{\mathbf{x}} \varphi^w(\mathbf{x})r(\mathbf{x})l(\mathbf{x}) & \sum_{\mathbf{x}} \varphi^w(\mathbf{x})l^2(\mathbf{x}) & \sum_{\mathbf{x}} \varphi^w(\mathbf{x})l(\mathbf{x})c(\mathbf{x}) & \sum_{\mathbf{x}} \varphi^w(\mathbf{x})l(\mathbf{x}) \\ \sum_{\mathbf{x}} \varphi^w(\mathbf{x})r(\mathbf{x})c(\mathbf{x}) & \sum_{\mathbf{x}} \varphi^w(\mathbf{x})l(\mathbf{x})c(\mathbf{x}) & \sum_{\mathbf{x}} \varphi^w(\mathbf{x})c^2(\mathbf{x}) & \sum_{\mathbf{x}} \varphi^w(\mathbf{x})c(\mathbf{x}) \\ \sum_{\mathbf{x}} \varphi^w(\mathbf{x})r(\mathbf{x}) & \sum_{\mathbf{x}} \varphi^w(\mathbf{x})l(\mathbf{x}) & \sum_{\mathbf{x}} \varphi^w(\mathbf{x})c(\mathbf{x}) & \sum_{\mathbf{x}} \varphi^w(\mathbf{x}) \end{pmatrix} \begin{pmatrix} A_{dr} \\ A_{dl} \\ A_{dc} \\ b_d \end{pmatrix}$$

$$= \begin{pmatrix} \sum_{\mathbf{x}} \varphi^w(\mathbf{x})(\mathbf{v}(\mathbf{x}) \cdot \hat{\mathbf{e}}_d(\mathbf{x}))r(\mathbf{x}) \\ \sum_{\mathbf{x}} \varphi^w(\mathbf{x})(\mathbf{v}(\mathbf{x}) \cdot \hat{\mathbf{e}}_d(\mathbf{x}))l(\mathbf{x}) \\ \sum_{\mathbf{x}} \varphi^w(\mathbf{x})(\mathbf{v}(\mathbf{x}) \cdot \hat{\mathbf{e}}_d(\mathbf{x}))c(\mathbf{x}) \\ \sum_{\mathbf{x}} \varphi^w(\mathbf{x})(\mathbf{v}(\mathbf{x}) \cdot \hat{\mathbf{e}}_d(\mathbf{x})) \end{pmatrix} \quad . \tag{4}$$

Linear transformations per window (Eq. 2) can then be converted to Cartesian coordinates by using the RLC basis vectors and then merged through the window functions $\varphi^w(\mathbf{x})$ to deform the whole ventricle.

Multi-scale Approach. To obtain a multi-scale representation, we first averaged all per-window systems and solved for a global linear transformation (global level). We then averaged the systems of basal, mid and apical windows and fitted the residual from the global reconstructed field (level 2). We finally proceeded similarly for the finest level (level 3, 17 AHA segments).

Healthy Shape Model. PCA was applied to the thickness and radius maps on the healthy population. Both thickness and radius maps were vectorized before PCA. Also, two normalization strategies were considered. The first was to normalize all input maps by their average value. The second normalized them by the cavity volume. Plotting the explained variance ratio revealed that the volume normalization led to a more compact representation than the average one. Volume normalization was therefore chosen in our experiments. The same explained variance curve were also used to set the number of PCA components used for reconstruction. A 90% threshold led to 5 components for radius and 7 for thickness.

Healthy Motion Model. All frame to frame linear regression results $\mathbf{A}^w$ and $\mathbf{b}^w$ were interpolated (and scaled) along time to a normalized heart cycle of 30 frames. We used mitral valve opening and the beginning/end of the cycle as temporal landmarks. For computing healthy motion modes of variations, we combined different concatenation strategies. First, the *global* transformations were vectorized as described in [4] and concatenated across all time points. Computing a PCA across all subjects encoded the temporal variability of global healthy motion. Then, for the next resolutions (levels 2 and 3), matrices were vectorized and concatenated across the different spatial regions (3 at level 2, 17 at level 3) for each frame separately. This set of PCAs thus encoded the spatial variability at each frame seen over healthy subjects. The number of components to be used for each PCA in the healthy motion model was selected with a 90% threshold at each level on the explained variance curves. For levels 2 and 3, we took the maximum number of components over all time points after applying the threshold. This gave 18 components for the global level, and 8/12 components (per frame) for the subsequent levels.

Reconstruction. For a given new subject, both shape and motion can be projected onto the healthy PCA output space and reconstructed. Computing the difference in the input spaces between the original and the reconstructed scalar/vector fields will indicate pathological deviations from the healthy model. For shape, as we treat two static scalar fields (radius and thickness), we defined the difference between original and healthy-based reconstructed maps as *absolute radius/thickness abnormality*. For motion, we computed the relative error between the displacement encoded by the original and reconstructed mesh points. This was done for the normalized cardiac cycle at the myocardial mid-layer. We defined a *relative motion abnormality* as

$$\mu_d(\mathbf{x}) = (\mathbf{u}(\mathbf{x}, \mathbf{t}) - \hat{\mathbf{u}}(\mathbf{x}, \mathbf{t})) \frac{\min(1, \alpha \| \hat{\mathbf{u}}(\mathbf{x}, \mathbf{t}) \|)}{\| \hat{\mathbf{u}}(\mathbf{x}, \mathbf{t}) \|} \quad , \tag{5}$$

where $\mathbf{u}(\mathbf{x}, \mathbf{t})$ is the displacement obtained by chaining the poly-affine estimate of Eq. 2 over time and $\hat{\mathbf{u}}(\mathbf{x}, \mathbf{t})$ is its reconstructed counterpart (from the healthy model). In Eq. 5, the numerator of the second factor neutralizes the denominator for small displacement values.

3   Results

Shape Reconstruction. Figure 2 shows the same input shapes as in Fig. 1 with the reconstructed radius map using the PCA normality model (overlaid as wireframe). The first case (healthy) was left out of the PCA training set. For the pathological case, the septal bulge is smoothed out by the normality model, resulting in negative and positive reconstruction errors along the septum, both at baseline and follow-up.

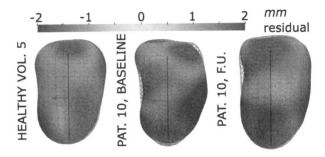

Fig. 2. Signed difference between the original radius map and its reconstruction (shown in wireframe) for a healthy subject and the patient shown in Fig. 1 (here at baseline and follow-up).

Motion Reconstruction. Figure 3 (see links in the caption) plots the mid-layer meshes tracked around end-systolic frames at baseline and follow-ups (blue color). A difference in LV shape is visible between baseline and follow-up. The LV moved back to a more ellipsoidal shape at follow-up, which is known as reverse remodeling. The reconstructed healthy motion for this patient is plotted in green in Fig. 3. Differences between the healthy reconstruction and the patient's motion appear more pronounced at baseline than follow-up. This indicates that similarly to shape, motion abnormalities also reduced for this patient at follow-up. For taking a closer look at motion abnormalities, Fig. 4 plots the Bull's eye diagrams of μ_d (Eq. 5) over time points close to the end of systole together with scar transmurality. The latter was segmented from late enhancement MR using the Segment (Medviso) software and resampled to the same Bull's eye as the motion abnormality maps. The μ_d maps return high abnormality values around the septal region, being reduced at follow-up. The extent and exact location of abnormalities differ between our index and the scar. This can first be explained by some potential misalignment when defining AHA segments in both modalities. Also, complex interaction mechanisms between stunned, infarcted and normal tissue (at baseline), or between normal, re-vascularized and scar tissue (at follow-up), can also explain the larger extent of the abnormality with our index.

Abnormality Spreads. Figure 5a plots the L^2 norm of radius, thickness, motion and deformation abnormalities as box plots for controls and infarcts. Controls shape and motion abnormalities were computed by excluding that subject from the normality model database (leave-one-out approach). For each abnormality index, values were centered and normalized for the healthy population to have zero mean and unit standard deviation. Also, a two-sided Mann-Whitney rank test between controls and infarcts returned the p-values showed over the curly brackets.

Joint Analysis. Finally, we selected one feature for shape and one for dynamics (radius and motion based on the p-values from Fig. 5a). These two indexes were rendered as a scatter plot in Fig. 5b. Similarly to Fig. 5(a), both axes show the L^2 norm of the radius and μ_d abnormality maps. Controls, infarcts at baseline and infarcts at follow-up are plotted using different symbols. Most patients at follow-up had reduced relative motion abnormality μ_d compared to the overall distribution of baseline infarcts. A more detailed analysis would be required to compare the evolution in shape and motion abnormality values to the clinical outcome of each patient. Although the values in Fig. 5 do not seem to offer the same classification accuracy as some of the submissions to the shape modelling challenge [9], it should be emphasized that our results were extracted from a database of 3D US and not cine MR images. Also, as the patient populations differ, results can not be directly compared in terms of accuracy.

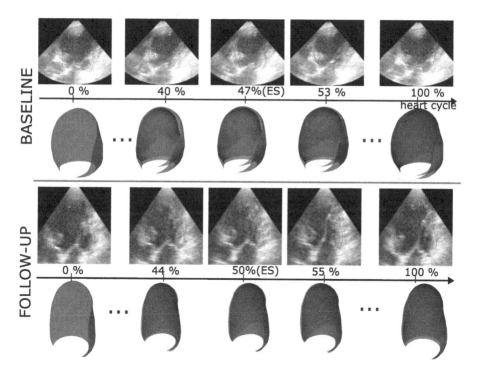

Fig. 3. LV meshes over the normalized cardiac cycle at baseline and follow-up for infarcted Patient #4. The reconstruction based on the healthy model is shown in green. For an animated version, click on the *following links*: baseline, follow-up. (Color figure online)

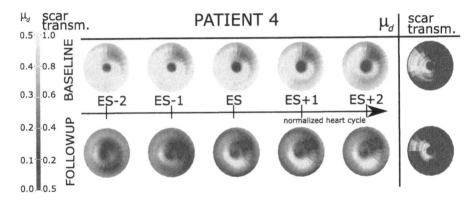

Fig. 4. Bull's eye plots of the motion (μ_d) abnormality computed over time for the same patient shown in Fig. 3 at baseline and follow-up. The scar transmurality (between 0 and 1) as measured from late enhancement images is plotted on the right.

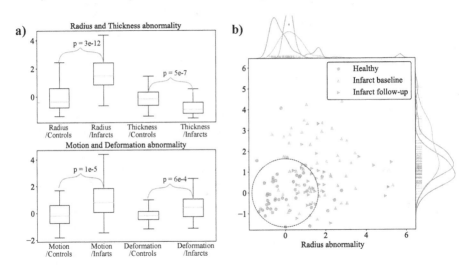

Fig. 5. (a) Abnormality spreads for radius, thickness, motion and deformation among controls and infarcts. (b) Distribution of the L^2 norm of μ_d and radius abnormality maps across healthy and infarcted populations, distinguishing between baseline and follow-up.

4   Conclusions

We proposed a generic way to quantify abnormalities from different input descriptors, here illustrated on shape and motion to quantify cardiac remodeling in a population of patients with infarcts, comparing it to the normality pattern observed from a population of healthy subjects. A simple PCA approach was used to capture the healthy variability of both shape and motion descriptors. This model was then used on unseen cases to compute a reconstruction error that is used as an abnormality signature for both shape and motion. Including more pathological groups could emphasize different remodeling patterns. Also, the complex relationship between the presence of fibrosis (as revealed by late enhancement MR) and deformation requires to be further analyzed for a better comparison of scar and shape/motion abnormality maps.

References

1. Duchateau, N., et al.: A spatiotemporal statistical atlas of motion for the quantification of abnormal myocardial tissue velocities. Med. Image Anal. **15**(3), 316–328 (2011)
2. Ecabert, O., et al.: Automatic model-based segmentation of the heart in CT images. IEEE Trans. Med. Imaging **27**(9), 1189–1201 (2008)
3. Fertin, M., et al.: Usefulness of serial assessment of b-type natriuretic peptide, troponin i, and c-reactive protein to predict left ventricular remodeling after acute myocardial infarction (from the REVE-2 Study). Am. J. Cardiol. **106**(10), 1410–1416 (2010)
4. McLeod, K., et al.: Spatio-temporal tensor decomposition of a polyaffine motion model for a better analysis of pathological left ventricular dynamics. IEEE Trans. Med. Imaging **34**(7), 1562–1575 (2015)
5. Peressutti, D., et al.: Prospective identification of CRT super responders using a motion atlas and random projection ensemble learning. In: Navab, N., Hornegger, J., Wells, W.M., Frangi, A.F. (eds.) MICCAI 2015. LNCS, vol. 9351, pp. 493–500. Springer, Cham (2015). https://doi.org/10.1007/978-3-319-24574-4_59
6. Puyol-Antón, E., et al.: A multimodal spatiotemporal cardiac motion atlas from mr and ultrasound data. Med. Image Anal. **40**, 96–110 (2017)
7. Sanchez-Martinez, S., et al.: Characterization of myocardial motion patterns by unsupervised multiple kernel learning. Med. Image Anal. **35**, 70–82 (2017)
8. Somphone, O., et al.: Fast myocardial motion and strain estimation in 3D cardiac ultrasound with sparse demons. In: 2013 IEEE 10th International Symposium on Biomedical Imaging, pp. 1182–1185 (2013)
9. Suinesiaputra, A., et al.: Statistical shape modeling of the left ventricle: myocardial infarct classification challenge. IEEE J. Biomed. Health Inform. **22**(2), 503–515 (2018)
10. Zhang, X., et al.: Atlas-based quantification of cardiac remodeling due to myocardial infarction. PloS ONE **9**(10), e110243 (2014)
11. Zhou, Y., et al.: 3D harmonic phase tracking with anatomical regularization. Med. Image Anal. **26**(1), 70–81 (2015)

Towards Automated Quantification of Atrial Fibrosis in Images from Catheterized Fiber-Optics Confocal Microscopy Using Convolutional Neural Networks

Chao Huang[1,2], Stephen L. Wasmund[1], Takanori Yamaguchi[1],
Nathan Knighton[3], Robert W. Hitchcock[3], Irina A. Polejaeva[4],
Kenneth L. White[4], Nassir F. Marrouche[1],
and Frank B. Sachse[1,2,3(✉)]

[1] Comprehensive Arrhythmia and Research Management (CARMA) Center,
Division of Cardiovascular Medicine, University of Utah, Salt Lake City, USA
[2] Nora Eccles Harrison Cardiovascular Research and Training Institute,
University of Utah, Salt Lake City, USA
frank.sachse@utah.edu
[3] Department of Biomedical Engineering, University of Utah,
Salt Lake City, USA
[4] Department of Animal, Dairy and Veterinary Sciences,
Utah State University, Logan, USA

Abstract. Clinical approaches for quantification of atrial fibrosis are currently based on digital image processing of magnetic resonance images. Here, we introduce and evaluate a comprehensive framework based on convolutional neural networks for quantifying atrial fibrosis from images acquired with catheterized fiber-optics confocal microscopy (FCM). FCM images in three regions of the atria were acquired in the beating heart in situ in an established transgenic animal model of atrial fibrosis. Fibrosis in the imaged regions was histologically assessed in excised tissue. FCM images and their corresponding histologically-assessed fibrosis levels were used for training of a convolutional neural network. We evaluated the utility and performance of the convolutional neural networks by varying parameters including image dimension and training batch size. In general, we observed that the root-mean square error (RMSE) of the predicted fibrosis was decreased with increasing image dimension. We achieved a RMSE of 2.6% and a Pearson correlation coefficient of 0.953 when applying a network trained on images with a dimension of 400×400 pixels and a batch size of 128 to our test image set. The findings indicate feasibility of our approach for fibrosis quantification from images acquired with catheterized FCM using convolutional neural networks. We suggest that the developed framework will facilitate translation of catheterized FCM into a clinical approach that complements current approaches for quantification of atrial fibrosis.

Keywords: Atrial fibrosis · Fiber-optics confocal microscopy · Machine learning

© Springer Nature Switzerland AG 2019
Y. Coudière et al. (Eds.): FIMH 2019, LNCS 11504, pp. 168–176, 2019.
https://doi.org/10.1007/978-3-030-21949-9_19

1   Introduction

Atrial fibrosis, defined as excessive formation of connective tissue in the atria, is associated with a variety of conditions including heart failure, ischemia, hypertension, and mitral valve disease [1]. Atrial fibrosis is also associated with an increased risk of atrial fibrillation. Recent studies showed that many atrial fibrillation patients exhibit atrial fibrosis, which is thought to be the major contributor to arrhythmogenic tissue remodeling in these patients [2, 3]. Also, atrial fibrosis is an important predictor of outcome in patients undergoing catheter ablation for the treatment of atrial fibrillation [4].

Several approaches for atrial fibrosis identification and quantification at the microscopic and macroscopic scale have been developed. Recently developed macroscopic approaches are based on magnetic resonance imaging (MRI) and include T1 mapping and late gadolinium enhancement MRI [5]. Further approaches for identification of fibrotic remodeling and scarring apply endocardial electrical measurements. In particular, fractionated electrograms and reduced voltage amplitudes are thought to identify fibrosis [6, 7]. Microscopic approaches for fibrosis quantification apply stains to assess collagen content in sections of excised tissue. For both macroscopic and microscopic approaches, image processing is required to quantify fibrosis from the image data.

Here, we present a methodology for evaluating microscopy-based fibrosis quantification in the beating heart in situ. Our studies are based on a transgenic goat model of fibrosis, which was found to exhibit a significant increase in the atrial collagen area fraction versus control [8]. We imaged the atria in this model applying previously introduced catheterized fiber-optics confocal microscopy (FCM) of the beating heart in-situ [9]. We histologically assessed fibrosis based on excised atrial tissues from these imaged regions. The histologically derived fibrosis was indexed to their corresponding FCM images, which were subsequently applied for training of convolutional neural networks. We explored this machine learning approach by varying parameters of the networks and training. We evaluated the trained networks in quantifying fibrosis from a test set of FCM images and assessed its utility using the root mean square error (RMSE) of predicted fibrosis levels.

2   Methods

2.1   Animal Model

All experimental procedures in this study conformed to the National Institutes of Health Guide for the Care, and Use of Laboratory Animals, and were approved by the Institutional Animal Care and Use Committee (IACUC) of the University of Utah (IACUC protocol #17-08008). We used a transgenic goat model of atrial fibrosis produced by cardiac specific overexpression of human transforming growth factor-β1 [8]. To further increase atrial fibrosis based on an approach previously applied in canine [10], we implanted a pacemaker in one animal, and paced the right ventricle at 240 bpm for 3 weeks prior to FCM imaging.

2.2   Catheterized Fiber-Optics Confocal Microscopy

Transgenic goats (n = 2) weighing 40 to 60 kg were anesthetized with Propofol and mechanically ventilated. We performed cardiac catheterization in these anesthetized animals using a percutaneous femoral approach. Under intracardiac echocardiographic and fluoroscopic guidance, a transseptal puncture was performed and a bi-directional steerable guiding sheath (14F Destino; Oscor Inc, Palm Harbor, FL, USA) was placed into the left atrium. Afterward, we performed cardiac mapping (EnSite Precision™, Abbott St. Jude Medical, Green Oaks, Illinois, USA) to generate 3D models of the atrial chambers. Following mapping, fluorescein sodium (AK-Fluor 10%, Akorn Inc., Lake Forest, IL) was intravenously injected at a dose of 7.7 mg/kg to label the extracellular space. The FCM imaging microprobe (UltraMiniO, CellVizio, Mauna Kea Technologies, Paris, France) was steered and positioned via the steerable sheath on the endocardial surfaces within atrial chambers. We acquired images at a field of view (xy) of 245 μm at a rate of 12 images/s. The lateral and axial resolution as well as depth of field of the image system was 1.4, 9, and 66 μm, respectively. The distal tip of our imaging microprobe was customized with electrodes to track its 3D position via cardiac mapping. Image sequences, tip positions, and 3D anatomical models were annotated, indexed and stored on a file server for subsequent analyses. Following the imaging study, the animals were sacrificed and their hearts excised and formalin fixed.

2.3   Histological Assessment of Fibrosis

Full-thickness biopsies from the imaged left atrial anterior and right atrial free wall were paraffin embedded, transmurally sectioned, and Masson's trichrome stained. The slide sections were imaged with a digital slide scanner (Axio Scan.Z1, Zeiss, Jena, Germany) equipped with a 40x lens at a lateral resolution of 0.89 μm (Fig. 1A).

Quantitative fibrosis assessment was performed on these acquired images using the image processing software Fiji [11]. In short, we selected a rectangular region of interest through the full myocardial thickness excluding endo- and epicardium, as well as greater vessels (Fig. 1A). We applied color deconvolution for Masson's trichrome stains to separate out the contribution of the blue and red dyes based on their RGB absorption within the selected region (Figs. 1B and C). The deconvolved blue and red images corresponding to their respective dyes were subsequently histogram-derived thresholded (Figs. 1D and E). The thresholded red component image was subtracted from the blue component image. The resulting binary image displayed only the blue component pixels that correspond to collagen (Fig. 1F). A selection was created of all dark pixels in the original blue component image and applied to the collagen-only image. The area fraction of the collagen component within the selected blue component was used as our measure of fibrosis.

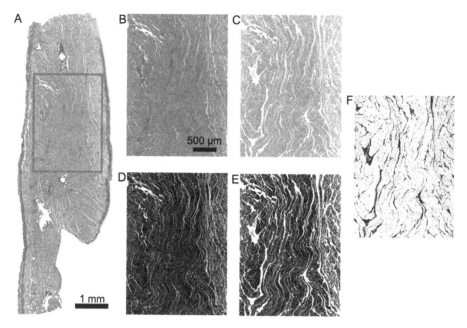

Fig. 1. Image processing for fibrosis assessment. (A) Representative Masson's trichrome cross-section of the right atrial free wall. (B–F) Processed region of interest outlined in green. Color deconvolved region separated into (B) blue, (C) red components and after (D, E) histogram-derived thresholding. (F) Visualization of fibrosis from subtracting (E) from (D). Scale bar in (B) applies to (C–F). (Color figure online)

Table 1. Convolutional neural network configuration.

Layer	Layer type	Parameters and range
1	Input	Dimensions (pixels): [50 50], [100 100], ... , [400 400]
2, 6, ... , 18	Convolution	Filter size: 2; Number of filters: 2, 4, ... , 32
3, 7, ... , 19	Normalization	
4, 8, ... , 20	Rectified linear unit	
5, 9, ... , 17	Maxima pooling	Pool size: 2; Stride: 2
21	Fully connected	
22	Regression	

2.4 Fibrosis Quantification Using Convolutional Neural Networks

We configured a convolutional neural network (Table 1) for quantification of fibrosis in FCM images from 3 atrial regions using histological assessment as ground truth. In short, we inspected all FCM image sequences and selected images free of artifacts. We avoided inclusion of replicates. Through this process we were able to stratify these images into three groups based on imaged region and their histologically-derived fibrosis level. The 3 imaged regions were of the right atrial free wall in both the paced and non-paced transgenic goat as well as the left atrial anterior wall in the paced transgenic goat. Eight FCM images from each of the three atrial regions were selected and subsequently rotated 35 times in $10°$ increments to expand the image pool (n = 864). The original images from each of the 3 atrial regions and their corresponding 35 rotated variants were randomly split into a training and test set. A split of 75% and 25% for the training (n = 648) and test (n = 216) sets, respectively was determined to adequately reduce variance and overfitting in the training process as well as eliminating overlap of images and their variants across the training and test groups. Initially, we trained the network on full resolution images, but also investigated image dimensions of 50, 100, and 200 pixels. We compared a stochastic gradient descent and the Adam optimizer for training of the network. Training was terminated after a maximal number of epochs of 1000 or when the training RMSE did not decrease for 500 iterations. Batches of training data were used in each epoch and shuffled after each epoch. We varied batch sizes between 64, 128, and 256. Each combination of batch size and image dimension was applied 10 times for training of a network.

3   Results

Example Masson's trichome images for the 3 different regions with low, intermediate and high fibrosis in the atria of transgenic animals are presented in Fig. 2A, D and G, respectively. Corresponding images of the detected fibrosis are shown in Fig. 2B, E and H. The measured degree of fibrosis is presented in Table 2. We annotated images from catheterized FCM from these regions with the measured degree of fibrosis. Example collagen-only images from atrial regions with low, intermediate and high fibrosis are presented in Fig. 2C, F, and I, respectively.

Statistical information on RMSE for training using the stochastic gradient descent optimizer and with varying batch sizes and image dimensions is summarized in Fig. 3A. In general, decreasing image dimension or batch size increased training RMSE. The number of iterations for each training process ranged between 350 and 2200 (Fig. 3B). We calculated RMSE for application of trained neural networks to the test image set (Fig. 3C). In general, test RMSE decreased with increasing image dimension. The stochastic gradient descent optimizer outperformed the Adam optimizer when applied to both the training and test image sets. The smallest training and test RMSE achieved with the Adam optimizer was 0.80 and 7.9%, respectively.

Table 2. Comparison of fibrosis measured in Masson's trichrome and predicted from FCM images and the convolutional neural network.

FCM image set	Transgenic model	Tissue region	Measured fibrosis (%)	Predicted fibrosis (%, mean ± stdev)
1	Paced	Left atrial anterior wall	16.5	16.7 ± 2.3
2	Non-paced	Right atrial free wall	21.8	23.0 ± 3.8
3	Paced	Right atrial free wall	39.7	37.7 ± 1.8

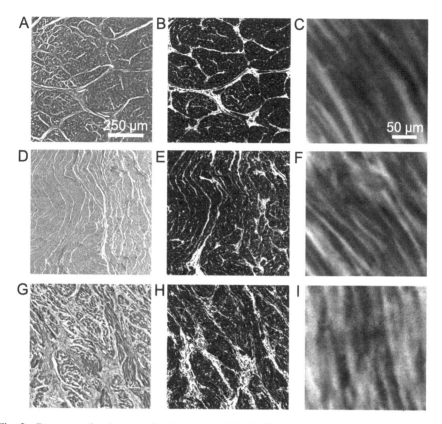

Fig. 2. Representative images of atrial tissue with (A–C) low, (D–F) intermediate, and (G–I) high fibrosis. Zoom-in of Masson's trichrome stained regions depicting fibrosis (blue pixels) at (A) low, (D) intermediate, and (G) high levels. (B, E, H) Visualization of fibrosis (bright pixels) in the Masson trichrome zoom-ins post-process. Representative images from FCM image sequences acquired from the (C) low, (F) intermediate, and (I) high fibrosis regions in situ. Scale bar in (A) applies to (B, D, E, G, H). Scale bar in (C) applies to (F, I). (Color figure online)

The stochastic gradient descent optimizer achieved a training and test RMSE of 0.50 and 2.6%, respectively, for a network trained on images with a dimension of 400×400 pixels and a batch size of 128. The relationship between batch size and test RMSE was weak.

The predicted degree of fibrosis for each region using the network with the smallest training RMSE is listed in Table 2. The Pearson correlation coefficient between predicted and measured fibrosis was high (0.953).

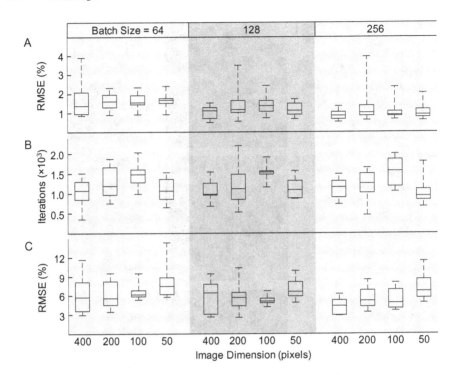

Fig. 3. Statistical analysis of convolutional neural network trained with varying batch sizes and image dimensions. (A) RMSE and (B) number of iterations upon termination for training. (C) RMSE after applying the trained neural network to the test image set.

4   Discussion and Conclusion

In this study, we introduced and evaluated a framework for quantification of atrial fibrosis in the beating heart in situ based on microscopic imaging and convolutional neural networks. We explored our approach in a transgenic animal model of atrial fibrosis and using histological analyses from excised atrial tissues as ground truth. An important finding was that reduction of image resolution reduced the accuracy of fibrosis quantification. The overall performance of the neural network for fibrosis prediction was promising. The predicted fibrosis values in our test image set was close to the ground truth values based on our RMSE and cross correlation analysis.

Our results suggest that catheterized FCM may translate into a clinical approach complementing current approaches for quantification of atrial fibrosis. Currently, assessment of fibrosis burden is based on MRI or endocardial voltage mapping. MRI and voltage mapping lack the spatial resolution to accurately and precisely localize atrial fibrosis. In addition, these macroscopic techniques have pigeonholed fibrosis into only two categories: patchy or diffuse. Microscopic approaches such as FCM may expand these definitions and etiologies of fibrosis based on insights that only becomes apparent at this scale. These opportunities motivate a need for a high-resolution imaging technology to accurately and precisely localize atrial fibrosis and assess fibrosis burden. The introduced framework takes steps towards translating catheterized FCM for clinical assessment of atrial fibrosis at a meaningful resolution.

We note several limitations of our study. We selected only images without notable motion artifacts for training and testing of the network. Future refinements to our framework could automatically detect such artifacts using the neural network and remove detected images from subsequent analysis. In addition, we focused our analyses on a small number of atrial regions and animals which limited the diversity in the images acquired. However, even with this small dataset there appeared to be learnable features that produced reliable quantification of fibrosis using our framework. Randomization of the images into their respective training and test sets was performed once. A more comprehensive evaluation of our framework would be to perform several iterations of the randomization process and how this parameter affects the overall performance of the neural network. We also evaluated only two of many numerical optimizers available for training. The two optimizers were limited in that they produced largely varying outputs due to the stochastic nature of the learning process. Alternative approaches such as particle swarm optimization could be more efficient.

Acknowledgments. We acknowledge support by the National Institutes of Health (R01HL 135077 and T32HL007576-31), American Heart Association (18POST34020052), the Nora Eccles Treadwell Foundation, and the Technology and Venture Commercialization, University of Utah.

References

1. Burstein, B., Nattel, S.: Atrial fibrosis: mechanisms and clinical relevance in atrial fibrillation. J. Am. Coll. Cardiol. **51**, 802–809 (2008)
2. Platonov, P.G., Mitrofanova, L.B., Orshanskaya, V., Ho, S.Y.: Structural abnormalities in atrial walls are associated with presence and persistency of atrial fibrillation but not with age. J. Am. Coll. Cardiol. **58**, 2225–2232 (2011)
3. Boldt, A., et al.: Fibrosis in left atrial tissue of patients with atrial fibrillation with and without underlying mitral valve disease. Heart **90**, 400–405 (2004)
4. Marrouche, N.F., et al.: Association of atrial tissue fibrosis identified by delayed enhancement MRI and atrial fibrillation catheter ablation: the DECAAF study. JAMA **311**, 498–506 (2014)
5. Oakes, R.S., et al.: Detection and quantification of left atrial structural remodeling with delayed-enhancement magnetic resonance imaging in patients with atrial fibrillation. Circulation **119**, 1758–1767 (2009)
6. Jadidi, A.S., et al.: Inverse relationship between fractionated electrograms and atrial fibrosis in persistent atrial fibrillation: combined magnetic resonance imaging and high-density mapping. J. Am. Coll. Cardiol. **62**, 802–812 (2013)
7. Yamaguchi, T., et al.: Impact of the extent of low-voltage zone on outcomes after voltage-based catheter ablation for persistent atrial fibrillation. J. Cardiol. **72**, 427–433 (2018)
8. Polejaeva, I.A., et al.: Increased susceptibility to atrial fibrillation secondary to atrial fibrosis in transgenic goats expressing transforming growth factor-beta1. J. Cardiovasc. Electrophysiol. **27**, 1220–1229 (2016)

9. Huang, C., Wasmund, S., Hitchcock, R., Marrouche, N.F., Sachse, F.B.: Catheterized fiber-optics confocal microscopy of the beating heart in situ. Circ. Cardiovasc. Imaging **10**, e006881 (2017)

10. Hanna, N., Cardin, S., Leung, T.K., Nattel, S.: Differences in atrial versus ventricular remodeling in dogs with ventricular tachypacing-induced congestive heart failure. Cardiovasc. Res. **63**, 236–244 (2004)

11. Schindelin, J., et al.: Fiji: an open-source platform for biological-image analysis. Nat. Methods **9**, 676–682 (2012)

High-Resolution *Ex Vivo* Microstructural MRI After Restoring Ventricular Geometry via 3D Printing

Tyler E. Cork[1,2] , Luigi E. Perotti[3] , Ilya A. Verzhbinsky[1] ,
Michael Loecher[1] , and Daniel B. Ennis[1(✉)]

[1] Department of Radiology, Stanford University, Stanford, CA 94305, USA
dbe@stanford.edu
[2] Department of Bioengineering, Stanford University, Stanford, CA 94305, USA
[3] Department of Mechanical and Aerospace Engineering,
University of Central Florida, Orlando, FL 32816, USA

Abstract. Computational modeling of the heart requires accurately incorporating both gross anatomical detail and local microstructural information. Together, these provide the necessary data to build 3D meshes for simulation of cardiac mechanics and electrophysiology. Recent MRI advances make it possible to measure detailed heart motion *in vivo*, but *in vivo* microstructural imaging of the heart remains challenging. Consequently, the most detailed measurements of microstructural organization and microanatomical infarct details are obtained *ex vivo*. The objective of this work was to develop and evaluate a new method for restoring *ex vivo* ventricular geometry to match the *in vivo* configuration. This approach aids the integration of high-resolution *ex vivo* microstructural information with *in vivo* motion measurements. The method uses *in vivo* cine imaging to generate surface meshes, then creates a 3D printed left ventricular (LV) blood pool cast and a pericardial mold to restore the *ex vivo* cardiac geometry to a mid-diastasis reference configuration. The method was evaluated in healthy (N = 7) and infarcted (N = 3) swine. Dice similarity coefficients were calculated between *in vivo* and *ex vivo* images for the LV cavity (0.93 ± 0.01), right ventricle (RV) cavity (0.80 ± 0.05), and the myocardium (0.72 ± 0.04). The R^2 coefficient between *in vivo* and *ex vivo* LV and RV cavity volumes were 0.95 and 0.91, respectively. These results suggest that this method adequately restores *ex vivo* geometry to match *in vivo* geometry. This approach permits a more precise incorporation of high-resolution *ex vivo* anatomical and microstructural data into computational models that use *in vivo* data for simulation of cardiac mechanics and electrophysiology.

Keywords: 3D printing · Magnetic resonance imaging ·
Cardiac electromechanics · Computational modeling

This work was supported by NIH/NHLBI K25-HL135408 and R01-HL131823. The content is solely the responsibility of the authors and does not necessarily represent the official views of the National Institutes of Health.

© Springer Nature Switzerland AG 2019
Y. Coudière et al. (Eds.): FIMH 2019, LNCS 11504, pp. 177–186, 2019.
https://doi.org/10.1007/978-3-030-21949-9_20

1   Introduction

Computational modeling can be used to better understand the mechanisms of cardiac electromechanics in both healthy and infarcted hearts. Advanced MRI methods enable quantifying both functional and microstructural properties of the heart for incorporation into computational models. For example, MRI can be used to: (1) measure tissue displacements with the cine DENSE technique, thereby enabling the estimation of cardiac strain [16]; (2) characterize microstructural orientation and compartment shape through diffusion tensor imaging (DTI) [1]; and (3) characterize myocardial infarct microanatomy using late gadolinium enhancement (LGE) [4].

Several groups have recently developed *in vivo* cardiac DTI (cDTI) techniques [1,9] that have already proven valuable for describing microstructural differences in several forms of cardiomyopathy [6]. However, *in vivo* cDTI still presents major challenges, including the relatively large voxel size needed to acquire sufficient signal-to-noise, compensating for bulk cardiac motion in the preferred reference configuration (i.e., mid-diastasis), and exam time constraints. In order to overcome these limitations, high-resolution *ex vivo* cDTI imaging can be used to build subject-specific computational models [7,12].

In addition, *in vivo* LGE images are a proven technique for estimating myocardial infarct volume, transmurality, and location. Infarct detection based on LGE, however, is limited by inherently low spatial resolution and the sensitivity to contrast agent dosing and washout. A large voxel size makes quantifying scar volume prone to error, especially in the border zones where low signal intensity in healthy tissue is averaged with high signal intensity due to fibrotic tissue [5]. Although LGE provides diagnostic clinical data, this method is generally inadequate for computational model building. Therefore, one approach is to use high-resolution *ex vivo* LGE to overcome the limitations of low resolution *in vivo* data.

Consequently, while meaningful cine DENSE data can only be acquired *in vivo*, both cardiac DTI and LGE scar imaging are better performed *ex vivo*. There remains, however, an open problem of acquiring *ex vivo* MRI data in hearts that accurately conform to *in vivo* geometry. Current *ex vivo* cDTI methods typically use an injection of rubber silicone to fill the four heart chambers [5,8]. Although this method approximately restores the geometry of the compliant heart, it is very difficult to control cavity shape and volume, which limits conformance of the *ex vivo* data to the *in vivo* data. Post-processing techniques such as elastic registrations are able to partially overcome the mismatch, but can distort the quantitative information stored in the registered image voxels. Indeed, there is no guarantee that the elastic registration correctly accounts for the underlying tissue deformation between the *in vivo* and *ex vivo* configurations. For example, through plane motion and intra-slice torsion cannot be accurately estimated with image-based elastic registration techniques [11] and the resulting warping field may not transform microstructural orientation information accurately. In other organs, 3D printing has proven to be a feasible approach to provide external shape constraints that maintain and enforce geometric accuracy, thereby permitting registration to *in vivo* targets with a simpler rigid registration [15].

Herein, we propose a new method for *ex vivo* heart preparation that uses *in vivo* cardiac MRI to build 3D printed molds. The proposed method: (1) accurately restores *ex vivo* ventricular cardiac geometry to an *in vivo* mid-diastasis reference configuration; and (2) enables rigid registration of high-resolution *ex vivo* anatomical scar and cDTI data to the corresponding reference configuration derived from the *in vivo* images.

2 Methods

2.1 Study Design

Healthy (N = 7) and infarcted (N = 3) swine hearts were included in this study. All animal experiments were conducted according to a research protocol approved by the UCLA Institutional Animal Care and Use Committee (ARC protocol # 2015-124). Infarct induction was achieved by fluoroscopic guided injection of microspheres into a secondary branch of the left anterior descending (N = 2) and left circumflex (N = 1) coronary arteries to occlude blood flow. Infarcted subjects were then imaged 6–10 weeks after induction to allow for complete scar formation.

2.2 *In Vivo* Acquisition and 3D Printing of Cardiac Scaffolds

All subjects were imaged *in vivo* using a 3T MRI scanner (Prisma, Siemens) and a 32-channel chest and spine coil. Balanced steady state free precession (bSSFP) 2D cine cardiac scans were pulse-oximetry gated with single breath holds (base matrix, 160×288; field of view, $236\,\mathrm{mm} \times 340\,\mathrm{mm}$; spatial resolution, $1.18 \times 1.18 \times 8.0\,\mathrm{mm}^3$; 60 phases per cardiac cycle). For infarct subjects, LGE was performed by injecting a gadolinium-based contrast agent ($0.3\,\mathrm{ml/kg}$ diluted in 5 ml of saline and injected at $0.5\,\mathrm{ml/s}$) 10 min prior to imaging. A phase sensitive inversion recovery (PSIR) LGE sequence was used to image the entire heart volume (base matrix, 154×256; field of view, $255\,\mathrm{mm} \times 340\,\mathrm{mm}$; flip angle, $20°$; Inversion Time = 260 ms; spatial resolution, $1.33 \times 1.33 \times 8.0\,\mathrm{mm}^3$).

Cine images were manually segmented (Horos, v3.1.1) at a mid-diastolic time frame. Masks of the left ventricular (LV) blood pool and pericardium were drawn for all mid-diastolic images. Segmentation time for the entire heart volume for both the LV blood pool and pericardium was approximately 30 min. The LV blood pool and pericardial masks were converted into surface meshes with 3Matics (Materialise NV). Local smoothing was used to connect the short-axis contours corresponding to slices acquired 8 mm apart. Meshes of the pericardial molds and LV blood casts were then imported and printed (Ultimaker 3 extended, Ultimaker) with polylactic acid (PLA). Print times were approximately 45 min for the LV blood pool casts and approximately 210 min for the pericardial molds (Fig. 1).

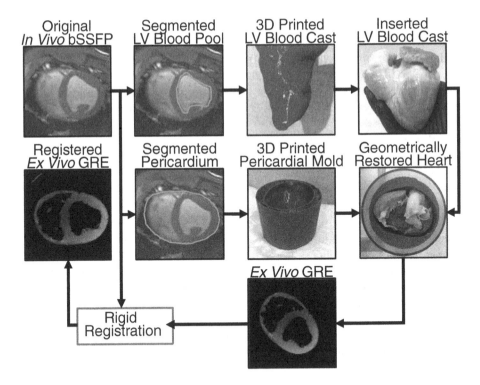

Fig. 1. Workflow for 3D printing restoration of ventricular geometry. *In vivo* images were acquired followed by manual left ventricle (LV) and pericardial segmentation. Segmentations were then converted to surface meshes and 3D printed. The LV blood pool cast was inserted into the excised heart, the heart was placed into the pericardial mold, and the right ventricle (RV) was filled with a rubber silicone compound. *Ex vivo* images were acquired and rigidly registered with *in vivo* images.

2.3 *Ex Vivo* Extraction and Acquisition

For infarcted subjects, a double dose of gadolinium (gadobenate dimeglumine or gadopentetic acid, 0.6 ml/kg) was injected intravenously ten minutes prior to euthanasia (following institutional guidelines) to achieve LGE when performing *ex vivo* imaging.

After excision of the heart, the LV blood cast was inserted into the LV, using the papillary muscles and position of the aorta as reference to correctly orient the 3D print. The heart was then placed inside the customized pericardial mold to provide an external boundary for the epicardium (Fig. 1). The right ventricle (RV) was then filled with silicone rubber compound (Polyvinylsiloxane, Microsonic Inc.). The geometrically restored heart was placed in a cylindrical container and filled with a perfluoropolyether solution (Fomblin, Solvay) that produces no MRI signal.

Ex vivo MRI was performed on the same 3T MRI scanner used during the *in vivo* studies. A 15-channel transmit/receive knee coil was used to acquire T1-weighted gradient echo (GRE) images (echo time(TE)/repetition time (TR), 3.15 ms/12 ms; flip angle, 25°; base matrix, 160 × 160; field of view, 160 mm × 160 mm; spatial resolution, 1.0 × 1.0 × 1.0 mm³; averages, 6) and DTI (TE/TR, 62 ms/15560 ms; flip angle, 180°; base matrix, 150 × 150; field of view, 150 mm × 150 mm; spatial resolution, 1.0 × 1.0 × 1.0 mm³; b-values, 0 s/mm² and 1000 s/mm²; number of directions, 30; averages, 5).

2.4 Image Processing and Analysis

The workflow of image processing and analysis included five steps:

1. *Landmark Registration* – *In vivo* images (bSSFP and LGE) were registered to *ex vivo* T1-weighted images at their native voxel resolution using the MITK Workbench [14]. The registration method used landmark-based rigid registration, which allowed for qualitative image overlays between *in vivo* and *ex vivo* images and provided a more accurate basis for the following optimized semi-automatic 3D rigid registration. Landmarks consisted of four points in the most basal slice (center of the septal wall, anterior papillary muscle, center of the free wall, and posterior papillary muscle) and four points in similar locations near the apex.
2. *Diffusion Tensor Reconstruction* – *Ex vivo* DTI tensors were reconstructed using linear least squares. *Ex vivo* high-resolution reconstructed diffusion tensors were overlaid on the registered *in vivo* images and shown as superquadric tensor glyphs [3], which completely characterize the tensor shape and orientation (Fig. 2).
3. *Segmentation* – Manual segmentation of the landmark based *in vivo* and *ex vivo* LV blood pool, RV blood pool, and the myocardial tissue were performed in MITK Workbench and exported as binary masks. The apical and basal boundaries for each segmentation were defined from the *in vivo* cine images.
4. *Semi-Automated Registration* – *In vivo* and *ex vivo* LV blood pool binary images were registered with an automated 3D rigid regular-step gradient descent algorithm (Matlab, MathWorks), which scaled *ex vivo* voxels to match *in vivo* voxels. The transformation generated from this registration was also applied directly to the RV blood pool and the myocardial tissue binary images. The registered binary masks were constructed using the *in vivo* voxel size in order to avoid interpolating data from low to high-resolution.
5. *Quantitative Analysis* – *In vivo* and *ex vivo* registered LV blood pool, RV blood pool, and myocardial tissue binary images were used to compute volumes and Dice similarity coefficients (DSC). LV and RV volumes were also compared between *in vivo* and *ex vivo*. Linear regressions were computed for each of the different measurements displaying the line of best fit and 95% confidence intervals. Pearson's correlation coefficients (R^2) were calculated to determine the strength of correlation between *in vivo* and *ex vivo* volumetric data.

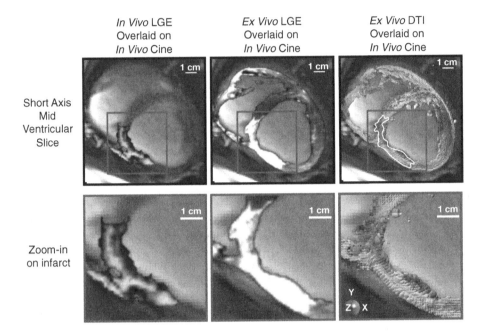

| *In Vivo* LGE Overlaid on *In Vivo* Cine | *Ex Vivo* LGE Overlaid on *In Vivo* Cine | *Ex Vivo* DTI Overlaid on *In Vivo* Cine |

Fig. 2. Image overlays of a mid-ventricular slice of an infarcted swine heart with a magnified view of the infarcted tissue. *In vivo* cine images were registered with *in vivo* LGE (left), *ex vivo* LGE (middle), and *ex vivo* cDTI tensor glyphs (right). The epicardial and endocardial boundaries between healthy and infarcted tissue agree well in each images. The manual segmentation (white contour) of the infarct as seen on the *ex vivo* LGE is superimposed on the *ex vivo* DTI.

3   Results

Image registration was qualitatively evaluated by overlaying different combinations of acquired data for all subjects. One infarcted subject is shown in Fig. 2. For healthy subjects, native resolution *ex vivo* T1-weighted images and *ex vivo* DTI glyphs were overlaid with *in vivo* bSSFP images at a short-axis mid-ventricular slice location. For infarcted subjects, an additional overlay of the *in vivo* LGE with *in vivo* bSSFP was also displayed to show concordance of the scar tissue with anatomy. In general, the images were all well registered.

DSC for each subject is shown in Table 1. Across all subjects, mean DSC $\pm$ standard deviation (SD) for the LV cavity was 0.93 ± 0.01, for the RV cavity was 0.80 ± 0.05, and for the myocardial tissue was 0.72 ± 0.04. In Fig. 3, Pearson correlation coefficients for linear regressions are $R^2 = 0.95$ and $R^2 = 0.91$ for the LV cavity and RV cavity, respectively.

Table 1. DSC for the LV cavity, RV cavity, and myocardial tissue in all subjects.

Heart #	Condition	LV cavity	RV cavity	Myocardium
1	Healthy	0.93	0.75	0.73
2	Healthy	0.94	0.81	0.72
3	Healthy	0.94	0.77	0.72
4	Healthy	0.94	0.79	0.71
5	Healthy	0.94	0.72	0.62
6	Healthy	0.93	0.81	0.75
7	Healthy	0.92	0.81	0.77
8	Infarct	0.90	0.84	0.71
9	Infarct	0.95	0.89	0.78
10	Infarct	0.92	0.83	0.67
mean ± SD		**0.93 ± 0.01**	**0.80 ± 0.05**	**0.72 ± 0.04**

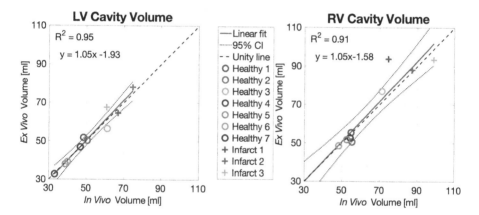

Fig. 3. Scatter plots of *in vivo* and *ex vivo* LV and RV cavity volumes showing both healthy hearts (o) and infarct subjects (+). Linear regression fits and 95% confidence interval (CI) are reported.

4   Discussion

Several computational studies of cardiac contraction and electrophysiology incorporate rule-based myofibers [2] into the models to account for cardiac microstructure. However, rule-based myofiber models do not take into account the regional variation of aggregate cardiomyocyte orientation, are not subject specific, and the effect of including rule-based models in lieu of subject specific myofibers has not been thoroughly evaluated. These factors highlight the necessity to build computational models based, when possible, on subject specific microstructural data.

Previous *ex vivo* DTI heart studies used several methods to restore ventricular cavity shape. Teh *et al.* [10] arrested rat hearts with potassium, perfused them with Karnovsky's fixative, and immediately fixed the tissue. Potassium arrest results in a relaxed, diastasis-like state, but is susceptible to tissue shrinkage from the fixative [13] and deformation due to lack of intracavity pressure or surface constraint. Kung *et al.* [5] and Sack *et al.* [8] fixed swine hearts with buffered formalin followed by an injection of rubber silicone compound to fill the four heart chambers. Using this method, the heart cavities may be over- or under-filled since there are no constraints to preserve the *in vivo* anatomy.

Using 3D printed molds to restore the shape of fresh myocardium eliminates tissue shrinkage artifacts, minimally impacts the imaged geometry, and enables high-resolution *ex vivo* LGE imaging. The reported high DSC for the LV cavity across subjects and good DSC for the RV cavity and myocardial tissue, in addition to highly correlated cavity volumes, suggest that using 3D printed scaffolds is an accurate approach to restore *ex vivo* geometry to a reference *in vivo* configuration. Subsequent alignment via rigid registration permits fusing *in vivo* and *ex vivo* data. From here, computational models of the heart can be built directly from the high-resolution *ex vivo* cDTI data in the preferred reference configuration without any additional pre-processing, while also accommodating direct incorporation of well-registered *ex vivo* infarct information and *in vivo* motion data.

One limitation of this approach is the unclear *in vivo* boundaries of the RV cavity and RV tissue during segmentation. While the cavity and tissue contrast is well defined in *ex vivo* scans, *in vivo* RV myocardium and blood pool boundaries are not as clearly defined, therefore limiting DSC to lower values for both the myocardial tissue and the RV cavity. Our approach also used rigid registration methods. Non-rigid registration methods might provide higher DSC and higher correlations of cavity volumes, but our method allows for microstructure reconstructed from the *ex vivo* DTI to be aligned without non-rigidly deforming the tensor orientation information, which would need to be validated. Another limitation of this approach arises from the relatively short time interval between *in vivo* and *ex vivo* imaging. There were approximately eight hours between the acquisition of *in vivo* cardiac cine MRI and the extraction of the heart. Approximately five of these hours were required for 3D printing the LV blood cast and pericardial mold. The remaining time allowed for only one segmentation, limiting the evaluation of intra- and inter-observer variations in image segmentation. After *in vivo* cine segmentation, the rest of the printing preparation time was used to generate smooth LV blood pool and pericardial surfaces from image contours. Finally, additional time was included into the preparation workflow as a precaution in case of print failure.

Although the results shown here are promising, further work is still needed. Separate segmentations of the LV and RV myocardium would permit computing DSC and tissue volumes comparison between *in vivo* and *ex vivo* LV and RV independently. Due to a lower image contrast in the RV, segmentation of the RV tissue is prone to larger errors and uncertainty, which need to be evaluated. Additionally, segmentation of the *in vivo* and *ex vivo* infarct regions would provide further quantitative measures such as scar DSC and volumes.

This work demonstrates a new method to restore *ex vivo* ventricular geometry to an *in vivo* reference configuration through 3D printing and rigid registration. Quantitative validation through volumetric and DSC comparisons confirms the accuracy of the approach and makes available anatomical and microstructural data more amenable to computational electromechanical modeling studies of healthy and infarcted hearts.

References

1. Aliotta, E., Wu, H.H., Ennis, D.B.: Convex optimized diffusion encoding (CODE) gradient waveforms for minimum echo time and bulk motion-compensated diffusion-weighted MRI. Magn. Reson. Med. **77**(2), 717–729 (2017)
2. Bayer, J.D., Blake, R.C., Plank, G., Trayanova, N.A.: A novel rule-based algorithm for assigning myocardial fiber orientation to computational heart models. Ann. Biomed. Eng. **40**(10), 2243–2254 (2012)
3. Ennis, D.B., Kindlmann, G.: Orthogonal tensor invariants and the analysis of diffusion tensor magnetic resonance images. Magn. Reson. Med. Off. J. Int. Soc. Magn. Reson. Med. **55**(1), 136–146 (2006)
4. Kellman, P., Arai, A.E., McVeigh, E.R., Aletras, A.H.: Phase-sensitive inversion recovery for detecting myocardial infarction using gadolinium-delayed hyperenhancement. Magn. Reson. Med. Off. J. Int. Soc. Magn. Reson. Med. **47**(2), 372–383 (2002)
5. Kung, G.L., et al.: Microstructural infarct border zone remodeling in the postinfarct swine heart measured by diffusion tensor MRI. Front. Physiol. **9** (2018)
6. Nielles-Vallespin, S., et al.: Assessment of myocardial microstructural dynamics by in vivo diffusion tensor cardiac magnetic resonance. J. Am. Coll. Cardiol. **69**(6), 661–676 (2017)
7. Ponnaluri, A.V., et al.: Electrophysiology of heart failure using a rabbit model: from the failing myocyte to ventricular fibrillation. PLoS Comput. Biol. **12**(6), e1004968 (2016)
8. Sack, K., et al.: Effect of intra-myocardial Algisyl-LVR^{TM} injectates on fibre structure in porcine heart failure. J. Mech. Behav. Biomed. Mater. **87**, 172–179 (2018)
9. Stoeck, C.T., et al.: Dual-phase cardiac diffusion tensor imaging with strain correction. PLoS One **9**(9), e107159 (2014)
10. Teh, I., et al.: Resolving fine cardiac structures in rats with high-resolution diffusion tensor imaging. Sci. Rep. **6**, 30573 (2016)
11. Verzhbinsky, I.A., Magrath, P., Aliotta, E., Ennis, D.B., Perotti, L.E.: Time resolved displacement-based registration of in vivo cDTI cardiomyocyte orientations. In: 2018 IEEE 15th International Symposium on Biomedical Imaging (ISBI 2018), pp. 474–478. IEEE (2018)
12. Wang, V.Y., Lam, H., Ennis, D.B., Cowan, B.R., Young, A.A., Nash, M.P.: Modelling passive diastolic mechanics with quantitative MRI of cardiac structure and function. Med. Image Anal. **13**(5), 773–784 (2009)
13. Wehrl, H.F., et al.: Assessment of murine brain tissue shrinkage caused by different histological fixatives using magnetic resonance and computed tomography imaging. Histol. Histopathol. **30**(5), 601–613 (2015)

14. Wolf, I., et al.: The medical imaging interaction toolkit. Med. Image Anal. **9**(6), 594–604 (2005)
15. Wu, H.H., et al.: A system using patient-specific 3D-printed molds to spatially align in vivo MRI with ex vivo MRI and whole-mount histopathology for prostate cancer research. J. Magn. Reson. Imaging **49**(1), 270–279 (2019)
16. Zhong, X., Spottiswoode, B.S., Meyer, C.H., Kramer, C.M., Epstein, F.H.: Imaging three-dimensional myocardial mechanics using navigator-gated volumetric spiral cine DENSE MRI. Magn. Reson. Med. **64**(4), 1089–1097 (2010)

Synchrotron X-Ray Phase Contrast Imaging and Deep Neural Networks for Cardiac Collagen Quantification in Hypertensive Rat Model

Hector Dejea[1,2(✉)], Christine Tanner[3], Radhakrishna Achanta[4],
Marco Stampanoni[2], Fernando Perez-Cruz[4], Ender Konukoglu[3],
and Anne Bonnin[1]

[1] Paul Scherrer Institut, Villigen PSI, Villigen, Switzerland
`hector.dejea@psi.ch`
[2] Institute for Biomedical Engineering, University and ETH Zürich,
Zurich, Switzerland
[3] Computer Vision Laboratory, ETH Zürich, Zurich, Switzerland
[4] Swiss Data Science Center, Lausanne, Switzerland

Abstract. An excessive deposition of collagen matrix in the myocardium has been clearly identified as an indication of the progression towards heart failure. Nevertheless, few studies have been performed for its quantification and most of them use 2D histological images, thus losing valuable encoded 3D information. In this study, several biopsies of areas of the left ventricle from age-matched spontaneously hypertensive rats and Wistar Kyoto rats were imaged using synchrotron radiation-based X-ray phase contrast imaging. Then, an optimized deep neural network was used for automatic image segmentation in order to assess collagen fraction differences between models as well as its age dependency. The results show a general increase in the collagen percentage in the hypertensive model and for older rats. Such tendency is comparable with the reports found in the literature. Therefore, this proof of concept shows that synchrotron imaging in combination with deep neural networks is a powerful tool for the investigation and quantification of cardiac microstructures.

Keywords: Myocardial collagen fraction · Synchrotron imaging ·
Deep neural network · Image segmentation

1 Introduction

Cardiovascular diseases (CVDs) are the leading cause of mortality, being responsible for 31% of the deaths worldwide [1]. Particularly, remodelling CVDs cause structural changes in the heart and vasculature at all scales, from organ- to cell-level.

H. Dejea, C. Tanner, E. Konukoglu and A. Bonnin—These authors contributed equally to this work.

Y. Coudière et al. (Eds.): FIMH 2019, LNCS 11504, pp. 187–195, 2019.
https://doi.org/10.1007/978-3-030-21949-9_21

The cardiac tissue (myocardium) is primarily composed of muscle cells called cardiomyocytes. They are arranged in spatially-oriented fibre-like structures (myofibers) and provide the contractile function of the heart under the electrical stimulus transmitted by pacemaker cells [2].

In the myocardial extracellular space, a collagenous matrix serves as a scaffold, helping the alignment of the cardiomyocytes. This matrix also avoids overstretching of cardiomyocytes and thus, ensures their integrity, among other functions. It is mainly composed of type 1 fibrillar collagen and maintained by specialized cells called myofibroblasts. Such matrix is organized in three different layers: endomysial collagen is found around single cardiomyocytes, perymysial collagen surrounds myofibers and epimysial collagen is located around groups of myofibres [3].

The collagen matrix is known to be affected by aging and remodelling disorders such as hypertension, myocardial infarction or transplant rejection. In these situations, the activity of the myofibroblast is increased and, consequently, an increased matrix deposition and fibrosis formation is observed. This situation causes myocardial stiffening and contractile dysfunction, thus eventually leading to heart failure syndrome [4].

Even if the relationship between collagen matrix remodelling and development of heart failure is known, the direct quantification of collagen amounts and its association with disease progression has not been extensively reported. Mainly, several quantification methods have been developed for 2D histological slices [5], which lack the spatial information and field of view that can be achieved with 3D imaging techniques, and suffer from artifacts due to slice preparation, such as tearing.

The main contribution in this topic was given by LeGrice et al. [6], who compared spontaneously hypertensive rats (SHR) and control Wistar Kyoto (WKY) to quantify the age-dependent deposition of collagen in the left ventricular lateral wall. Tissue blocks were prepared (including tissue processing, embedding and staining) and imaged in 3D using confocal microscopy, and collagen fraction was computed using dedicated image processing.

Alternatively, a recent preliminary study [7] proposed synchrotron radiation-based X-ray phase contrast imaging (X-PCI) as a powerful methodology for the quantification of collagen amounts, thanks to its 3D high resolution capabilities and the exploitation of the tissue's change of index of refraction for contrast generation. This study quantified the collagen fraction in the left ventricular basal septum, lateral wall and apex for 3 months old control WKY rats using a semi-automatic machine learning method.

In this paper, we extend this study by including the SHR model and 12 months old specimens, so that differences between models as well as age dependency can be assessed. In addition, we make use of an optimized deep neural network for automatic collagen fraction calculation, such that datasets without any user inputs can be included in our analysis.

2   Materials and Methods

2.1   Sample Preparation

Animal care and experimentation was performed in accordance with the European Union (Directive 2010/63/UE) and Spanish guidelines (RD 53/2013) for the use of experimental animals. Approval was received from the local animal experimentation ethics committee (CEEA 68/5435 and CEEA OB533/16).

The presented study was performed on endomyocardial biopsies ($\sim 2 \times 2 \times 4$ mm^3) of the basal septum, lateral wall and apex of the left ventricle in four different rats – two 3 months old and two 12 months old WKY, and four age-matching SHR.

The samples were fixed in 4% paraformaldehyde and placed in specifically designed sample holders, mainly consisting of a thin-walled borosilicate glass tubes with an inner diameter of 2 mm (Fig. 1a).

2.2   Data Acquisition and Reconstruction

The synchrotron radiation-based X-ray tomography experiments were performed at the TOMCAT beamline (X02DA) of the Swiss Light Source (Paul Scherrer Institut, Villigen, Switzerland). Propagation-based X-PCI at an energy of 20 keV, propagation (sample-to-detector) distance of 20 cm and 0.65 μm voxel size was used for image acquisition of the previously described cardiac biopsies. The biopsies were larger than the available field of view (1.6 $\times$ 1.44 mm^2), so three overlapping volumes from each biopsy were obtained. The volumes were acquired by a PCO.Edge 5.5 sCMOS Camera detector after conversion of X-rays to visible light with a LuAG:Ce 20 μm scintillator (Fig. 1b). For each volume, 2501 projections over 180°, 20 darks and 50 flats were measured with an exposure time of 120 ms, leading to ~ 7 min acquisition time per volume. The dark projections are acquired without beam exposure to correct for the electronic noise of the detector, whereas flat projections are acquired without sample, to compensate for non-uniformities in beam intensity and the optical components.

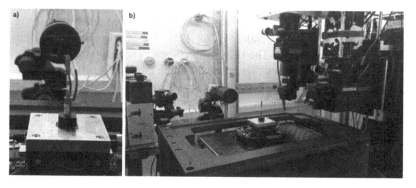

Fig. 1.   (a) Cardiac biopsy in a dedicated holder on the sample stage for acquisition. (b) Overview of the synchrotron imaging setup.

After acquisition, the sinograms for each volume were computed. Then, the tomograms were reconstructed using the Gridrec algorithm [8], which is a modified fast version of the typical filtered back projection algorithm thanks to a regridding process to resample the Fourier space from polar to Cartesian coordinates. Such reconstruction procedure was performed with and without application of the Paganin's single-distance phase retrieval method [9], which allows to retrieve the phase information of the image thanks to a projection filtering process determined by the δ/β ratio from the index of refraction, which was set to 56.9. Reconstruction times were approximately 2 and 6.5 min, without and with Paganin method, respectively.

2.3 Tissue Segmentation

Manual segmentation of the datasets was prohibitively time-consuming. Therefore, we first generated ground truth data from a subset of volumes using a well-established machine-learning based tool, which required manual interaction, then used it to train a neural network, which is later employed to segment the rest of the datasets.

Ground truth generation. As demonstrated in [7], ground truth data was generated with the open-source software Ilastik [10] and its interactive pixel classification module. Briefly, several regions of interest (of $460 \times 460 \times 460$ voxels representing regions of $300 \times 300 \times 300$ μm^3) from the acquired datasets were cropped and loaded on Ilastik. Iterative supervised labelling for background, cells and collagen was performed in 3D. The phase retrieved images were used to segment cells and background due to an increased contrast resulting from the Paganin method, while the collagen was segmented from the non-retrieved images due to increased sharpness, which can be slightly lost by the application of the Paganin method. Later, objects smaller than 150 voxels were removed for noise reduction purposes and both masks were fused to generate the final segmentation. We could have used the Ilastik tool to segment all datasets, however, empirically we observed that this tool required many user interactions for a satisfactory segmentation. Instead, we used Ilastik to carefully segment a few datasets with user-interactions and used it as ground truth to train the algorithms described next.

U-Net for Collagen Quantification. We used a well-studied and widely-used deep convolutional neural network, called U-Net [11], for segmenting the images into background, collagen and cells. The U-Net was chosen as it has previously shown good performance in several segmentation challenges. We built a 2D network that takes as input images of size 240×240 pixels and outputs $240 \times 240 \times 3$ images, where the three channels at the output are the class probabilities for each pixel, corresponding to the three classes: background, collagen and cells. Based on experience and previously proposed architectures, we designed a U-Net with 4 resolution levels in the encoding path that results in a bottleneck layer of size 15×15 with 512 channels (the architecture: input >> $240 \times 240 \times 32$ >> $120 \times 120 \times 64$ >> $60 \times 60 \times 128$ >> $30 \times 30 \times 256$ >> $15 \times 15 \times 512$). Each resolution level had two convolutional layers (kernel size of 2x2 with stride 1, followed by ReLU activation) and a max-pooling layer. The decoding path was formed of 4 resolution levels, which increased the size to 240×240 and reduced the channels to 3 in a symmetric way using transposed convolution with 4×4 kernels and stride 2 followed by ReLU activation. At the output, no non-linear activation was used, instead a soft-max layer converted the outputs to pixel-wise probabilities. Skip connections concatenated activations in the encoding path to the corresponding blocks in the decoding path.

Seven hyper-parameters (a-g) were optimized via an ablation study, where each time a single parameter is changed with respect to a reference setting. Next, we describe per parameter the two options, with the underlined first option being the reference setting. The input consisted of either (a0) a single slice or (a1) 3 neighbouring slices from (b0) only the non-retrieved image or (b1) both the non- and phase-retrieved image. The images were (c0) not augmented or (c1) augmented by 90, 180, 270° rotations, horizontal or vertical flips. Optimization was based on (d0) the mean Dice coefficient or (d1) the logarithm of the mean Dice coefficient to cope with class imbalance [12], where Dice coefficient is defined as $D = 2|S_{GT} \cap S_{UNet}|/(|S_{GT}| + |S_{UNet}|)$ with $|S|$ denoting the number of foreground (=1) pixels of binary segmentation S. Regularization via setting weights to zero (dropout) was (e0) not enabled or (e1) with a probability of 50% performed. The U-Net consisted of (f0) 4 or (f1) 3 resolution levels. The U-Net decoder used (g0) transposed convolutions or (g1) bilinear upsampling followed by convolution. Furthermore, we compared the performance when increasing from 2.5% of the annotated data to 80%.

The annotated data, which consisted of ten volumes of size $460 \times 460 \times 460$ as described in the previous section, were randomly split into 6 training, 2 validation and 2 test volumes. A certain number of slices, e.g. 2.5% (11.5 slices), was then randomly selected from the training volumes for training the U-Net and from the validation volumes for determining training convergence (defined as no performance improvement over 1000 iterations for the validation data). To segment a volume, the trained U-Net was applied patch-wise with a stride length of 60 pixels and predicted class-probabilities were averaged. Ten U-Nets were trained based on different initial random initializations and randomly selected data splits to study gains from fusing their results and for getting results for all test cases. Fusion of the 1–4 results per manually segmented test case and of the 10 results for the other cases, was done by averaging per case, class and pixel the predicted probabilities from the U-Nets.

A supervised Gaussian Mixture Model (GMM) was used as baseline method. It clustered pixel intensities into 3 classes, based on model parameters (mean, standard deviation, mixture coefficients) learned from the ground truth segmentations of 8 volumes. It achieved a mean Dice coefficient of 66.3% in leave-2-volume-out tests. The correlation between the ground truth and GMM collagen fraction, measured by Pearson's correlation coefficient, was 0.17.

The ablation study of the U-Net hyper-parameters, listed in (a-g) above, showed that options c1, d1, e1, f1, g1 improved mean Dice coefficient by 0.3 to 1.4% over the reference configuration and lead to a positive correlation between the ground truth and U-Net collagen fraction (0.17–0.43). Combining these 5 changes provided the highest correlation (0.60). Using much more data (80%) improved the mean Dice coefficient by 0.9% to 82.8% and the correlation to 0.73. As the latter is a substantial improvement, we report here results for the U-Nets trained on 80% of the data.

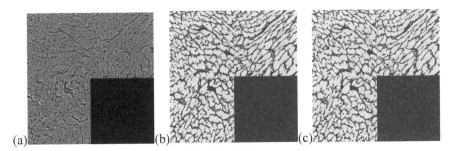

Fig. 2. Orthogonal central slices of (a) non-retrieved image volume (b) U-Net segmentation, and (c) ground truth for worst test result (Dice coefficient 75.2%). Yellow represents myocytes, blue shows background and cyan depicts collagen. Slices are 460 × 460 pixels. (Color figure online)

Figure 2 shows the segmentation result with the largest difference in collagen fraction to the ground truth (3.4% vs. 1.0%). It can be observed that even for this worst result the U-Net segmentation is consistent across slices and appears reasonable.

3   Results

The U-Net results of the collagen fraction in the different regions of WKY and SHR of 3 and 12 months old are summarized in Fig. 3. A general tendency towards higher collagen fractions can be observed for older rats as well as for the SHR case. Nevertheless, differences are observed when looking individually at each of the areas. Most of 3 months old measurements are located in a range between 2 and 7%, while most 12 months old fractions are found between 5 and 9%.

In the apex, there is an increased amount of collagen ($\sim 1\%$) in the SHR case both for 3 months old and 12 months old. In addition, the collagen fraction appears higher for older rats, being $\sim 3\%$ higher.

In the basal septum, very similar values are observed for WKY and SHR for the same age rats. Nevertheless, again an increase of $\sim 4\%$ can be observed for the older group of animals.

Finally, in the lateral wall large variations are observed. The value for WKY rats does not seem to change significantly between 3 and 12 months and the mean is even found to be $\sim 0.5\%$ lower. The same applies to SHR, which, even if higher than WKY for both age points ($\sim 1\%$ and $\sim 1.5\%$), has a mean value that seems to also decrease by $\sim 0.5\%$ in older rats.

4   Discussion

In this study we assess the deposition of collagen in WKY and SHR rats at two different time points, corresponding to 3 and 12 months old, and three different regions of the left ventricle (apex, basal septum and lateral wall). For this purpose, X-PCI tomograms (0.65 μm voxel size) of rat heart biopsies were acquired, and segmented using an optimized deep neural network.

In this way, this manuscript presents an extension of preliminary work in this field [7] using the same imaging technique, but an improved analysis method that allows automatic collagen segmentation and fraction computation thanks to trained deep neural networks.

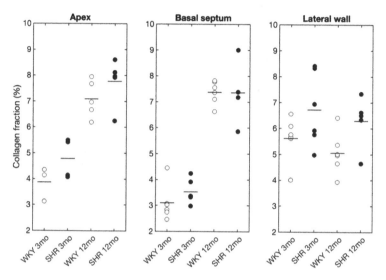

Fig. 3. Results for collagen fraction in the apex, basal septum and lateral wall of the left ventricle, respectively. Circles and full dots represent data points of control (WKY) and hypertensive model (SHR) respectively. Lines correspond to the mean value.

The results presented in Fig. 3 show a general increased tendency in collagen with age as well as in the hypertensive case. This is in accordance with the literature, since it is well-known that the collagen deposition in the extracellular matrix increases with aging. It is also known that the response of hypertensive hearts is to produce a stiffening of the myocardium (by collagen deposition) to generate higher forces and improve its pumping capabilities [4, 6, 13].

With regards to aging, we can see that in both the apex and basal septum, the mean collagen fraction increases by ∼3–4% in the 12 months old cases. Nevertheless, in the lateral wall the values for the WKY remain similar or even slightly lower in the older specimens.

Looking at the differences between models, we see that in the apex there is an increase of ∼1% in SHR at both time points. Nevertheless, the basal septum appears to show no or very small difference in the collagen values between the models, and the lateral wall only shows an increase of ∼1–1.5% in SHR at both ages.

Even if lower, the presented values show a similar tendency as observed in the most recent related publication [6]. There, quantification of collagen in the lateral wall of left ventricle of these two models by using confocal microscopy and a dedicate image processing pipeline was performed. For collagen fraction, they find close values in 3 months old WKY and SHR, which, as similarly shown in Fig. 2, do not change either for 12 months old WKY. Nevertheless, they show an increase with high variability for 12 months old SHR, which is not so clear in our case even if we also observe high variance.

The differences in values, as reviewed elsewhere [5], can be caused by many factors, mostly related to setup and sample preparation, such as imaging technique, resolution, use of staining, differences in tissue processing or collagen fraction calculation method.

The results presented in this paper have been obtained from a low number of specimens (2 rats per model and time point) and a few regions of interest analyzed per biopsy (3–6), which limits this study from reaching significant clinical conclusions. Moreover, during the segmentation procedure, another limitation that arises is the fact the blood cells and collagen appear to have similar characteristics, which may introduce some unwanted variations. For this reason, the selected regions of interest to analyze were chosen to have the least possible amount of visible vasculature possible.

The fact that the presented values match the well-known evolution of collagen amount as well as the most recent literature, demonstrates that X-PCI and the use of deep neural network architectures such as the optimized U-net shown here are an alternative powerful tool for the quantitative investigation of cardiac collagen matrix.

5 Conclusion

This study extends the assessment of collagen fraction by X-PCI to the comparison between WKY and SHR rats at 3 and 12 months old. The analysis has been performed using an optimized deep neural network, which enables automatic segmentation for unlabeled regions of interest. The results obtained follow the tendencies observed in the literature, therefore posing X-PCI as a valuable tool for the assessment and quantification of cardiac micro-structure. Nevertheless, higher statistics would be needed to attain clinical significance.

Acknowledgments. We acknowledge Monica Zamora, Fatima Crispi and Eduard Guasch for animal handling, and Xavier Buyse Sánchez for his support in the labelling task. In addition, we acknowledge the Paul Scherrer Institut, Villigen, Switzerland for provision of synchrotron radiation beamtime at the beamline TOMCAT (X02DA) of the Swiss Light Source. This project was supported by the grant #2017-303 of the Strategic Focal Area "Personalized Health and Related Technologies (PHRT)", and the grant C17-04 of the Strategic Focal Area "Swiss Data Science Center (SDSC)" of the ETH Domain.

References

1. World Health Organization: Cardiovascular Diseases (CVDs) Fact Sheet (2017). http://www.who.int/en/news-room/fact-sheets/detail/cardiovascular-diseases-(cvds). Accessed 18 Jan 2019
2. Filipoiu, F.M.: Atlas of Heart Anatomy and Development. Springer, London (2013). https://doi.org/10.1007/978-1-4471-5382-5

3. Weber, K.T., Sun, Y., Bhattacharya, S.K., et al.: Myofibroblast-mediated mechanisms of pathological remodelling of the heart. Nat. Rev. Cardiol. **10**(1), 15 (2013)
4. Burchfield, J.S., Xie, M., Hill, J.A.: Pathological ventricular remodeling: mechanisms: Part 1 of 2. Circulation (2013). https://doi.org/10.1161/CIRCULATIONAHA.113.001878
5. Schipke, J., Brandenberger, C., Rajces, A., et al.: Assessment of cardiac fibrosis: a morphometric method comparison for collagen quantification. J. Appl. Physiol. (2017) https://doi.org/10.1152/japplphysiol.00987.2016
6. LeGrice, I.J., Pope, A.J., Sands, G.B., et al.: Progression of myocardial remodeling and mechanical dysfunction in the spontaneously hypertensive rat. AJP Hear Circ Physiol. (2012). https://doi.org/10.1152/ajpheart.00748.2011
7. Dejea, H., et al.: Microstructural analysis of cardiac endomyocardial biopsies with synchrotron radiation-based x-ray phase contrast imaging. In: Pop, M., Wright, Graham A. (eds.) FIMH 2017. LNCS, vol. 10263, pp. 23–31. Springer, Cham (2017). https://doi.org/10.1007/978-3-319-59448-4_3
8. Marone, F., Stampanoni, M.: Regridding reconstruction algorithm for real-time tomographic imaging. J. Synchrotron. Radiat. (2012). https://doi.org/10.1107/S0909049512032864
9. Paganin, D., Mayo, S.C., Gureyev, T.E., et al.: Simultaneous phase and amplitude extraction from a single defocused image of a homogeneous object. J. Microsc. (2002). https://doi.org/10.1046/j.1365-2818.2002.01010.x
10. Sommer, C., Straehle, C., Ullrich, K., Hamprecht, F.A.: Ilastik: interactive learning and segmentation toolkit. In: Eighth IEEE International Symposium Biomedical Imaging, Heidelberg Collaboratory for Image Processing (HCI), University of Heidelberg (2011). https://doi.org/10.1109/ISBI.2011.5872394
11. Ronneberger, O., Fischer, P., Brox, T.: U-Net: convolutional networks for biomedical image segmentation. In: Navab, N., Hornegger, J., Wells, William M., Frangi, Alejandro F. (eds.) MICCAI 2015. LNCS, vol. 9351, pp. 234–241. Springer, Cham (2015). https://doi.org/10.1007/978-3-319-24574-4_28
12. Wong, K.C.L., Moradi, M., Tang, H., Syeda-Mahmood, T.: 3D segmentation with exponential logarithmic loss for highly unbalanced object sizes. In: Frangi, Alejandro F., Schnabel, Julia A., Davatzikos, C., Alberola-López, C., Fichtinger, G. (eds.) MICCAI 2018. LNCS, vol. 11072, pp. 612–619. Springer, Cham (2018). https://doi.org/10.1007/978-3-030-00931-1_70
13. Horn, M.A., Trafford, A.W.: Aging and the cardiac collagen matrix: novel mediators of fibrotic remodelling. J. Mol. Cell. Cardiol. **93**, 175–185 (2016)

3D High Resolution Imaging of Human Heart for Visualization of the Cardiac Structure

Kylian Haliot[1,2,3(✉)], Julie Magat[1,2,3], Valéry Ozenne[1,2,3],
Emma Abell[1,2,3], Virginie Dubes[1,2,3], Laura Bear[1,2,3],
Stephen H. Gilbert[4], Mark L. Trew[5], Michel Haissaguerre[1,2,6],
Bruno Quesson[1,2,3], and Olivier Bernus[1,2,3]

[1] IHU Liryc, Electrophysiology and Heart Modeling Institute,
Foundation Bordeaux Université, 33600 Pessac-Bordeaux, France
kylian.haliot@ihu-liryc.fr
[2] Univ. Bordeaux, Centre de recherche Cardio-Thoracique de Bordeaux, U1045,
33000 Bordeaux, France
[3] INSERM, Centre de recherche Cardio-Thoracique de Bordeaux, U1045,
33000 Bordeaux, France
[4] Mathematical Cell Physiology, Max Delbrück Centre for Molecular Medicine
in the Helmholtz Association, 13125 Berlin-Buch, Germany
[5] Bioengineering Institute, Department of Physiology, University of Auckland,
Auckland, New Zealand
[6] Bordeaux University Hospital (CHU), Electrophysiology and Ablation Unit,
33600 Pessac, France

Abstract. Imaging of cardiac structure is thus essential for understanding both electrical propagation and efficient contraction in human models. The processing pipeline of diffusion tensor imaging (DTI) and structure tensor imaging (STI) is described and the first *ex vivo* demonstration of this approach in a human heart is provided at 9.4T.

A human heart was fixed in formaldehyde with gadolinium then immersed in Fomblin. MRI acquisitions were performed at 9.4T/30 cm with a 7 elements transmit/receive coil. 3D spin-echo DTI at $600 \times 600 \times 600$ μm^3 and 3D FLASH echo image at $150 \times 150 \times 150$ μm^3 were produced. Tensor extraction and analysis were performed on both volumes.

3D gradient echo at $150 \times 150 \times 150$ μm^3 allows direct visualization of detailed structure of LV. Abrupt change in sheetlet orientation is observed in the LV and is confirmed with STI. The DTI helix angle has a smooth transmural change from endocardium to epicardium. Both the helix and transverse angles are shown to be similar between DTI and STI. The sheetlet organization between both acquisitions displays the same pattern even though local angle differences are demonstrated.

In conclusion, these preliminary results are promising for investigating 3D structural characterization of normal/pathologic cardiac organization in human. It opens new perspectives to better understand the links between structural remodeling and electrical disorders of the heart.

Keywords: Cardiac microstructure · Sheetlet organization · High field MRI · Helix angles · Structure tensor · Diffusion tensor

© Springer Nature Switzerland AG 2019
Y. Coudière et al. (Eds.): FIMH 2019, LNCS 11504, pp. 196–207, 2019.
https://doi.org/10.1007/978-3-030-21949-9_22

1   Introduction

Efficient contraction and electrical propagation are key parameters in the normal function of the heart, and they are influenced by the myocardial structure [1]. Visualizing and quantifying myocardial arrangement is important for assessing impact on cardiac function.

Two cardiac organizations have been described [2, 3]:

- The myocyte orientation defined by the long axis of the myocytes, shows a regular low order organization in the ventricular wall often described as a helical transmural arrangement [4].
- The orientation of the myolaminae, i.e. the laminar architecture, is more complex, and comprises sheetlets disconnected by collagenous bundle interstices [5].

Both these architectures help defining an orthotropic myocardial basis along three orthogonal directions: normal to the sheetlet plane, normal to the fiber axis in the sheetlet plane and collinear to the fiber axis.

Magnetic Resonance Imaging (MRI) technique can provide information on the myocardial myocyte orientation and the myolaminar structure covering the whole heart. Diffusion Tensor Imaging (DTI) has been applied to investigate cardiac myocyte orientation *in vivo* [6, 7] and *ex vivo* in 3D on several species [7–11] including human [3]. DTI was also validated by comparison to histology [12]. Following from this earlier work, 3D high resolution MRI (HR-MRI) acquisition with at $50 \times 50 \times 50$ μm^3 isotropic resolution was proposed and validated by Gilbert et al. [13] to visualize and quantify cardiac microstructure in rats. They demonstrated that sheetlet orientation found using structure tensor imaging (STI) was proved superior to DTI [14]. However, thus far, this technique on human has been unexplored.

In this study, we apply these approaches to whole *ex vivo* human hearts from patients not eligible to cardiac transplantation. 3D FLASH/ST and diffusion images were acquired at a high spatial resolution using a 9.4T/30 cm magnet and a dedicated transmit/receive MR coil designed for this application. Quantitative structural information are computed from the resulting images using structure tensor analysis and leading to the STI pipeline. The fiber and the sheetlet orientations are extracted from both DTI and STI. The first results on fiber and sheetlet orientations in an intact human heart using a combination of DTI and STI are reported.

2   Materials and Methods

2.1   Sample Preparation

The heart was obtained through the human donor research project approved by the French Biomedicine Agency. A healthy heart (74 y.o. woman) of dimension $11 \times 10 \times 12$ cm^3 (Fig. 1) was fixed for at least 2 h with formalin containing 2 ml Dotarem (Gadoterate Meglumine, Guerbet, France). Imaging was carried out with the heart removed from formalin and immersed in Fomblin (Solvay Solexis, Inc) oil to reduce susceptibility artifacts at the interface with the tissue [13].

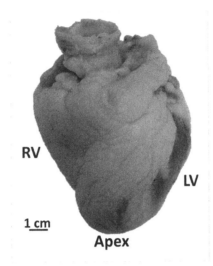

Fig. 1. Human donor heart (74 y.o woman) of dimension $11 \times 10 \times 12$ cm^3. Apex, right ventricle (RV) and left ventricle (LV) are identified. Yellow tissue is corresponding to pericardial fat and brown tissue to cardiac tissue. (Color figure online)

2.2   Magnet Set Up

MRI. All experiments were performed at 9.4T/30 cm (Bruker Biospin MRI, Ettlingen Germany). A cylindrical (165 mm inner diameter) 7 channels volume array Tx/Rx was used for ex vivo imaging. After acquiring scout images, a standard B1 map using Bloch-Siegert method in 3D (matrix size of $100 \times 100 \times 32$ and resolution of $1 \times 1 \times 2$ mm) was acquired [15].

3D DTI. DT-MRI was carried out using a 3D diffusion-weighted spin-echo sequence with TE = 23 ms, TR = 500 ms, at an isotropic resolution of $600 \times 600 \times 600$ μm^3. The diffusion gradient had 4.35 ms duration each and with an 11 ms delay between them. Six gradient directions were applied with a b-value of 1000 s/mm^2, as described previously [14], and a partial Fourier factor of 1.8 was used, for a total acquisition time of 23 h. Raw diffusion weighted images were processed using Paravison 6.0 to compute the diffusion tensor.

For DTI post processing pipeline in Fig. 2, an N4 bias correction [16] was applied on b = 0 s/mm^2 maps to segment the cardiac ventricles, thus avoiding cutting off regions with B$_1$ inhomogeneity. Binary masks were created using low and high cut-off thresholds based on FA, Trace and diffusion weighted image to remove the background noise. The first DT eigenvector has been shown to correspond to the myocyte orientation [3, 9–11]. The second and third eigenvectors have been associated with the sheetlet in-plane and normal directions respectively,

As described in previous studies [14, 17, 18], all structural information was obtained using a specific cardiac reference system, with an apex-base left ventricle (LV) axis running through the center of the left ventricular cavity. For each voxel in the segmented

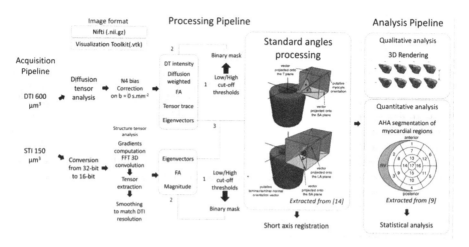

Fig. 2. DTI and STI data processing pipeline of the human donor heart of dimension $11 \times 10 \times 12$ cm^3. Diffusion tensor (DT) spin-echo and structure tensor (ST) gradient-echo data are processed through two different pipeline resulting in the helix angle (HA), transverse angle (TA), sheetlet elevation angle (SE), sheetlet azimuth angle (SA), sheetlet normal elevation angle (NE) and sheetlet normal azimuth angle (NA). The angle maps are then registered in short axis to apply the 17-segment AHA model for quantitative analysis.

datasets, the orientation of myocytes (fibers) and sheetlets were computed in this coordinate system.

3D STI. The whole human heart volume was imaged with a 3D FLASH/ST sequence with the following parameters: 12 averages and TE = 7 ms; TR = 30 ms; matrix size = $800 \times 731 \times 665$; voxel dimensions = 150 µm isotropic, flip angle = 32°, GRAPPA = 2 in phase encoding direction, for a total acquisition time of 27 h 36 min.

The image-processing pipeline (Fig. 2) was implemented [14] using Matlab 8.6 (The MathWorks Inc., Natick, MA, USA) and VTK libraries: images were converted into a stack of 16-bit images. Then, a structure tensor analysis was carried out: gradients, FFT-based convolution was computed as described in [13, 14] and smoothed at progressive resolution (from $150 \times 150 \times 150$ µm^3 to $600 \times 600 \times 600$ µm^3 isotropic ($800 \times 731 \times 665$ tensors to $200 \times 183 \times 166$ tensors). Extraction of eigenvalues and eigenvectors was performed. The $600 \times 600 \times 600$ µm^3 smoothed structure tensor dataset was used to best match the expected diffusion tensor (DT) resolution.

The first eigenvector (largest magnitude eigenvalue) corresponds to the sheetlet/laminae normal direction, the second to the sheetlet/laminae in-plane direction and the third to the myocyte orientation (smallest magnitude eigenvalue) [13, 14].

Binary masks were created using low and high cut-off thresholds based on FA where we chose 0.11 and 1 (meanly between 0.4 and 0.8 in the *ex vivo* heart), Trace and image intensity to remove the background noise.

Comparison Between STI and DTI. The first DTI eigenvector was compared to the third STI eigenvector for myocyte orientation and the fiber helix angle (HA) was then computed. The fiber helix angle is the angle between the short axis plane (i.e. the transverse plane) and the projection of the fiber vector onto the wall tangent plane (i.e. the longitudinal plane). The fiber transverse angle (TA) is the angle between the longitudinal plane and the projection of the fiber vector onto the transverse plane as shown on the standard angles processing scheme on Fig. 2.

For sheetlet orientation, the first STI eigenvector was compared to the third DTI eigenvector, which is assumed to be positioned normal to the sheet. The sheetlet elevation angle (SE) is the angle between the short axis plane and the projection of the vector onto the radial plane. The sheetlet azimuth angle (SA) is the angle between the local radial plane and the projection of the vector onto the short axis plane.

Images Visualization and Quantification. Image reconstruction was performed using ParaVision 6.0 on a workstation equipped with 512 GB of RAM to process the very large data matrices.

Short axis registration of the parametric volume maps was made using 3D Slicer (www.slicer.org) [19].

3D volume renderings and parametric volume maps were displayed using Paraview (Kitware, Clifton Park, NY) software.

Quantitative transmural maps were sub-divided into regions defined by the 17-segment American Heart Association (AHA) model [20] and post processed using custom written software in Matlab 8.6 (MathWorks Inc., Natic Massachusetts, USA). Descriptive statistics with linear regression (R^2 linearity and slope) and mean curve fitting were performed. A pair wise difference between DTI and STI measurements was performed as described [13]. All quantitative treatments were obtained using GraphPad Prism 8 (GraphPad Software, San Diego, CA) on the segment 7 to 12 of AHA model, corresponding to the mid ventricular region of the heart. Each segment contains a mean of 19600 ± 1419.3 voxels for a total of 117600 voxels in all the considered segments.

3   Results

3.1   Structural Characterization: A DTI and STI Comparison

Myocyte Orientation. Quantitative maps of standard angles describing the myocyte orientation (HA, TA) for both DTI and STI are presented (Fig. 3a–d). Based on DTI eigenvectors, HA map (Fig. 3a) shows a smooth transmural profile myocyte orientation from the LV sub-epicardium ($-40°$) to sub-endocardium ($90°$). In comparison with STI, HA map (Fig. 3b) preserves the same left to right helical orientation (from $-40°$ to almost $90°$) but with more abrupt transitions. The DTI TA map (Fig. 3c) displays a mean close to $0°$, indicating that myocytes orientation stays in the short axis plane. STI TA map (Fig. 3d) is noisier than the DTI, but the overall trend is at values close to $20°$.

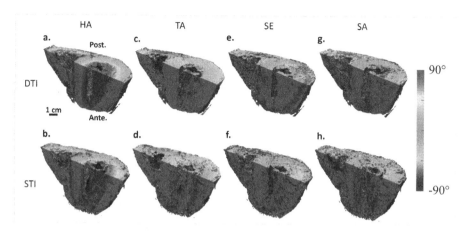

Fig. 3. Quantitative angle maps comparison between DTI and STI. Helix angle (HA), transverse angle (TA), sheetlet elevation angle (SE) and sheetlet azimuth angle (SA) in degrees are displayed at $600 \times 600 \times 600 \ \mu m^3$ for both DTI and STI.

Moreover, we observe that both HA and TA maps present an abrupt separation between the LV and RV. The separation between the LV and RV is more pronounced in HA map than in TA map.

Transmural profiles of HA and TA of DTI and STI in the mid ventricular region as measured in segment 7 to 12 based on the AHA heart model (AHA7-12) are presented (Fig. 4a–l). Their linearity and slope are summarized (Table 1) and the pair wise difference between the DTI and STI profiles is presented (Fig. 5a–l). DTI transmural profiles of HA on AHA7-12 (Fig. 4a–f) is found to be mostly linear ($R^2 \geq 0.8$) even though there is a negative slope range of between $-120.1°$ in the mid anteroseptal region (Fig. 4b, segment 8 on Table 1) to $-178.3°$ in the mid anterior region (Fig. 4a, segment 7 on Table 1). The pair wise difference of HA (Fig. 5a–f) shows a globally constant variation in the transmural profile with an exception for the segment 7 (Fig. 5a). There is also an offset of about 10–20° between DT and ST The pair wise

Table 1. Linearity (R2) and coefficient slope of DTI and STI myocyte orientation (helix angle and transverse angle). Linearity and slope were measured from LV endocardium to epicardium.

	DTI Helix angle		STI Helix angle		DTI Transverse angle		STI Transverse angle	
	Linearity	Slope	Linearity	Slope	Linearity	Slope	Linearity	Slope
7	0.80	-178.3 ± 1.3	0.09	-63.4 ± 3.0	0.04	23.1 ± 1.5	0.00	8.7 ± 3.2
8	0.71	-120.1 ± 1.0	0.55	-136.2 ± 1.7	0.25	38.4 ± 0.9	0.34	111.7 ± 2.1
9	0.77	-126.5 ± 1.0	0.56	-139.9 ± 1.9	0.23	44.6 ± 1.2	0.21	97.2 ± 2.7
10	0.62	-148.5 ± 1.7	0.37	-112.8 ± 2.1	0.00	5.7 ± 1.8	0.02	-31.0 ± 2.9
11	0.90	-153.1 ± 0.7	0.77	-150.3 ± 1.1	0.01	-4.8 ± 0.7	0.26	-61.5 ± 1.4
12	0.79	-153.9 ± 1.2	0.52	-149.6 ± 2.0	0.00	-1.3 ± 1.0	0.00	-4.7 ± 1.9

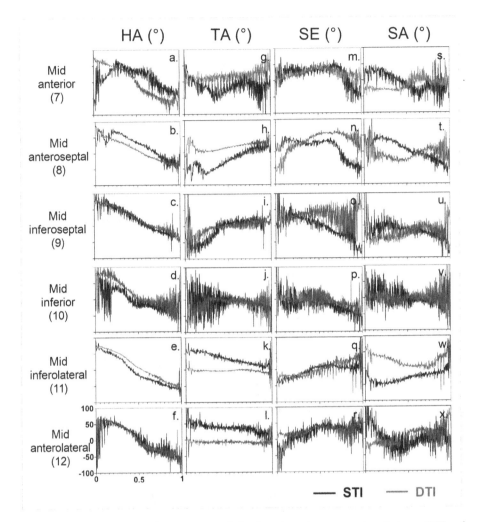

Fig. 4. Transmural variation in myocytes and sheetlet orientations comparison between DTI and STI. The 17-segment AHA model defines regions and AHA7-12 of STI (black line) and DTI (gray line) helix angle (HA), transverse angle (TA), sheet elevation (SE) and sheet azimuth (SA) are displayed. Means of angles normalized to wall thickness ae given.

difference of TA (Fig. 5g–l) shows a less consistent variation between STI and DTI for segment 8 and 11 (Fig. 5i and k) which goes from 80° in the sub-endocardium to 10° in the sub-epicardium.

Angles profiles of TA (Fig. 4g–l) have a low or absent linearity ($R^2 \leq 0.25$) with a low slope (mean of 17.7° for DTI) and indicates that the orientation is in the short axis plane. STI transmural profiles of HA (Fig. 4a–f) are weakly linear ($0.55 \leq R^2 \leq 0.8$), which represent a higher discrepancy on 3D volume (Fig. 3d), and the slope range is of between $-150.3°$ in the mid inferolateral (Fig. 4e, segment 11 on Table 1) to $-63.4°$ in the mid anterior (Fig. 4a, segment 7 on Table 1). The transmural profiles of TA using

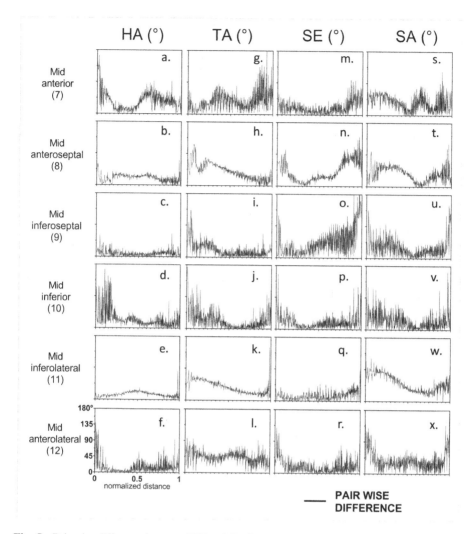

Fig. 5. Pair wise difference between DTI and STI in myocytes and sheetlet orientations. The 17-segment AHA model defines regions and AHA7-12 of pair wise difference (black line) of helix angle (HA), transverse angle (TA), sheet elevation (SE) and sheet azimuth (SA) are displayed.

STI (Fig. 4g–l) have a low linearity ($R^2 \leq 0.34$). We can notice that the slope of 111.7° in the mid anteroseptal (Fig. 4b, segment 7 on Table 1), 97.2° in the mid inferoseptal (Fig. 4c, segment 9 on Table 1) and −61.5° in the mid inferolateral (Fig. 4e, segment 9 on Table 1) appears to be not close to 0°.

Laminar Orientation. Quantitative maps of standard angles describing the laminar orientation (SE, SA) for both DTI and STI are presented (Fig. 3e–h). Based on DTI eigenvectors, SE map (Fig. 3e) is heterogeneous in the lateral regions of the LV from the sub-epicardium (80–90°) to the sub-endocardium (close to 0°) while the septal region is uniform and close to 0°. In comparison with STI, SE map (Fig. 3f) reveals the same trend as the DTI but with a more moderate elevation (45°). For SA map using DTI (Fig. 3g) we observe homogeneous values, and the angle values are close to 0°. Both SA and SE maps present an abrupt separation between the LV and RV orientation. The STI SA map (Fig. 3h) shows values close to 0° in the sub-epicardium and about −90° in the sub-endocardium. The intersection between the LV and RV are more homogeneous than with DTI SA and SE maps.

Transmural profiles of SE and SA of DTI and STI in the mid ventricular region as measured in segment 7 to 12 based are presented (Fig. 4m–x), their linearity and slope summarized (Table 2) and the pair wise difference between the profiles of DTI and STI is presented (Fig. 5m–x).

The DTI transmural profiles of SE (Fig. 4m–r) are not linear ($R^2 \leq 0.13$) and the slope range is broad with −19.8° in the mid inferior region (Fig. 4p, segment 10 on Table 2) and 73.8° in the mid inferolateral region (Fig. 4q, segment 11 on Table 2). Moreover, the pair wise difference of SE (Fig. 5m–r) shows a globally constant variation between DT and ST in the transmural profile for the segments 7 and 10 to 12 (Fig. 5m and p–r). The pair wise difference of SA (Fig. 5s–x) displays a less consistent variation between STI and DTI for segment 7 and 11 (Fig. 5q and w), which decreases from 70–90° in the sub-endocardium to 10–20° in the middle of the wall and then increases to 30–40° in the sub-epicardium.

In comparison, STI transmural profiles of SE (Fig. 4m–r) have a lower linearity, than DTI ($0.01 \leq R^2 \leq 0.25$) and the slope of 44.1° in the mid inferolateral (Fig. 4q, segment 11 on Table 2) is small in comparison with DTI slope of 73.8° in the same region (Fig. 4q, segment 11 on Table 1). The DTI SA transmural profiles (Fig. 4s–x) displays the non-linearity ($R^2 \leq 0.22$) and a broad slope range even if the septal region (Fig. 4n–o, segment 8–9 on Table 2) is of a greater homogeneity. For STI, Transmural profiles of SA (Fig. 4s–x) have the overall same linearity and slope range as the DTI (Table 2).

Table 2. Linearity (R2) and coefficient slope of DTI and STI laminar orientation (sheet elevation angle and sheet azimuth angle). Linearity and slope were measured from LV endocardium to epicardium.

	DTI Sheet elevation		STI Sheet elevation		DTI Sheet azimuth		STI Sheet azimuth	
	Linearity	Slope	Linearity	Slope	Linearity	Slope	Linearity	Slope
7	0.01	12.2 ± 1.5	0.01	−21.5 ± 3.0	0.16	48.3 ± 1.6	0.12	−78.2 ± 3.1
8	0.10	44.8 ± 1.8	0.23	−95.0 ± 2.4	0.08	40.5 ± 1.9	0.33	−111.9 ± 2.2
9	0.00	−2.6 ± 3.0	0.25	−108.2 ± 2.8	0.1	51.2 ± 2.3	0.09	−58.7 ± 2.8
10	0.02	−19.8 ± 2.0	0.03	−34.1 ± 2.8	0.01	−13.7 ± 1.9	0.02	−26.5 ± 3.0
11	0.13	73.8 ± 2.6	0.09	44.1 ± 1.9	0.02	−21.2 ± 2.0	0.12	46.9 ± 1.7
12	0.06	29.0 ± 1.7	0.05	49.0 ± 2.9	0.22	65.8 ± 1.8	0.00	−2.5 ± 3.5

4 Discussion and Conclusion

This study present a comparison between 3D DTI and STI using a processing pipeline allowing the investigation of cardiac microstructure in the whole human heart. The STI data process using a FLASH/ST sequence are acquired at an isotropic resolution of $150 \times 150 \times 150$ μm^3, and DTI images are obtained on a volume at $600 \times 600 \times 600$ μm^3. Both acquisitions are post processed and compared using tensor imaging with an isotropic resolution of $600 \times 600 \times 600$ μm^3. Here, we show for the first time full 3D reconstructions of myocardial myocytes and sheetlet orientation in a human heart using Structure Tensor Image Analysis.

Myocyte orientation characterized by STI demonstrated that transmural profiles of HA and TA (Fig. 4a–l) are consistent with a previous study on rodent heart [13].

Cardiac laminar organization is more complex. Indeed, Gilbert et al. [13] and Bernus et al. [14] demonstrated that laminar structure has many regions of localized structural complexity. Thus, that it is not possible to label unique laminae and unequivocally determine their dimensions using one eigenvector since they form a densely branching three-dimensional network with more than one preferred direction and high spatial change of the laminar orientation [5]. The strong non-linearity of SE and SA in both DTI and STI (Table 2) is consistent with the discontinuous nature of myolaminae organization. Nonetheless, we observed strong similarities in 3D on SE and SA maps (Fig. 3e–h) between STI and DTI.

The pair wise difference between DTI and STI transmural profiles (Fig. 5a–x) highlights the consistency of HA and TA calculations in all regions even though there is an offset (more than 20°) between DTI and STI measurements as demonstrated by Bernus et al. [14]. Laminar organization is more complex and the comparison between SA and SE is more challenging particularly in the septal region (segments 7 to 9), and the junction of the two ventricles. We also need to improve our STI pipeline, especially concerning the angles unwrapping on the edges.

Bernus et al. [14] have shown a bias between DTI and STI (more than 20°) on rat hearts and emphasized the fact that the assignment of eigenvectors for DTI could provide limited assessment of laminae directions. DTI has an important role for describing myocyte orientation and STI is reliable for myolaminar measurements, we can hypothesis that combining both techniques in one acquisition could lead to the exploitation of the best of each one.

STI pipeline has been developed for rat hearts and applied to a human heart in our study. The first STI at $150 \times 150 \times 150$ μm^3 allows us to perform a 3D volume giving a large amount of structure information (size of ventricles, vessels, etc.) and the consequence in the case of a human heart is a loss of SNR (SNR = 90 for FLASH/ST at $150 \times 150 \times 150$ μm^3 against 479 for DTI at $600 \times 600 \times 600$ μm^3 in the same region). Moreover, some spatial heterogeneities have been observed, which could be reduced by adapting the STI pipeline to large volumes. There are discrepancies in different segments but also similarities. We can identify good similarities between segments. Nevertheless, we identify noisier results on segments 7 and 10, which correspond to the junction between RV and LV cardiac muscle and a specific organization of tissues in these areas. More investigation in several areas, especially in base apex region, LV/RV junction have to be studied.

In conclusion, these preliminary results are a promising foundation for investigating the 3D structural characterization of normal/pathologic cardiac organization. This work could be useful for simulation especially if we applied the pipeline on pathological heart. It could be integrate to provide information to better understand the inverse problem of electrocardiography, as it will serve to ameliorate the regularization techniques that help solve ill-posed problem

References

1. Hooks, D.A., Trew, M.L., Caldwell, B.J., Sands, G.B., LeGrice, I.J., Smaill, B.H.: Laminar arrangement of ventricular myocytes influences electrical behavior of the heart. Circ. Res. **101**(10), e103–e112 (2007)
2. Streeter Jr., D.D., Spotnitz, H.M., Patel, D.P., Ross Jr., J., Sonnenblick, E.H.: Fiber orientation in the canine left ventricle during diastole and systole. Circ. Res. **24**(3), 339–347 (1969)
3. Rohmer, D., Sitek, A., Gullberg, G.T.: Reconstruction and Visualization of Fiber and Sheet Structure with Regularized Tensor Diffusion MRI in the Human Heart, Lawrence Berkeley National Laboratory (2006)
4. Gilbert, S.H., Benson, A.P., Li, P., Holden, A.V.: Regional localization of left ventricular sheet structure: integration with current models of cardiac fiber, sheet and band structure. Eur. J. Cardiothorac. Surg. **32**(2), 231–249 (2007)
5. LeGrice, I.J., Smaill, B.H., Chai, L.Z., Edgar, S.G., Gavin, J.B., Hunter, P.J.: Laminar structure of the heart: ventricular myocyte arrangement and connective tissue architecture in the dog. Am. J. Physiol. **269**(2 Pt 2), H571–H582 (1995)
6. Toussaint, N., Stoeck, C.T., Schaeffter, T., Kozerke, S., Sermesant, M., Batchelor, P.G.: In vivo human cardiac fiber architecture estimation using shape-based diffusion tensor processing. Med. Image Anal. **17**(8), 1243–1255 (2013)
7. Nielles-Vallespin, S., et al.: Assessment of myocardial microstructural dynamics by in vivo diffusion tensor cardiac magnetic resonance. J. Am. Coll. Cardiol. **69**(6), 661–676 (2017)
8. Holmes, A.A., Scollan, D.F., Winslow, R.L.: Direct histological validation of diffusion tensor MRI in formaldehyde-fixed myocardium. Magn. Reson. Med. **44**(1), 157–161 (2000)
9. Teh, I., et al.: Resolving fine cardiac structures in rats with high-resolution diffusion tensor imaging. Sci. Rep. **6**, 30573 (2016)
10. Helm, P.A., Tseng, H.J., Younes, L., McVeigh, E.R., Winslow, R.L.: Ex vivo 3D diffusion tensor imaging and quantification of cardiac laminar structure. Magn. Reson. Med. **54**(4), 850–859 (2005)
11. Healy, L.J., Jiang, Y., Hsu, E.W.: Quantitative comparison of myocardial fiber structure between mice, rabbit, and sheep using diffusion tensor cardiovascular magnetic resonance. J. Cardiovasc. Magn. Reson. **13**, 74 (2011)
12. Köhler, S., Hiller, K.H., Waller, C., Jakob, P.M., Bauer, W.R., Haase, A.: Visualization of myocardial microstructure using high-resolution T_2^* imaging at high magnetic field. Magn. Reson. Med. **49**(2), 371–375 (2003)
13. Gilbert, S.H., et al.: Visualization and quantification of whole rat heart laminar structure using high-spatial resolution contrast-enhanced MRI. Am. J. Physiol. Heart Circ. Physiol. **302**(1), H287–H298 (2012)

14. Bernus, O., et al.: Comparison of diffusion tensor imaging by cardiovascular magnetic resonance and gadolinium-enhanced 3D image intensity approaches to investigation of structural anisotropy in explanted rat hearts. J. Cardiovasc. Magn. Reson. **17**, 31 (2015)
15. Sacolick, L.I., Wiesinger, F., Hancu, I., Vogel, M.W.: B_1 mapping by Bloch-Siegert shift. Magn. Reson. Med. **63**(5), 1315–1322 (2010)
16. Tustison, N.J., et al.: N4ITK: improved N3 bias correction. IEEE Trans. Med. Imaging **29**(6), 1310–1320 (2010)
17. Gilbert, S.H., Smaill, B.H., Walton, R.D., Trew, M.L., Bernus, O.: DT-MRI measurement of myolaminar structure: accuracy and sensitivity to time post-fixation, b-value and number of directions. In: Conference Proceedings of IEEE Engineering in Medicine and Biology Society, pp. 699–702 (2013)
18. Gilbert, S., Trew, M., Smaill, B., Radjenovic, A., Bernus, O.: Measurement of myocardial structure: 3D structure tensor analysis of high resolution MRI quantitatively compared to DT-MRI. In: Camara, O., Mansi, T., Pop, M., Rhode, K., Sermesant, M., Young, A. (eds.) STACOM 2012. LNCS, vol. 7746, pp. 207–214. Springer, Heidelberg (2013). https://doi.org/10.1007/978-3-642-36961-2_24
19. Fedorov, A., et al.: 3D slicer as an image computing platform for the quantitative imaging network. Magn. Reson. Imaging **30**(9), 1323–1341 (2012)
20. Cerqueira, M.D., et al.: Standardized myocardial segmentation and nomenclature for tomographic imaging of the heart. A statement for healthcare professionals from the Cardiac Imaging Committee of the Council on Clinical Cardiology of the American Heart Association. Circulation **105**, 539–542 (2002)

Investigating the 3D Local Myocytes Arrangement in the Human LV Mid-Wall with the Transverse Angle

Shunli Wang[1,2,3], Iulia Mirea[1], François Varray[1(✉)], Wan-Yu Liu[2,3], and Isabelle E. Magnin[1,2,3]

[1] Lyon University, CREATIS, INSA-Lyon, UCBL Univ., UJM Univ., CNRS, Inserm, Lyon, France
francois.varray@creatis.insa-lyon.fr
[2] Metislab, Harbin Institute of Technology, Harbin, China
[3] Metislab, Shanghai University, Shanghai, China

Abstract. Myolaminar Layer Arrangement plays an essential role in cardiac biomechanics. In this preliminary study, we investigate the local 3D arrangement of the myocytes inside the sheets (layers) in three LV human heart transparietal samples imaged by X-ray phase contrast micro-tomography. We extract the large cleavage planes (CPs) of the extracellular matrix, manually select the middle wall region within each sample and compute the skeleton surface (chamfer distance and nonwitness-points selection) of the layers containing the myocytes. We compute the transverse angles of the myocytes in windows ($32 \times 32 \times 32$ voxels i.e. $112 \times 112 \times 112$ μm^3) centered on the 3D skeleton surface. Our results show that the myocytes are organized (i) in two populations in a LV samples close to the base with an angular distribution alternatively changing from one layer to the next and (ii) in a continuous angular evolution in samples located close to the apex. We find a mean angular difference between the two populations of about 8° in the two LV posterior samples and about 13° in the LV anterior sample. It is too early to statistically confirm that values as "universal" therefore we currently pursue our analysis of other available human LV samples to assess those first results.

Keywords: Human left ventricular middle wall · Myocytes arrangement · Laminar structure · Transverse angle · X-ray phase contrast micro-tomography · Image analysis

1 Introduction

Left ventricular (LV) myocytes are organized into branching myolaminar 'sheets' about four cells thick by an extensive extracellular collagen matrix [1]. The cardiac endomysium collagen provides tight coupling within each sheet. And, the perimysium appears as cleavage planes, providing looser connections and potential spaces for relative slippage between adjacent sheets. During the cardiac systole and diastole, the role of the laminar architecture was proposed to distinctly facilitate the wall thickening

© Springer Nature Switzerland AG 2019
Y. Coudière et al. (Eds.): FIMH 2019, LNCS 11504, pp. 208–216, 2019.
https://doi.org/10.1007/978-3-030-21949-9_23

deformation by the sheet extension, shear and thinning (or thickening), thus enhancing the ventricular ejection [2–6].

Morphology observations and feature measurements about the sheets have been carried out in several species at different phases of the cardiac cycle with various imaging techniques such as dissection and histology [6], polarised light imaging [7], electron microscopy [8], Magnetic Resonance imaging (MRI) [9], and X-ray phase contrast micro-tomography [10]. In Ref. [8], the sheet morphology was described locally and precisely. In Ref. [6], the sheet angle was measured regionally. In Ref. [10], the thickness of the cleavage planes and the distance between the cleavage planes were statistically analyzed. Besides, the existence of herring-like patterns of laminar populations (i.e. sheet angles are around $\pm 45°$) was demonstrated [6, 11, 12] and the related strains were measured regionally [3, 9].

Myocytes are the main structural component of the sheets and their orientation notably affects the strain as shown in Ref. [13]. Past studies focus on the LV transmural orientation of the myocytes measured with the helix angle. They reveal that the myocytes are locally approximately parallel and the myocytes orientation continuously varies with the depth of the wall [14]. When the myocyte orientation is quantified using the transverse angle alone, it clearly appears that the myocytes are wrapped between endocardium and epicardium [15]. The important contribution of the myocytes distribution (fibers) to the LV wall stress has also been demonstrated [16].

In our study, we extract and analyses the 3D local orientation of the myocytes within the layers from X-ray phase contrast micro-tomography using the local transverse angle distribution.

2   Material and Methods

2.1   Data Acquisition

The Medico-Legal Institute of Lyon IML HCL (N°DC-2012-1588) supplied a series of fresh human LV transmural heart samples. We performed the image acquisition campaign at the European Synchrotron Radiation Facility located in Grenoble (ESRF). Details on the data and acquisition procedure are given in Ref. [17]. The reconstructed tissue samples have an isotropic spatial resolution with voxel's edges equal to 3.5 μm. The locations of three samples used in this paper are presented in Fig. 1. Samples 1 and 2 are from the LV posterior part. Sample 3 is from LV anterior part. Sample 2 and 3 are near the apex.

2.2   Data Segmentation

We pre-process the data with a Gaussian filter $G(0, 44)$ to suppress the very low frequency components induced by the acquisition procedure. We binarise them using the method of maximum variance between classes (i.e. Otsu algorithm) [18], eliminate the small extracellular structures and select the main cleavage planes (CPs) by connected components filtering (Fig. 2). We analyze the layers located between the binary CPs.

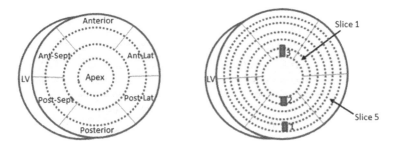

Fig. 1. The samples' locations in LV wall.

2.3 Skeletonization of the Main Layers

The myocytes of the LV wall are locally organized in layers lying between two sheetlets (or Cleavage planes (CP)) as a sandwich. We consider each layer as a local active unit (LAU). The two sheetlets slide relatively to each other to contribute to the thickening of the wall during the heart systole. We measure the local orientation of the myocytes within each layer to understand their 3D arrangement and their contribution to the local motion/deformation. To have it, we compute the chamfer distance from the CPs to get the distance field within each layer and select the 'nonwitness points' as the centers of maximal balls to compose each 3D layer's skeleton [19, 20].

2.4 Local Extraction of the Myocytes Transverse Angle

The left ventricular wall comprises three types of micro-architecture, which distinguishes the sub-endocardial, the middle wall and the sub-epicardial regions (Fig. 2a). Based on CP appearance, we manually select the three regions and extract the middle wall, as shown in Fig. 2b.

We use the right-hand coordinate system (Fig. 2c), where x corresponds to the radial direction, y to the circumferential direction and z to the longitudinal direction. The transverse angle $\varnothing$ is the angle between the y axis and the projection of the myocytes' local orientation onto the (x, y) circumferential-radial plane.

We compute the local transverse angle of the myocytes at each voxel of the layer's skeleton within a 32^3-voxel (i.e. 112^3 μm^3) size window. The method involves a 3D FFT followed by a Principle Component Analysis of the binarised spectrum as described in Ref. [17]. As depicted in Fig. 3a, the selected stack of myocytes can be regarded as a bundle of parallel cylinders. We note the (Fig. 3b) oblate spheroidal shape of the binary FFT spectrum which short axis (smallest eigenvectors) coincides with the myocytes orientation.

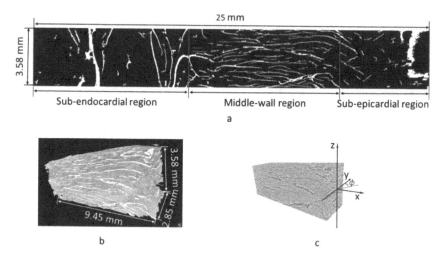

Fig. 2. (a) Longitudinal-radial section of a binarized left ventricular wall sample (sample 1, LV posterior) manually divided into three regions; (b) The sample 1 middle wall region with the main cleavage planes; (c) The coordinate system with x, y, z respectively corresponding to the radial, circumferential and longitudinal directions and $\varnothing$ the transverse angle.

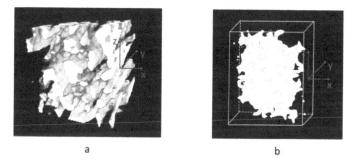

Fig. 3. Extraction of the 3D local orientation of myocytes. (a) Data sample in a 32^3 voxel size window centered on a layer's skeleton. (b) Related binarised 3D Fourier spectrum of (a) with the 3 eigenvectors.

3   Results

Figure 4 depicts a distance field map and the skeletons extracted in the mid-wall longitudinal-radial section of sample 1. The deep red parts are the skeletons of the layers of myocytes located between the sheetlets (cleavage planes) in deep blue. The surface skeletons (and cleavage planes) are branching and merging. They are transmurally oriented and relatively locally parallel. The thickness of layers (or sheets) varies.

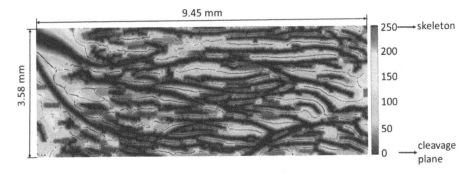

Fig. 4. Sample 1 (LV posterior). A longitudinal-radial section of the 3D distance field map with the skeletons of the layers of myocytes in deep red. The deep blue corresponds to the cleavage planes. (Color figure online)

Figure 5a presents the transverse angle distribution of sample 1 (LV posterior, heart 1) and Fig. 5b that of four ROIs. The four ROIs are longitudinal-radial sections 51–100 (i.e. from section 51 to section 100), sections 251–300, sections 451–500, and sections 651–700. We observe, especially in Fig. 5b, two shifted and superimposed Gaussian distributions with different means in each distribution that we can automatically separate using the Otsu thresholding method [18]. The obtained transverse angle threshold values are around 17°.

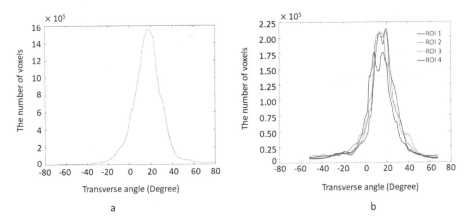

Fig. 5. Transverse angle distribution of sample 1 (LV posterior, heart 1) (a) and that of four ROIs in it (b). The four ROIs are longitudinal-radial sections 51–100 (i.e. from section 51 to section 100), sections 251–300, sections 451–500, and sections 651–700.

The arrangement of the two transverse angle populations in sample 1 are presented in Fig. 6. The c, d and e are circumferential-longitudinal sections respectively lying in the inner, middle and external parts of the LV mid-wall sample. In the circumferential-longitudinal sections c, d, e, the orientation of the two angle populations is parallel.

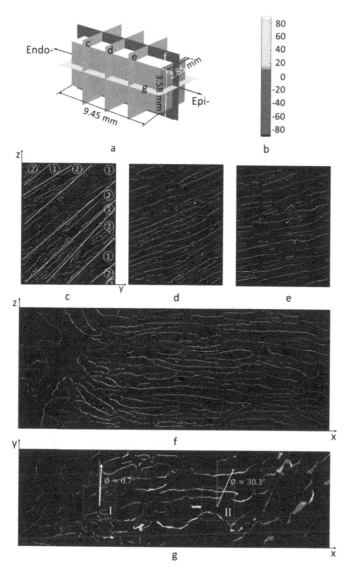

Fig. 6. Transverse angle. Display of two transverse-angle populations in sample 1 mid-wall. (a) Schematic mid-wall region with the location of three circumferential-longitudinal sections c, d, e, one longitudinal-radial section f and one radial-circumferential section g. (b) Color scale in degrees dividing the 2 populations (red and yellow). (c) ① denotes the red population and ② denotes the yellow population. (Color figure online)

The bundles of myocytes filling the layers are alternatively oriented in both orientations from one layer (to the next. In the longitudinal-radial section f and radial-circumferential section g, the two populations are interlaced. The mean transverse angles (Fig. 6g) of ROI I (in red) is about $0.7°$, that of ROI II (in yellow) is about $30.3°$.

214   S. Wang et al.

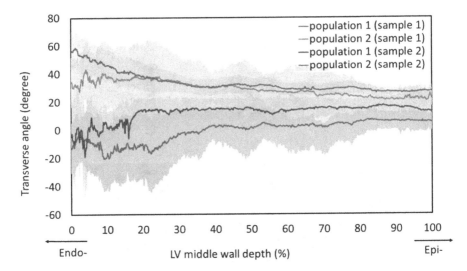

Fig. 7. Transverse angle of the two populations along the radial axis in LV posterior samples 1 and 2

The two populations' transverse angles distributions are computed section by section in the circumferential-longitudinal plane (Fig. 7). The mean value in each population remains nearly constant from the inner to the external part of the region. But their standard deviations (SD) are large.

Table 1. Statistical results of the transverse angle of the myocytes in each samples

LV	Population 1	Population 2	Population 1 + Population 2
Sample 1 (LV posterior)	$0.7° \pm 16.2°$	$30.2° \pm 13.1°$	$13.3° \pm 20.3°$
Sample 2 (LV posterior)	$13.4° \pm 10.1°$	$35.2° \pm 10.3°$	$23.1° \pm 13.9°$
Sample 3 (LV anterior)	$-8.3° \pm 6.2°$	$4.9° \pm 6.0°$	$-1.7° \pm 8.7°$

The Table 1 lists the statistical results of the transverse angle in three samples, presented as mean $\pm$ SD. The mean transverse angle value and SD of each population varies the in different samples. It is larger in the LV posterior samples than in the LV anterior sample. The SDs of the populations in each sample are similar.

4   Discussion and Conclusion

We investigated the transverse angle of myocytes located in the main layers in three human LV mid-wall samples using X-ray phase contrast micro-tomography.

The transverse angle was calculated, in a 32^3 voxels size window roughly corresponding both to the average length of myocytes and to the average thickness of layer.

Our preliminary results show that the myocytes transverse angle is not uniform in the LV mid-wall layers of all samples. As shown in Fig. 5b, the transverse angles in each ROI belong to two Gaussians and these two Gaussians are separable. While enlarging the sample size, the transverse angle distribution reduces to one main peak as shown in Fig. 5a, which accounts for the transmural transition of transverse angle.

The Fig. 7 and Table 1 present statistical results of the two transverse angle populations in LV mid-wall samples. The results show that the distribution of the two transverse angle populations is regional. We automatically separated the two populations and observed that the mean angle between those two populations in the LV posterior samples ranges from 22° to 29°. The first results on a LV anterior sample exhibit a smaller mean angle between the two populations of 13° (Table 1). The SDs in each sample's two populations are large, but similar, which may account for the structure noises.

Thought the results in this paper are interesting, the shortage is also apparent. The layers related study demonstrates that layer features are regional. The number and size of the samples used remain small. We observe several transverse angle distribution patterns and the pattern discussed here is only one of them. Besides, we manually selected the LV mid-walls based on their CPs appearance as in Fig. 2 where the CPs are relatively parallel. We note, Fig. 7, that the transverse angle distribution gradually varies trough the wall. So an adaptive segmentation method could be chosen to get the series of local thresholds separating two successive layers.

References

1. Weber, K.T.: Cardiac interstitium in health and disease: the fibrillar collagen network. J. Am. Coll. Cardiol. **13**, 1637–1652 (1989)
2. LeGrice, I.J., Takayama, Y., Covell, J.W.: Transverse shear along myocardial cleavage planes provides a mechanism for normal systolic wall thickening. Circ. Res. **77**, 182–193 (1995)
3. Costa, K.D., Takayama, Y., McCulloch, A.D., Covell, J.W.: Laminar fiber architecture and three-dimensional systolic mechanics in canine ventricular myocardium. Am. J. Physiol. **276**, H595–H607 (1999)
4. Spotnitz, H.M., Spotnitz, W.D., Cottrell, T.S., Spiro, D., Sonnenblick, E.H.: Cellular basis for volume related wall thickness changes in the rat left ventricle. J. Mol. Cell Cardiol. **6**, 317–331 (1974)
5. Takayama, Y., Costa, K.D., Covell, J.W.: Contribution of laminar myofiber architecture to load-dependent changes in mechanics of LV myocardium. Am. J. Physiol. Circ. Physiol. **51**, H1510 (2002)
6. Harrington, K.B., et al.: Direct measurement of transmural laminar architecture in the anterolateral wall of the ovine left ventricle: new implications for wall thickening mechanics. Am. J. Physiol. Circ. Physiol. **288**, H1324–H1330 (2005)
7. Jouk, P.S., et al.: Analysis of the fiber architecture of the heart by quantitative polarized light microscopy. Accuracy, limitations and contribution to the study of the fiber architecture of the ventricles during fetal and neonatal life. Eur. J. Cardio-Thoracic Surg. **31**, 916–922 (2007)
8. LeGrice, I.J., Smaill, B.H., Chai, L.Z., Edgar, S.G., Gavin, J.B., Hunter, P.J.: Laminar structure of the heart: ventricular myocyte arrangement and connective tissue architecture in the dog. Am. J. Physiol. Circ. Physiol. **269**, H571–H582 (1995)

9. Rademakers, F.E., et al.: Relation of regional cross-fiber shortening to wall thickening in the intact heart. Three-dimensional strain analysis by NMR tagging. Circulation **89**, 1174–1182 (1994)

10. Mirea, I., Wang, L., Varray, F., Zhu, Y.-M., Serrano, E.E.D., Magnin, I.E.: Statistical analysis of transmural laminar microarchitecture of the human left ventricle. In: 2016 IEEE 13th International Conference on Signal Processing (ICSP), pp. 53–56. IEEE (2016)

11. Kung, G.L., et al.: The presence of two local myocardial sheet populations confirmed by diffusion tensor MRI and histological validation. J. Magn. Reson. Imaging **34**, 1080–1091 (2011)

12. Gilbert, S.H., et al.: Visualization and quantification of whole rat heart laminar structure using high-spatial resolution contrast enhanced MRI. Am. J. Physiol. Circ. Physiol. **302**, H287–H298 (2011)

13. Ubbink, S., Bovendeerd, P., Delhaas, T., Arts, T., van de Vosse, F.: Left ventricular shear strain in model and experiment: the role of myofiber orientation. In: Frangi, A.F., Radeva, P. I., Santos, A., Hernandez, M. (eds.) FIMH 2005. LNCS, vol. 3504, pp. 314–324. Springer, Heidelberg (2005). https://doi.org/10.1007/11494621_32

14. Streeter Jr., D.D., Bassett, D.L.: An engineering analysis of myocardial fiber orientation in pig's left ventricle in systole. Anat. Rec. **155**(4), 503–511 (1966)

15. Streeter Jr., D.D.: Gross morphology and fiber geometry of the heart. In: Berne, R.M. (ed.) Handbook of Physiology-The Cardiovascular System I. The Heart, vol. 1, chap. 4, pp. 61–112. American Physiology Society, Bethesda (1979)

16. Bovendeerd, P.H.M., Huyghe, J.M., Arts, T., Van Campen, D.H., Reneman, R.S.: Influence of endocardial-epicardial crossover of muscle fibers on left ventricular wall mechanics. J. Biomech. **27**(7), 941–951 (1994)

17. Varray, F., Mirea, I., Langer, M., Peyrin, F., Fanton, L., Magnin, I.E.: Extraction of the 3D local orientation of myocytes in human cardiac tissue using X-ray phase-contrast micro-tomography and multi-scale analysis. Med. Image Anal. **38**, 117–132 (2017)

18. Otsu, N.: A threshold selection method from gray-level histograms. IEEE Trans. Syst. Man Cybern. **9**, 62–66 (1979)

19. Borgefors, G.: Distance transformations in arbitrary dimensions. Comput. Vis. Graph Image Process **27**, 321–345 (1984)

20. Pudney, C.: Distance-ordered homotopic thinning: a skeletonization algorithm for 3D digital images. Comput. Vis. Image Underst. **72**(404–413), 0680 (1998)

Biomechanics: Modelling and Tissue Property Measurements

Development of a Computational Fluid Dynamics (CFD)-Model of the Arterial Epicardial Vasculature

Johannes Martens[1,2(✉)], Sabine Panzer[1,2],
Jeroen P. H. M. van den Wijngaard[3,4], Maria Siebes[3],
and Laura M. Schreiber[1,2]

[1] Chair of Molecular and Cellular Imaging, Comprehensive Heart Failure Center
(CHFC), University Hospitals Wuerzburg, Würzburg, Germany
martens_j@ukw.de
[2] Department of Cardiovascular Imaging, CHFC, University Hospitals
Wuerzburg, Würzburg, Germany
[3] Department of Biomedical Engineering and Physics, Academic Medical Center,
Amsterdam, Netherlands
[4] Department of Clinical Chemistry and Hematology, Diakonessenhuis,
Utrecht, Netherlands

Abstract. Motivation of the project is the analysis of systematic errors
in contrast-enhanced cardiac perfusion imaging by CFD simulations
of contrast agent transport in the arterial epicardial vasculature. This
requires the realistic modeling of volume blood flow (VBF) in the coronary arteries to provide a physiologically relevant computational framework for the transport simulations. For this purpose, 3D-models of the
left and right coronary trees are extracted from high-resolution cardiovascular cryomicrotome imaging data and meshed with computational
grids. A dedicated model integrating characteristics of coronary blood
flow is used to generate boundary conditions (BCs). Subsequently, VBF
is analyzed in left and right ventricular myocardial regions (VBF_m) and
in dependence of the vessel sizes (VBF_v). Regarding the distribution of
VBF_m in the myocardial segments, good agreement with literature values is found. Partial compliance of the findings of the VBF_v-analysis
with results from other groups is promising, however, indicates room for
improvement.

Keywords: Computational fluid dynamics · In-silico modeling ·
Epicardial blood flow

1   Introduction

Long-term aim of the project is the CFD modeling of contrast agent transport in coronary arteries in order to investigate systematic errors in contrast-
enhanced Magnetic Resonance Imaging of myocardial perfusion [1–4]. To insure

© Springer Nature Switzerland AG 2019
Y. Coudière et al. (Eds.): FIMH 2019, LNCS 11504, pp. 219–229, 2019.
https://doi.org/10.1007/978-3-030-21949-9_24

the physiologic relevance of the transport simulations, a framework realistically describing blood flow in the coronary vasculature is essential.

Unlike systemic blood flow, cardiac circulation is increased during diastole in comparison to systole. This is due to the inherent embedding of the cardiovascular network directly within the myocardial tissue, which it supplies. The compression (systole) and relaxation (diastole) of the myocardium results in decreased and increased blood flow in cardiac vessels. This effect can be pronounced to an extent that even retrograde flow is observed in large epicardial arteries during early systole [5].

In order to correctly reflect this property characteristic of cardiovascular blood flow in CFD simulations, dedicated BCs need to be conceived and implemented. A rather sophisticated model makes use of the analogy of cardiac and electrical circuit. It was first proposed by Westerhof et al., [6] and enhanced and further developed in [7,8], among others.

In this work, an implementation of this dedicated BC is described and it is utilized to perform CFD simulations of blood flow in a coronary tree including vessels at the pre-arteriolar level. Since the realistic modeling of blood flow represents an indispensable prerequisite for future contrast agent transport simulations, the computations are validated with regard to several different aspects of coronary and blood flow in general.

2   Materials and Methods

2.1   Imaging Data and Computing Facilities

The used 3D models are extracted from a high-resolution (160 μm) cryomicrotome imaging dataset of an ex-vivo porcine heart [9]. The dedicated software package *SimVascular*[1] is used for model creation. Due to the high complexity

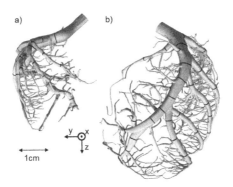

Fig. 1. Anterior view of the generated models of the right (a) and the left coronary tree (b). The cross sections indicated in black are used in the analysis of VBF in dependence of vessel diameter.

---

[1] SimTK, simvascular.github.io [10].

of the models, the complete model preparation is a manual endeavor of several months. The generated models are depicted in Fig. 1.

The vascular geometries are automatically discretized with the help of the software package *cfMesh*[2] taking approximately 1.5 h and 20 min for the models of the left (LCT) and the right coronary tree (RCT), respectively. The generated computational grids consist of $\sim$ 14 and 7 Mio largely hexahedral ($>$ 95%) cells, respectively. The used mesh resolution is based on [4]. The simulations are performed at the high performance computing cluster *CoolMUC-2* at the Leibniz Rechenzentrum Munich, parallelized on 56 (RCT) and 140 (LCT) processors.

2.2 The Electrical Analog of Coronary Circulation

In Fig. 2 the electrical analog of cardiac circulation is depicted, an RC-circuit. Accordingly, the pressure that is "stored" across a vessel's compliance C during a time interval Δt is given by:

$$p_C(t) = p_C(t - \Delta t) + \frac{1}{C} \int_{t-\Delta t}^{t} F(t')\mathrm{d}t' \tag{1}$$

$$\approx p_C(t - \Delta t) + \frac{\Delta t}{C} \cdot (F(t) - F(t - \Delta t)), \tag{2}$$

where $F(t), F(t-\Delta t)$ denote the flow through the vessel. In the case that $\Delta t \to 0$, Eq. 2 becomes exact. Depending on the initial "charge" of the myocardial compliance (capacitor), the system at some point reaches a stationary periodic state. This can be solved iteratively, where the parameter Δt must be chosen small enough to resolve the shortest timescales in which changes in blood flow occur. With the aortic pressure curve taken from [11] and an estimated myocardial pressure curve in a physiologic range according to [12], $\Delta t = 0.001$ s is sufficient. The cardiac cycle duration at rest is scaled to 0.7 s, yielding a typical heart rate for pigs weighing 25–35 kg [13], which corresponds to the size of the animals from which the used dataset was extracted [9].

In Fig. 2 p_{Cap} is to be understood as the pressure in the middle of the capillaries, thus, the resistances R_{Art} and R_{Ven} (cf. Fig. 2) comprise capillary contributions each. These contributions are chosen as follows: arteries ($\sim$ 0%), pre-/arterioles ($\sim$ 60%), capillaries ($\sim$ 25%), veins and venules ($\sim$ 15%) [14,15]. Depending on the 3D model's inlet radius, the total vascular resistance is estimated according to [16].

The second parameter from the analogy of cardiac circulation and electric current to be determined is the compliance C of the vessels within the myocardium. Due to the difficulty to measure this parameter [7], in this work, a sensitivity analysis is performed in order to make a reasonable estimation based on physiologically relevant quantities [8]. With the help of the open source

---

[2] cfmesh.com by *creativeFields* (London, United Kingdom).

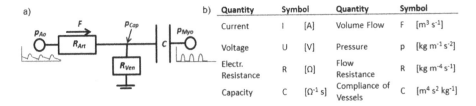

Fig. 2. Electrical circuit analog of cardiac blood flow. (a) p_{Ao} represents the pressure in the Aorta and p_{Myo} the pressure in the myocardium, which is approximated by an estimated ventricular pressure curve. The relations between the flow resistances in the arteries and veins, R_{Art}, R_{Ven}, and the compliance C of the vessels determine how blood flow F and the pressure in the capillaries p_{Cap} adjust. (b) The corresponding units of electrical circuit and cardiac circulation.

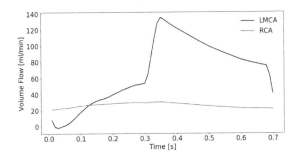

Fig. 3. Applied inlet volume flow curves for LMCA and RCA.

software package $QUCS^3$, the electrical circuit from Fig. 2 is set up and the influence of C is analyzed in a sensitivity study. In accordance with [8] the relationship $R_{Art} \cdot C = 1.14$ yields the physiologically most realistic ratio of mean right to mean left coronary flow of $\sim \frac{1}{4}$, in agreement with [12]. The obtained total volume flow curves are depicted in Fig. 3. In the LMCA, higher diastolic flow is obtained, whereas flow in the RCA hardly varies. Overall, the general shapes are in good agreement with normal coronary flow curves [12,17]; however, variations on shorter timescales as in [18,19] do not occur. These steps are performed before the actual CFD computations, where the obtained solutions for p_{Cap} and F_{total} are then used as BCs.

2.3   CFD Simulations

In order to map the considerations from Sect. 2.2 onto the models shown in Fig. 1 and provide the CFD simulations with a well-defined set of BCs, the following approach is chosen. The capillary pressure (p_{Cap} in Fig. 2 (a) is scaled depending on the depth of the considered outlet's position within the myocardium [20]. Subsequently, the arterial resistance R_{Art} is split up as sketched in Fig. 4.

---

[3] *Quite Unified Circuit Simulator*, http://qucs.sourceforge.net/.

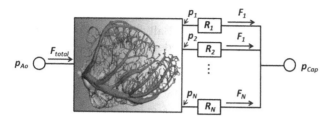

Fig. 4. Division of arterial resistance into a 3D-model with several parallel arranged outlets and associated downstream blood flow resistances. The arterial resistance R_{Art} from Fig. 2 is subdivided into a contribution from the model (CFD-simulations) and estimated corresponding outlet resistances R_i. Accordingly, total flow F splits into F_i, which distribute over all N model outlets. p_i denotes the pressure at outlet i.

At each timestep t of the simulation, the pressure at the outlets is thus given by

$$p_i(t) = F_i(t) \cdot R_i + p_i^{Cap}(t). \tag{3}$$

This approach guarantees that the flows to the different outlets $F_i(t)$ and, thus, the full cardiovascular model can adapt according to the conditions within the 3D model. At the same time, intramyocardial pressure $p_i^{Cap}(t)$ and flow resistance R_i of the remaining vasculature behind the outlets are taken into account, as well.

The required resistances R_i are approximated on the basis of the coronary arterial tree's self-similarity [16, 21] in dependence of the respective outlet diameter:

$$R_i = \text{const}_R \cdot \frac{L_{\text{crown}}}{D_{\text{outlet}}^4}, \tag{4}$$

where L_{crown} is the cumulative length of all vessels downstream of the outlet with diameter D_{outlet} and $\text{const}_R = R_{\max} \cdot \frac{D_{\max}^4}{L_{\text{total}}}$. The parameter $D_{\max}$ represents the most proximal stem diameter (i.e., inlet diameters of the left main coronary artery, LMCA, and the right coronary artery RCA), L_{total} the total cumulative crown length and $R_{\max}$ the total resistance of the full vascular tree starting at $D_{\max}$, respectively. Using the scaling laws of vascular volume [21] and hemodynamic flow resistance [16], the parameters in Eq. 4 can then be approximated.

The CFD simulations are performed with the software OpenFOAM[4] to solve the Navier-Stokes equations

$$\rho \left(\frac{\partial \mathbf{v}}{\partial t} + \mathbf{v} \cdot \nabla \mathbf{v} \right) = -\nabla p + \mu \Delta \mathbf{v} , \tag{5}$$

for blood flow, where $\rho, \mathbf{v}, p$ and μ denote density, velocity, pressure and viscosity of the fluid. This is performed until periodicity is reached. At the vessel walls, which are modeled to be rigid, a no-slip condition is applied (i.e., velocity

---

[4] www.openfoam.org, (Vs. 2.3.1).

U=0). Blood is assumed to be incompressible with density $\rho = 1060\frac{kg}{m^3}$. Non-Newtonian blood viscosity is given by Ballyk's generalized power law [22] and turbulences in blood flow are integrated using the large eddy simulation approach *Dynamic Smagorinsky* [23], a combined approach previously used in [2–4]. At the model inlets, the volume flow curves from Fig. 3 are applied. The pressure at the model outlets is computed by Eq. 3 (cf. Fig. 4) and the required flow resistances describing the downstream vasculature are calculated as outlined in Eq. 4.

2.4 Evaluation of VBF

The solutions of the blood flow simulations are evaluated with regard to the 17 myocardial segments of the left ventricular myocardium as defined in [24]. An analogous segmentation into four regions is defined for the right ventricle (RV). Since the imaging cryomicrotome dataset did not allow clear segmentation of distal vessels in the apical regions of the RV, these regions are neglected in the analysis. Based on their spatial coordinates, the model outlets can be associated with the defined myocardial segments allowing for a comparison of the amount of VBF per segment, VBF$_m$. To obtain physiologically realistic VBF$_m$ values in the myocardial segments, different rotations of the 3D models are tested. Since the used Imaging Cryomicrotome dataset does not contain information about the cardiac landmarks required to best fit the cardiac planes defined in [25], the orientation of the coronary trees is adapted to yield the most homogeneous distribution of VBF$_m$ across the 17 segments, as it would be expected in a healthy heart.

Moreover, to analyze if the simulations are in agreement with morphometric analyses on diameter and volume flow [21,29], the obtained VBFs F_i in vascular segments (VBF$_v$) are fitted with regard to the following relationship

$$\frac{F_i}{F_{max}} = A \cdot \left(\frac{D_i}{D_{max}}\right)^B. \tag{6}$$

Here D_i denotes the vessel diameter of the considered part of the vascular tree. The variable F_{max} is the volume flow through a larger upstream vessel, representing the beginning of the considered total vascular network and D_{max} the associated stem diameter. As in [21], B is fixed at $\frac{7}{3}$ and A is left free to vary in order to verify if $A = 1$, as expected.

3 Results

In Fig. 5 the obtained mean VBF$_m$s in the myocardial segments are shown. It becomes obvious that the orientation of the 3D dataset is critical for the analysis of VBF$_m$ heterogeneity. In particular, segments 3 and 19 show strong variability in dependence of the chosen orientation. The general distribution of the perfusion areas of the three large coronary arteries (RCA, LAD, LCX) is in accordance with what is expected [24]. However, Fig. 5 (d) shows the most homogeneous distribution across all segments. For all three orientations, both spread and regional variability of the obtained VBF_m across the LV are in the range of literature values for MBF measurements in pigs [26,27]. In accordance with [28], the inferior segments (except for segment 10) show lower VBF fractions than the anterior or lateral segments (with the exception of the apical lateral segment 16).

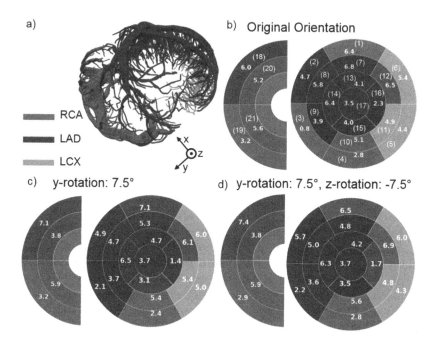

Fig. 5. Obtained VBF_m (in ml/min) in the left and right ventricular myocardial segments. The colors of the segments indicate the contributions from the large coronary branches (RCA, LAD, LCX). The numbers of the segments are given in brackets in panel (b) [24]. VBF_m in panels (c) and (d) are obtained by rotation of the 3D models relative to the original orientation (a, b). To examine the influence of erroneous assignment of an outlet to the perfusion of a particular myocardial segment, several rotations of the complete 3D dataset were tested. The chosen angles (c,d) yielded the most homogeneous overall distribution of VBF_m.

The results in Fig. 6 show slightly different behavior compared to what is found in the literature [21,29], $A < 1$ in all three coronary trees. This suggests that in the CFD simulations presented here, model outlets exhibit larger VBF_v with increasing diameter than assumed from the hypothesis (Eq. 6).

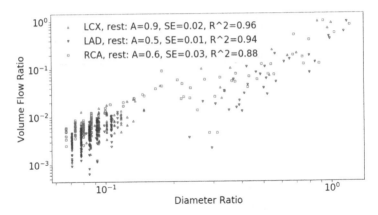

Fig. 6. Relative volume flow and diameter ratio at the model outlets and the cross sections marked in Fig. 1. The obtained fitting parameters are shown in the box.

4 Discussion

The examination of VBF_m in the different myocardial segments yields regional variations, which are in good accordance with literature values [24,26,27]. This applies both for the territories being supplied by the large coronary arteries (RCA, LAD and LCX) as well as the share of total VBF flowing into the segments. Even though distortions from segmentation errors can be recognized, overall, physiologically realistic results are obtained. However, it must be kept in mind that the analysis presented here is solely based upon the rheology of blood flow. Tissue demands and humoral regulation, which also play a decisive role in blood flow distribution, are not considered. Accordingly, the generated shapes of the utilized VBF curves in LMCA and RCA (Fig. 3) are still an area for improvement in future studies.

Regarding the dependence of VBF_v on vessel diameter, an acceptable agreement with the hypothesized volume scaling law from [21,29] is found. However, the CFD simulations cannot fully confirm the expected behavior. A major difference between this work and [21,29] is the relative diameter range that is considered in the analysis. While here only 2 orders of magnitude are included, the analyses in [21,29] are based on morphological data down to diameter ratios of 10^{-4}. The results from the hemodynamic analysis in [29] are based on a network flow analysis in which a simple symmetric model was used. All the vessel elements in any order were assumed to be of equal diameter and length, and they were arranged in parallel with equal blood pressures at the junctions between vessel orders. Moreover, vessel resistances in [29] are approximated by Poiseuille's law for laminar stationary flow. In comparison, the CFD approach used in this work allows for an analysis in a highly asymmetric (cardio-)vascular tree per se taking account of non-steady flow effects in the curved and tapering vessels. Considering this, the findings from this work are nonetheless highly relevant and the agreement with [21] is promising. However, in order to fully understand the underlying mechanisms of blood flow in the microvasculature, a more profound and dedicated analysis towards even smaller vessels is required.

Obviously, a substantial limitation of the results and the performed analysis lies in the fact that the vessels walls are assumed to be rigid in the simulations. In the beating heart, the coronary vasculature undergoes strong deformations both regarding the vessels' positions as well as their lumen. The implementation of the analog of cardiac circulation and the electrical RC-circuit (Fig. 2) presented in this work allows to include these effects in the outlet BCs (cf. Eq. 3); yet, the effects of myocardial pressure working on intramyocardial vessels, which are directly included in the 3D models are neglected due to the stiff modeling. Even though several authors only found small effects of vessel wall elasticity and periodic motion of the vasculature on blood flow modeling [30–32], the extent of their influence requires additional analysis. This could be done by including fluid-structure interactions in the simulations, which would, however, come with an increase of the already high computational costs and which was not feasible in this study.

5 Conclusion

In this work, an approach is presented to model blood flow as realistically as possible in the coronary arterial vasculature including smallest pre-arteriolar vessels with the help of a BC based on the electrical analog of cardiac circulation. Even though several simplifying assumptions are made, the physiological accuracy of the blood flow analysis considering the distribution of VBF_m in the myocardial segments shows acceptable agreement with the literature and can be seen as a confirmation of the proposed model and its high potential. However, the model still shows room for optimization with regard to the results from the dependence of VBF_v on the vessel diameter.

Although further validation particularly on the basis of real measurements is required, it represents a promising framework to analyze different aspects of cardiac circulation, such as the analysis of contrast agent dispersion in bolus-based quantitative myocardial perfusion measurements in medical settings, the initial objective of the implementation.

Acknowledgments. We acknowledge financial support of German Ministry of Education and Research (BMBF, grants: 01EO1004, 01EI01504). We acknowledge Leibniz Rechenzentrum Munich for access to the HPC-cluster *CoolMUC-2*. This manuscript is part of Johannes Martens' PhD-Thesis handed in at the Julius-Maximilians-University Wuerzburg in March 2019.

References

1. Graafen, D., Muennemann, K., Weber, S., Kreitner, K.-F., Schreiber, L.M.: Quantitative contrast-enhanced myocardial perfusion magnetic resonance imaging: simulation of bolus dispersion in constricted vessels. Med. Phys. **36**, 3099–3106 (2009)
2. Sommer, K., Schmidt, R., Graafen, D., Breit, H.-C., Schreiber, L.M.: Contrast agent bolus dispersion in a realistic coronary artery geometry: influence of outlet boundary conditions. Ann. Biomed. Eng. **42**, 787–796 (2013)
3. Sommer, K., Bernat, D., Schmidt, R., Breit, H.-C., Schreiber, L.M.: Resting myocardial blood flow quantification using contrast-enhanced magnetic resonance imaging in the presence of stenosis: A computational fluid dynamics study. Med. Phys. **42**, 4375–4384 (2015)
4. Martens, J., Panzer, S., van den Wijngaard, J.P.H.M., Siebes, M., Schreiber, L.M.: Analysis of coronary contrast agent transport in bolus-based quantitative myocardial perfusion mri measurements with computational fluid dynamics simulations. In: Pop, M., Wright, G.A. (eds.) FIMH 2017. LNCS, vol. 10263, pp. 369–380. Springer, Cham (2017). https://doi.org/10.1007/978-3-319-59448-4_35
5. Bender, S.B., van Houwelingen, M.J., Merkus, D., Duncker, D.J., Laughlin, M.H.: Quantitative analysis of exercise-induced enhancement of early- and late-systolic retrograde coronary blood flow. J. Appl. Physiol. **108**, 507–514 (2010)
6. Westerhof, N., Bosman, F., de Vries, C.J., Noordergraaf, A.: Analog studies of the human systemic arterial tree. J. Biomech. **2**, 121–143 (1969)
7. Burattini, R., Sipkema, P., van Huis, G.A., Westerhof, N.: Identification of canine coronary resistance and intramyocardial compliance on the basis of the waterfall model. Ann. Biomed. Eng. **13**, 385–404 (1985)
8. Kim, H.J., Vignon-Clementel, I.E., Coogan, J.S., Figueroa, C.A., Jansen, K.E., Taylor, C.A.: Patient-specific modeling of blood flow and pressure in human coronary arteries. Ann. Biomed. Eng. **38**, 3195–3209 (2010)
9. van den Wijngaard, J.P.H.M., et al.: Porcine Coronary Collateral Formation in the Absence of a pressure gradient remote of the iscemic border zone. Am. J. Physiol. Heart Circ. Physiol. **300**, H1930–H1937 (2010)
10. Updegrove, A., Wilson, N.M., Merkow, J., Lan, H., Marsden, A.L., Shadden, S.C.: SimVascular: an open source pipeline for cardiovascular simulation. Ann. Biomed. Eng. **45**, 525–541 (2017)
11. Kamoi, S., et al.: Continuous stroke volume estimation from aortic pressure using zero dimensional cardiovascular model: proof of concept study from porcine experiments. PLoS ONE **9**, e102476 (2014)
12. Guyton, A.C., Hall, J.E.: Textbook of Medical Physiology, 13th edn. Elsevier Saunders, Philadelphia (2005)
13. Lelovas, P.P., Kostomitsopoulos, N.G., Xanthos, T.T.: A comparative anatomic and physiologic overview of the porcine heart. J. Am. Assoc. Lab. Anim. Sci. **53**, 432–438 (2014)
14. Kaul, S., Jayaweera, A.R.: Determinants of microvascular flow. Eur. Heart J. **27**, 2272–2274 (2006)
15. Layland, J., Carrick, D., Lee, M., Oldroyd, K., Berry, C.: Adenosine: physiology, pharmacology, and clinical applications. JACC Cardiovasc. Interv. **7**, 581–591 (2014)
16. Huo, Y., Kassab, G.S.: The scaling of blood flow resistance: from a single vessel to the entire distal tree. Biophys. J. **96**, 339–346 (2009)

17. Itu, L., Sharma, P., Mihalef, V., Kamen, A., Suciu, C., Comaniciu, D.: A patient-specific reduced-order model for coronary circulation. In: 9th IEEE International Symposium on Biomedical Imaging, pp. 832–835. IEEE (2012)

18. Duanmu, Z., Yin, M., Fan, X., Yang, X., Luo, X.: A patient-specific lumped-parameter model of coronary circulation. Sci. Rep. **8**, 874 (2018)

19. Ge, X., Yin, Z., Fan, Y., Vassilevski, Y., Liang, F.: A multi-scale model of the coronary circulation applied to investigate transmural myocardial flow. Int. J. Numer. Method. Biomed. Eng. **34**, e3123 (2018)

20. Gerke, E., Juchelka, W., Mittmann, U., Schmier, J.: Der intramyokardiale Druck des Hundes in verschiedenen Tiefen bei Druckbelastung und Ischaemie des Herz-muskels. Basic Res. Cardiol. **70**, 537–546 (1975)

21. Huo, Y., Kassab, G.S.: A scaling law of vascular volume. Biophys. J. **96**, 347–353 (2009)

22. Ballyk, P.D., Steinman, D.A., Ethier, C.R.: Simulation of non-Newtonian blood flow in an end-to-side anastomosis. Biorheology **31**, 565–576 (1993)

23. Lesieur, M., Metais, O., Conte, P.: Large-Eddy Simulations of Turbulence. Cambridge University Press, New York (2005)

24. Cerqueira, M.D., et al.: Standardized myocardial segmentation and nomenclature for tomographic imaging of the heart. Ciculation **105**, 539–542 (2002)

25. No Authors Listed: Standardization of cardiac tomographic imaging. Circulation **86**, 338–339 (1992)

26. Rossi, A., et al.: Quantification of myocardial blood flow by adenosine-stress CT perfusion imaging in pigs during various degrees of stenosis correlates well with coronary artery blood flow and fractional flow reserve. Eur. Heart J. Cardiovasc. Imaging **14**, 331–338 (2013)

27. Fahmi, R.: Quantitative myocardial perfusion imaging in a porcine ischemia model using a prototype spectral detector CT system. Phys. Med. Biol. **61**, 2407–2431 (2016)

28. Chareonthaitawee, P., Kaufmann, P.A., Rimoldi, O., Gamici, P.G.: Heterogeneity of resting and hyperemic myocardial blood flow in healthy humans. Cardiovasc. Res. **50**, 151–161 (2001)

29. Kassab, G.S.: Scaling laws of vascular trees: of form and function. Am. J. Physiol. Heart Circ. Physiol. **290**, H894–H903 (2006)

30. Kolandavel, M.K., Fruend, E.T., Ringgaard, S., Walker, P.G.: The effects of time varying curvature on species transport in coronary arteries. Ann. Biomed. Eng. **34**, 1820–1832 (2006)

31. Theodorakakos, A., et al.: Simulation of cardiac motion on non-Newtonian, pulsating flow development in the human left anterior descending coronary artery. Phys. Med. Biol. **53**, 4875–4892 (2008)

32. Zeng, D., Boutsianis, E., Ammann, M., Boomsma, K., Wildermuth, S., Poulikakos, D.: A study on the compliance of a right coronary artery and its impact on wall shear stress. J. Biomech. Eng. **130**, 1–11 (2008)

Mesh Based Approximation of the Left Ventricle Using a Controlled Shrinkwrap Algorithm

Faniry H. Razafindrazaka[1,2(✉)], Katharina Vellguth[2], Franziska Degener[2],
Simon Suendermann[2], and Titus Kühne[2]

[1] Freie Universität Berlin, Berlin, Germany
[2] Institute for Imaging Science
and Computational Modelling in Cardiovascular Medicine,
Charité-Universitätsmedizin Berlin, Berlin, Germany
faniry.razafindrazaka@charite.de

Abstract. This research paper introduces an adaptive algorithm to reconstruct the left ventricle from computer tomographic images (CT). Often, manual image segmentation gives the right geometry but produces too smooth shapes while automatic segmentation algorithms incorporate too much details including unwanted noise with complicated topology. The correct topology is most important in the ventricle motion approximation gained for example from 4D computed tomography and applied in computational fluid dynamics (CFD) of the heart. We propose to bridge the two extremes i.e. producing models with common spherical topologies and smoothness level while preserving important details as much as possible. We propose a controlled shrinkwrap algorithm together with a framework where level of details can be chosen based on a single parameter. We demonstrate the results on a patient specific left ventricle to understand the effect of papillary muscles in computational fluid dynamic progressively from a smooth to a detailed surface.

Keywords: Shrinkwrap algorithm · Left ventricle segmentation ·
Computational modelling · Computational fluid dynamics

1 Introduction

Numerical simulation of the human cardiac blood flow is an appreciated method to gain further understanding in healthy and pathological hemodynamic processes inside the heart. Furthermore, it provides opportunities to support diagnostics and treatment planning in cardiology and cardiac surgery. The focus is thereby set on flow structures, vortex propagation, velocity distribution and pressure drop over the heart valves. In many cases the ventricular geometry is strongly simplified, such as e.g. performed in [11,12]. On the one hand, the imaging data is often not sufficiently detailed enough, on the other hand, a costly

© Springer Nature Switzerland AG 2019
Y. Coudière et al. (Eds.): FIMH 2019, LNCS 11504, pp. 230–239, 2019.
https://doi.org/10.1007/978-3-030-21949-9_25

tracking algorithm is needed to detect fine ventricular structures over an entire heart cycle, such as performed in [8,10].

The commonly used geometric approximation of the left ventricle (LV) has a spherical topology. The segmented geometry is hole free with neither cylindrical region (*frayings*) nor small spikes. These geometric configurations are considered as artifacts and are problematic in various applications such as shape morphing, displacement field computation, or numerical simulation. Figure 1 is an illustration of these undesired regions. To resolve these issues while maintaining an accurate representation of the LV, we focus on a progressive approximation of the details without breaking the spherical topology requirement. Naturally the topology of the LV is invariant during the cardiac cycle. We consider holes and frayings as artifacts either from low quality CT scans or from naive segmentation algorithm. We propose a mesh based controlled shrinkwrap approach to resolve these geometrical issues. The main advantage of this approach is that the resulting geometry is guaranteed to be manifold and preserve its initial topology. Shrinkwrap based remeshing is widely used in various area of science. From remeshing in computer graphics [5] to computer aided design (CAD) simulation [7,9] and segmentation in medical imaging [4]. However, all proposed methods either aim at a rough representation of the shape i.e ignores fine details or require a specific input such as spherical topology.

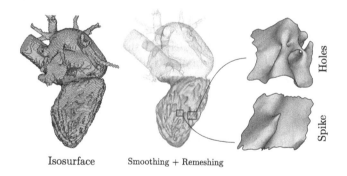

Isosurface   Smoothing + Remeshing

Fig. 1. Undesired artifacts from automatically segmented LV which cannot be removed from classical remeshing and smoothing algorithms. These are small holes (or tunnels) and frayings (or spikes).

Contribution. We propose a controlled shrinkwrap algorithm to approximate the LV detailed morphology. The method is independent of the input geometry topology which may also have holes, boundary, or frayings. The resulting geometry is guaranteed to be smooth with spherical topology. The flexibility of the proposed approach allows a control over the approximation quality via a single parameter balancing smoothness and papillary muscles awareness. The user can then decide if the final ventricle will have more trabeculae and papillary muscles or only a rough approximation. We show the result of the method on a set of patient specific 4D CT scans and use the derived displacement vectors in CFD.

Notation. We denote by $\mathcal{T}$ an initial representation of a given LV, usually non-smooth with complicated topology. We denote by $\mathcal{S}_\gamma$ a triangulated surface approximating $\mathcal{T}$. The parameter γ together with $\gamma_{\min} = \frac{4}{5}\gamma, \gamma_{\max} = \frac{4}{3}\gamma$ define the level of details of the approximation related respectively to the target edge length, minimum edge length, and maximum edge length of the final mesh. The ratios are derived according to the analysis given in [1]. Initially γ is the average edge length $\gamma = \frac{1}{|E(\mathcal{S})|} \sum_{e \in E(\mathcal{S})} \mathbf{length}(e)$ where $E(\mathcal{S})$ is the set of edges of $\mathcal{S}$ and **length** computes the length of the edge. A node or a vertex of $\mathcal{S}_\gamma$ is denoted by v_i which will denote at the same time its 3D coordinates. The normal vector at v_i pointing outwards is denoted by $\mathbf{n}_i$. The set of all nodes adjacent to v_i is denoted by $\mathcal{N}(v_i)$. Finally, we denote by $\mathcal{C}$ a region of $\mathcal{T}$ covered by $\mathcal{S}_\gamma$ during the approximation process (yellow region in Fig. 2).

2 The Shrinkwrap Model

The shrinkwrap algorithm is a diffusion process applied to a surface consisting of a node displacement and a smoothing. Physically it is a simulation of an elastic fabric wrapped around a given 3D object. The shrinking behaviour of a wrapping surface as an elastic fabric is discretized and simulated in order to approximate complicated surfaces. Each node movement aims at minimizing the Hausdorff distance $d_{\mathrm{H}} = \mathrm{distance}(\mathcal{S}, \mathcal{T})$ while a smoothing keeps the elasticity and smoothness quality of the surface. The difficulties in shrinkwrap based approximation are slow convergence, fold-overs and uncontrollable behaviour close to the holes. Traditional shrinkwarp approaches accept these bad configuration by projecting naively $\mathcal{S}$ onto $\mathcal{T}$. However, if the projection produces strong fold-over, it cannot be resolved by smoothing only. A more elegant approach is to control the motion of each node during the approximation process. It is also important to know when a given node reaches the surface $\mathcal{T}$. All these aspects and the choice of the node dynamics can be controlled via γ.

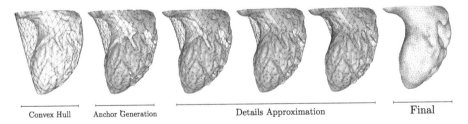

| Convex Hull | Anchor Generation | Details Approximation | Final |

Fig. 2. Approximation algorithm starting from a convex Hull to anchored nodes generation to details approximation and finally to a smooth surface. (Color figure online)

2.1   Controlled Shrinkwrap Algorithm

Our algorithm is subdivided into three fundamental steps. We generate S_γ and apply standard shrinkwrap algorithm in the form of anchored nodes generation. An efficient approximation of the shape details is then applied as illustrated in Fig. 2.

Generation of S_γ. We choose the convex Hull of T as a starting surface. In [7], a Cartesian grid is intersected with T to generate the wrapping surface. This is not suitable in our case since the isosurface itself is produced by a Cartesian marching cube. To obtain initial regular triangles and good starting elastic energy, we apply a moderate remeshing to S_γ as detailed in Sect. 2.2.

Anchored Nodes Generation. As a physical material, an elastic fabric tends to come back to its rest states unless some points of the surface are anchored to the target surface. We generate a set of anchored points, denoted by $\mathcal{A}$. The elements of $\mathcal{A}$ are points of S_γ filtered based on their distance to T. The anchored node generation is summarized by Algorithm 1.

Algorithm 1. Anchored node generation

1: **for** vertex $v_i \in S_\gamma$ **do**
2:   Compute $\mathbf{d_i} = d(v_i, T)$ #Hausdorff distance
3: **end for**
4: **for** vertex $v_i \in S_\gamma$ **do**
5:   **if** $\|\mathbf{d}_i\| < \frac{1}{2}\gamma_{\min}$ **then**
6:     $\mathcal{A} = \mathcal{A} \cup v_i$
7:     $v_i^{\mathrm{new}} = v_i + \mathbf{d}_i$
8:   **end if**
9: **end for**
10: Remesh(S_γ)
11: Smooth($S_\gamma \setminus \mathcal{A}$)

Notice that the existence of anchored nodes requires the initial surface S to be close to T. This observation motivated our choice of the convex Hull as starting surface. The choice of the upper-bound in Algorithm 1, line 5 is to prevent the mesh to self-intersect. Here, lower scale of $\gamma_{\min}$ could also be used. More the remeshing used as initialization guarantees that no edge has edge-length less than $\gamma_{\min}$. Lines 10 and 11 in Algorithm 1 maintained the regularity and elasticity of the mesh. The smoothing does not move anchored points.

Details Approximation. The execution of Algorithm 2 already provides a coarse approximation of T. If more details are required, then an external force similar to a "push" is needed to reach concave regions. This is a delicate process since information about details are a priory unknown. Iterating Algorithm 1 many times produces wrinkles and may never converge to the desired region. In Fig. 3

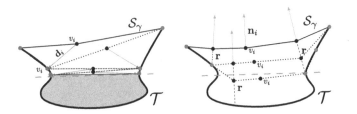

Fig. 3. Left: approximating concave regions using shortest distance only. Right: evolving along the normals by prioritizing nodes adjacent to element in $\mathcal{A}$ (in red). (Color figure online)

Algorithm 2. Details approximation

1: **while** $S_\gamma \neq \mathcal{A}$ **do**
2:   **for** vertex $v_i \in S_\gamma \setminus \mathcal{A}$ **do**
3:     #$\mathcal{N}(v_i)$ is the set of neighboring vertices of v_i
4:     **if** $\mathcal{N}(v_i) \cap \mathcal{A} \neq \emptyset$ **then**
5:       $\mathbf{r}_i = \mathcal{T} \cap \mathbf{n}_i$ #$\mathbf{n}_i$ is the normal vector at v_i
6:       **if** $\|\mathbf{r}_i\| < \frac{1}{2}\gamma_{\min}$ **then**
7:         $\mathcal{A} = \mathcal{A} \cup v_i$
8:         $v_i^{\text{new}} = v_i + \mathbf{d}_i$
9:       **else if** $\mathbf{r}_i \cap \mathcal{C} = \emptyset$ **then**
10:        $v_i^{\text{new}} = v_i + \frac{1}{4}\gamma_{\min}\mathbf{r}_i$
11:       **end if**
12:     **end if**
13:     Remesh(S_γ)
14:     Smooth(S_γ $\mathcal{A}$)
15:   **end for**
16:   **for** vertex $v_i \in S_\gamma$ **do**
17:     {Anchor if it is close enough to $\mathcal{T}$}
18:     **if** $\|\mathbf{d}_i\| < \frac{1}{8}\gamma_{\min}$ **then**
19:       $\mathcal{A} = \mathcal{A} \cup v_i$
20:       $v_i^{\text{new}} = v_i + \mathbf{d}_i$
21:     **end if**
22:   **end for**
23: **end while**

(left) is an example of this situation where the surface S_γ will never evolve in the green region. Several nodes will oscillate at the green dotted line. Moving along the normal approximate better the concave regions but is unfortunately unstable. The strategy is to prioritize nodes adjacent to nodes in $\mathcal{A}$ and move the rest using the elasticity property of the mesh (Fig. 3 right). A ray-triangle intersection is necessary to determine the amount of displacement along the normal and to avoid fold-overs with nodes movement not exceeding $\frac{1}{2}\gamma_{\min}$. The steps are summarized in Algorithm 2. In general, the while loop in Algorithm 2 may never be satisfied since some vertices may oscillate. We then also track a total change in node displacements. If it is inferior to a small epsilon threshold, we stop the wrapping. The condition $\mathbf{r}_i \cap \mathcal{C} = \emptyset$ in line 9 assures that $\mathcal{S}_\gamma$ is not evolving in areas of $\mathcal{T}$ which have been already covered ($\mathcal{C}$ is the set of triangle already covered by $\mathcal{S}_\gamma$). This prevents self-intersection especially in regions dominated by holes.

2.2   Remeshing

We use a well known surface remeshing from computer graphics combining local mesh operations, tangential smoothing and back-projection as described in [1]. In our case, we only need the back-projection in the anchored node generation. The remeshing assures that the final edge lengths of $\mathcal{S}_\gamma$ lies in the interval $[\gamma_{\min}, \gamma_{\max}]$. The local mesh modifications consist of edge flips, edge collapses and edge splits. An edge flip improves the regularity of each node i.e close to six. An edge collapse merges two near by vertices if their distance is less than $\gamma_{\min}$. A split subdivides an edge longer than $\gamma_{\max}$ as illustrated in Fig. 4.

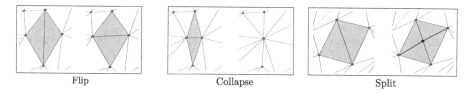

Flip   Collapse   Split

Fig. 4. Local mesh modifications used as basis for remeshing consisting of edge flips, edge collapse, and edge splits.

2.3   Smoothing

The elasticity property of the mesh $\mathcal{S}_\gamma$ is simulated by a uniform Laplacian smoothing, similar in some aspect to a spring attached to each edge. The smoothing tries to move each nodes to the centroid of the polygon by its adjacent nodes. The displacement vector is defined as $\mathbf{q}_i = \frac{1}{|\mathcal{N}(v_i)|} \sum_{j=0}^{|\mathcal{N}(v_i)|} (w_j - v_i)$ where $w_j \in \mathcal{N}(v_i)$ is a node adjacent to v_i. The tangential part is obtained by removing the normal component of $\mathbf{q}_i$. This minimizes oscillating motions. The new position is then updated as $v_i^{\mathrm{new}} = v_i + \beta(\mathbf{q} - \langle \mathbf{q}, \mathbf{n}_i \rangle \mathbf{n}_i)$.

3   Results and Outlook

The controlled shrinkwrap algorithm was implemented in Java using the JavaView (www.javaview.de) mesh processing library. We tested the algorithm on patient specific 4D CT scans taking two frames close to the diastolic case from ten captures with acquired resolution $0.7 \times 0.7 \times 0.5\,\mathrm{mm}^3$. The initial surface $\mathcal{T}$ is generated using the isosurface module with a connected component analysis (CCA) available in MevisLab (www.mevislab.de). The CCA returns the LV together with the left atrium and aorta as the largest component and is extracted. The LV is cut manually and approximated externally. We use Qhull (www.qhull.org) to generate the convex Hull. All the remaining operations such as remeshing, smoothing or Hausdorff distances are our own implementation. To generate a CFD ready model, we reattach the generated $\mathcal{S}_\gamma$'s using Meshmixer (www.meshmixer.com). No further remeshing/smoothing is required on the LV. For the CFD simulation, we use StarCCM+ (version 12.06.011, Siemens).

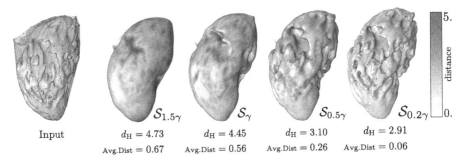

Input	$\mathcal{S}_{1.5\gamma}$	$\mathcal{S}_\gamma$	$\mathcal{S}_{0.5\gamma}$	$\mathcal{S}_{0.2\gamma}$
	$d_{\mathrm{H}} = 4.73$	$d_{\mathrm{H}} = 4.45$	$d_{\mathrm{H}} = 3.10$	$d_{\mathrm{H}} = 2.91$
	Avg.Dist $= 0.67$	Avg.Dist $= 0.56$	Avg.Dist $= 0.26$	Avg.Dist $= 0.06$

Fig. 5. Hausdorff distance (d_H) analysis of several approximations. Small scaling of γ corresponds to increase of details and small average distances in mm.

Progressive Approximation. In order to speed up the convergence rate, we apply a hierarchical multi-resolution strategy in a coarse to fine manner. For a given $a\gamma$ where $a \ll 1$ the size of $\mathcal{S}_{a\gamma}$ increases significantly such that the diffusion process may never converge in a reasonable amount of time. To solve this issue, we compute a coarser approximation $\mathcal{S}_{2a\gamma}$ which is then used as starting surface to generate $\mathcal{S}_{a\gamma}$. This is a recursive approach which forms the base of the hierarchy. Since most nodes of $\mathcal{S}_{2a\gamma}$ are anchored at the end of the diffusion, only a moderate amount of new vertices needs to be processed in $\mathcal{S}_{a\gamma}$. This approach also enables the diffusion process to ignore tiny holes and spikes at early stage. We use this technique to produce the models shown in Fig. 5.

Approximation Quality. In Fig. 5 is a quantitative comparison of several approximations of the ventricle. As expected, high scalings of γ correspond to rough approximations with high Hausdorff distances. If more details are required small scalings (arround 0.5) are more appropriate. Using $a\gamma$ with $a < 0.5$ produces too much details and may even capture unwanted configurations.

Cross Parameterization. The main advantage of having a common ventricle topology between several ventricle frames is information about morphing and displacement vectors. In Fig. 6 is an example where two near diastole isosurfaces have been approximated and cross parameterized using [6]. The correspondence points are prescribed by a user based on the frame motion from the corresponding 4D CT. The resulting displacement vector is used as a velocity boundary condition in the CFD simulation (see Fig. 7).

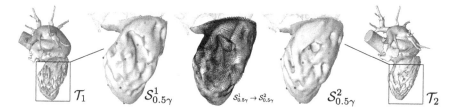

Fig. 6. Deriving a displacement field between two approximated isosurfaces from two frames of 4D CT scans. The colored points are correspondence points prescribed by the user. (Color figure online)

Computational Fluid Dynamics. In order to analyze the effect of increasing details on the LV hemodynamics, we exemplarily performed a CFD simulation of the LV filling phase on $\mathcal{S}_{1.5\gamma}, \mathcal{S}_{\gamma}$, and $\mathcal{S}_{0.5\gamma}$. A patient specific mitral valve is incorporated in the ventricle geometry. For simplicity, we only regard a short moment of early diastole, allowing the usage of rigid walls. The derived displacement field is used as a scaled velocity boundary condition such that the blood flow is 700 ml/s. Figure 7 shows the velocity distribution in different cross sections of the ventricle in each degree of detail. Maximum velocities (v_{max}) are very similar in all simulations, while jet orientation and localization of v_{max} differ with increasing detail of structures like trabeculae and papillary muscles. Comparing the horizontal cross sections, the velocity distribution and flow shape is rather similar. In contrast, the velocity distribution in the frontal section differs with increasing detail such that the jet is directed further towards the apex on the long axis and towards the center along the short axis. Thereby, also the localization of v_{max} shifts downwards.

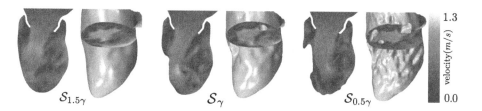

Fig. 7. Velocity distribution in two cross sections of the geometry with $\mathcal{S}_{1.5\gamma}$ (left), $\mathcal{S}_{\gamma}$ (middle), $\mathcal{S}_{0.5\gamma}$ (right).

3.1   Conclusion

In this paper we introduced a controlled shrinkwrap algorithm to approximate the LV. The use of well known remeshing strategy enables the algorithm to produce a spherical topology and fold-over free geometry. Applying the algorithm to 4D CT scans, we showed the advantage and easy use of the generated models in CFD. The employment of the algorithm on geometries for prescribed motion in fluid structure interaction simulations is of great interest. A profound analysis of the influence of ventricular geometry details on hemodynamics has to be performed in further work. A comparison to a ground truth MRI flow will be a good validation of our morphing strategy. Shrinkwrap based algorithms are not limited to these types of models, it can be used for other biological shapes with known spherical topology. However, for more articulated geometries an adaptive local remeshing might be necessary or applying a spherical parameterization based schrinkwrap approach is more appropriate by closing holes and tunnels explicitly in the target geometry.

Acknowledgement. This project is partly funded by the bundesministerium für bildung und forschung (Project VIP+, DSSMitral), Berlin, Germany. And partlty funded by the Deutsche Forschungsgemeinschaft (DFG, German Research Foundation) under Germany's Excellence Strategy – The Berlin Mathematics Research Center MATH+ (EXC-2046/1, project ID: 390685689).

References

1. Botsch, M., Kobbelt, L., Pauly, M., Alliez, P., Levy, B.: Polygon Mesh Processing, pp. 100–103. AK Peters, New York (2010)
2. Caballero, A., et al.: Modeling left ventricular blood flow using smoothed particle hydrodynamics. Cardiovasc. Eng. Technol. **8**, 465–479 (2017)
3. Domenichini, F., Pedrizzetti, G., Baccani, B.: Three-dimensional filling flow into a model left ventricle. J. Fluid Mech. **539**, 179–198 (2005)
4. Huang, Z., Dai, N., Liu, H.: Heuristically semi-automated segmentation of femur from 3-D CT images. Int. J. Comput. Theory Eng. **8**, 240–243 (2016)
5. Kobbelt, L., Vorsatz, J., Labsik, U., Seidel, H.-P.: A Shrink wrapping approach to remeshing Polygonal Surfaces. Comput. Graph. Forum **18**, 119–130 (1999)
6. Kwok, T.H., Zhang, Y., Wang, C.C.L.: Efficient optimization of common base domains for cross parameterization. IEEE Trans. Vis. Comput. Graph. **18**(10), 1678–1692 (2012)
7. Lee, Y.K., Lim, C.K., Ghazialam, H., Vardhan, H., Eklund, E.: Surface mesh generation for dirty geometries by the Cartesian shrink-wrapping technique. Eng. Comput. **26**, 377–390 (2010)
8. Lantz, J., Henriksson, L., Persson, A., Karlsson, M., Ebbers, T.: Patient-specific simulation of cardiac blood flow from high-resolution computed tomography. J. Biomech. Eng. **138**(12), 121004 (2016)
9. Martineau, D., Gould, J., Papper, J.: An integrated framework for wrapping and mesh generation of complex geometries. In: European Congress on Computational Methods in Applied Sciences and Engineering, pp. 6938–6954 (2016)
10. Mittal, R., et al.: Computational modeling of cardiac hemodynamics: current status and future outlook. J. Comput. Phys. **305**, 1065–1082 (2016)
11. Pedrizzetti, G., Domenichini, F.: Left ventricular fluid mechanics: the long way from theoretical models to clinical applications. Ann. Biomed. Eng. **43**(1), 26–40 (2015)

12. Schenkel, T., Malve, M., Reik, M., Markl, M., Jung, B., Oertel, H.: MRI-based CFD analysis of flow in a human left ventricle: methodology and application to a healthy heart. Ann. Biomed. Eng. **37**(3), 503–515 (2009)
13. Vedula, V., George, R., Younes, L., Mittal, R.: Hemodynamics in the left atrium and its effect on ventricular flow patterns. J. Biomech. Eng. **137**(11), 111003 (2015)

A Computational Approach on Sensitivity of Left Ventricular Wall Strains to Geometry

Luca Barbarotta$^{(\boxtimes)}$ and Peter Bovendeerd

Eindhoven University of Technology, Eindhoven, Netherlands
l.barbarotta@tue.nl

Abstract. In this work we use a Finite Element model of the left ventricular (LV) mechanics to assess the sensitivity of strains to geometry. Six principal shape modes extracted from an atlas of LV geometries using principal component analysis, have been used to model the variability of the geometry of a population of 1991 asymptomatic volunteers. We observed that shear strains are more sensitive than normal strains to geometry. For all the strains, shape mode 1, related with variation in size within the population, plays an major role, but none of the six principal modes can be considered non influential.

1 Introduction

Personalised cardiac medicine aims at assisting clinicians with model based interpretation of the patient's data. However, since the invasivity of clinical measurements must be limited as much as possible and not every quantity can be measured, the patient's data included in a model are typically affected by errors and uncertainty. It is therefore important to assess the sensitivity of a model to its input to understand whether a pathology can be observed notwithstanding the output is affected by uncertainty. In the last decades, several attempts were made to try estimating cardiac tissue properties such as contractility from patients [4,9,12–14]. Recently, a variational approach has been used to perfom high resolution data assimilation on strains computed from 4D echocardiographic measurements to estimate tissue properties in in-vivo patients [1,6]. These studies included in their analyses patient specific left ventricular (LV) [1,6,12–14] or bi-ventricular [4,9] geometries. Besides being affected by geometrical approximations, such as exclusions of the papillary muscles and trabeculae, patient specific geometries are also affected by measurement and reconstruction error. In this work we use a simulation of LV mechanics to assess sensitivity of 3D strain distributions on variations in model geometry, based on data from the Cardiac Atlas project [15].

© Springer Nature Switzerland AG 2019
Y. Coudière et al. (Eds.): FIMH 2019, LNCS 11504, pp. 240–248, 2019.
https://doi.org/10.1007/978-3-030-21949-9_26

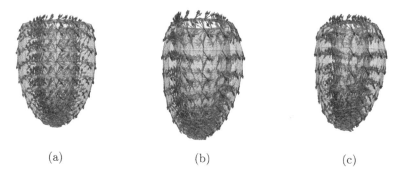

(a)   (b)   (c)

Fig. 1. View of meshes and fibers for the template geometry (left), the population-averaged end diastolic geometry (middle), and the reconstructed unloaded geometry (right).

2   Material and Methods

2.1   Geometry

We used data from the Cardiac Atlas Project (http://www.cardiacatlas.org, [15]) in which left ventricular surface meshes of a population of 1991 MESA models were represented by six principal shape modes. We interpolated the end diastolic surface meshes of the atlas onto the surfaces of a template volume mesh, after alignin the two basal planes. Then, using the transmural coordinate, numerically integrated from the ellipsoidal coordinates defining the template mesh, the internal nodes of the mesh were moved in order to obtain transmurally even elements. This method preserves the topology of the template mesh and allowed the generation of high quality meshes for arbitrary end diastolic configurations, defined by combinations of shape modes. This mesh generation algorithm also allows to compute the geometric mapping from the template mesh to the deformed mesh which we use to compute the deformation gradient to reorient the vector fields defined over the template mesh. An example of mesh and fiber generated using this technique can be seen in Fig. 1.

2.2   Material Properties

We simulated LV mechanics using the model in [2]. Briefly, the myocardium is described as an hyperelastic fiber-reinforced transversely isotropic active stress material [2]. The active stress is dependent on sarcomere length, time and velocity, and acts along fiber direction. Material incompressibility is enforced weakly using a modified volumetric part $\mathcal{W}_v$ in the strain energy density function

$$\mathcal{W}_v = 4k \left(J^2 - 1 - 2\ln J \right),\tag{1}$$

where J is the determinant of the deformation gradient F and k is a constitutive parameter. The formulation in (1) satisfies the requirements of being positive, convex, null in $J = 1$, and unbounded at zero and infinity [11]. The orientation of fibers inside the myocardium is described by high order Legendre polynomials of normalized transmural and longitudinal coordinates [2].

2.3   Computation of Left Ventricular Wall Mechanics

Left ventricular wall mechanics was computed by solving for equilibrium between forces related to active stress, passive stress and cavity pressure:

$$
\begin{cases}
\text{Div}\,(\mathsf{P}\,(\mathsf{F})) = \mathbf{0} & \text{in } \Omega_0, \\
\mathsf{P}\boldsymbol{n}_0 = -p J \mathsf{F}^{-T} \boldsymbol{n}_0 & \text{on } \Gamma_{0,Endo}, \\
\mathsf{P}\boldsymbol{n}_0 = \mathbf{0} & \text{on } \Gamma_{0,Epi},
\end{cases}
\tag{2}
$$

where P is the first Piola-Kirchhoff stress tensor; F is the deformation gradient; $\boldsymbol{u}$ is the displacement; and $\partial\Omega_0 = \Gamma_{0,Base} \cup \Gamma_{0,Endo} \cup \Gamma_{0,Epi}$ represents the decomposition of the unloaded boundary in basal, endocardial and epicardial surfaces, respectively; and $\boldsymbol{n}_0$ is the normal vector defined over those surfaces. To prevent rigid body motion we suppressed displacement normal to the basal plane and circumferential displacement of the basal endocardium. We did not couple our LV model to the closed circulation model described in [2]. Instead, we set a fixed preload of 1.5 kPa and a physiological afterload of 12 kPa, reached after the isovolumic contraction dictated by the fiber active stress. From those experiments, end systolic strains with respect of the end diastolic configuration were computed using the Green-Lagrange strain tensor along the wall-bound basis vectors e_c, e_l, and e_t (namely the circumferential, the longitudinal and the transmural directions, respectively).

2.4   Computation of the Unloaded Configuration

To perform a finite element simulation of LV mechanics we need to recover a virtually unloaded configuration. In order to estimate this unknown configuration, we deflated the end diastolic configuration. We assumed the cavity volume of the unloaded configuration to be equal to that of the end systolic configuration, while assuming a physiological ejection fraction of 60%. We reached this configuration by applying an endocardial pressure load. Since we have information about a deformed configuration Ω, we set there problem (2) but using the Cauchy stress tensor according to the principle of Inverse Design [8]

$$
\text{div}\,(\sigma\,(\mathsf{F})) = \mathbf{0} \quad \text{in } \Omega, \tag{3}
$$
$$
\sigma\boldsymbol{n}\cdot\boldsymbol{n} = -p \quad \text{on } \Gamma_{Endo}, \tag{4}
$$
$$
\sigma\boldsymbol{n} = \mathbf{0} \quad \text{on } \Gamma_{Epi}, \tag{5}
$$

and the same kinematic boundary conditions of (2). Note that taking advantage of the knowledge of the end diastolic deformed surfaces we can write the Natural boundary conditions at endocardium and epicardium using σ and the normal vector $\boldsymbol{n}$ in the deformed configuration. However, since F is unknown, we substitute it in the definition of σ with the equivalent known f^{-1}, as follows

$$\sigma\left(F\right) = \sigma\left(f^{-1}\right) = \sigma\left(\left(I - \frac{\partial \boldsymbol{u}}{\partial \boldsymbol{x}}\right)^{-1}\right). \tag{6}$$

Now the problem defined in (3) can be seen as the search of the displacement $\boldsymbol{u}$ that determines the unloaded configuration Ω_0, given Ω and the boundary conditions. Note that, since the solution of (3) modifies the domain on which the material properties of the body are defined, those properties must be changed accordingly. We therefore reorient the fibers accordingly during the deflation process.

2.5 Finite Element Implementation

We implemented our Finite Element model in a python package based on the FEniCS library (fenicsproject.org). The spatial discretisation of the problem has been studied in order to achieve an accurate prediction of strains at an affordable computational cost. In this respect, we use a sub-parametric finite element approach where the geometry is discretised using linear tetrahedra and the solution is discretised using quadratic Lagrangian polynomials. The spatial discretisation resulted in 5760 elements and 24519 degrees of freedom. The nonlinear problem has been solved using the Newton-Raphson algorithm and the resulting linear systems were solved using the LU decomposition implemented in the MUMPS library.

2.6 Sensitivity Analysis Approach

We applied the elementary effects method from [3]. This method relies on the definition of an elementary effect, EE_i which gives an estimation of the effect that a perturbation of a parameter x_i in a specific point $(x_1, ..., x_n)$ of the parameter space X by an amount of Δ introduces in an output variable y.

$$EE_i = \frac{y\left(x_1, ..., x_i + \Delta, ..., x_n\right) - y\left(x_1, ..., x_i, ..., x_n\right)}{\Delta}. \tag{7}$$

As model parameter x, we considered 6 shape modes ($n = 6$), each spanning a range of 1 SD around the mean. Using an optimized sampling strategy for the variations Δ, we built a 4 level mesh over the parameter space and generated 500 trajectories. As output y we considered the all six end systolic strain components, referred to the state at end diastole.

Elementary effects EE were computed from the per-node differences in all six end systolic strain components. By collecting all the elementary effects over the trajectories, three sensitivity indices can be computed as follows

$$\mu_i = \frac{1}{r}\sum_{j=1}^{r} EE_i^j, \quad \mu_i^* = \frac{1}{r}\sum_{j=1}^{r} |EE_i^j|, \quad \sigma_i = \frac{1}{r-1}\sum_{j=1}^{r}\left(EE_i^j - \mu_i\right)^2 , \tag{8}$$

where r is the number of trajectories, μ_i gives an overall estimation of the effect a perturbation on parameter x_i has, μ_i^* is used to detect whether variations on x_i are non influential, and σ_i gives information about how much the effect of a perturbation on x_i depends on the sampling point in the parameter space, namely the interactions between x_i and the other parameters.

3   Results

Figure 2 shows the coefficient of variation (CV) of normal strains. This coefficient consists in the standard deviation of a strain computed over all the nodes of the parameter space normalized with respect of the norm of the average respective strain. These norms were $\|E_{cc}^{avg}\|_\infty = 0.31$, $\|E_{ll}^{avg}\|_\infty = 0.22$, and $\|E_{tt}^{avg}\|_\infty = 1.20$. For the circumferential, the longitudinal and the transmural strains the coefficient of variation ranges 1.1%–18%, 2.6%–25%, and 2.3%–30%, respectively. The maximum values for E_{cc} and E_{ll} occur in the basal plane region, and for E_{tt} on the endocardial surface, while the minimum values occur on the epicardial surface for E_{cc} and E_{tt}, and in the basal and apical region of the endocardial

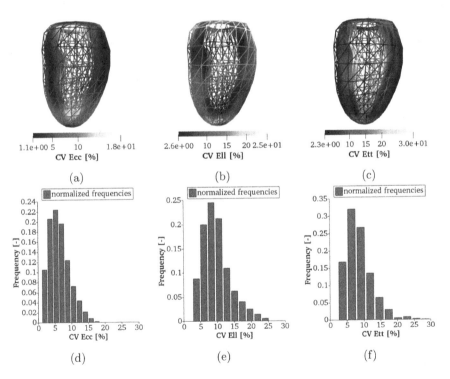

Fig. 2. Coefficient of variation of normal strains. From left to right the coefficient of variation of circumferential, longitudinal, and transmural strains. The top row shows the spatial distribution, the histograms in the bottom row show the frequency within the geometry.

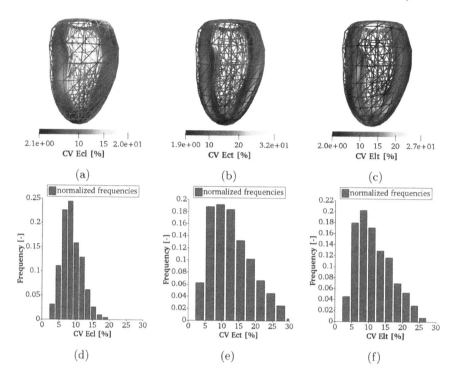

Fig. 3. Coefficient of variation of shear strains. From left to right the coefficient of variation of circumferential-longitudinal, circumferential-transmural, and longitudinal-transmural shear strains. The top row shows the spatial distribution, the histograms in the bottom row show the frequency within the geometry.

surface for E_{ll}. The bottom row of Fig. 2 better shows the distribution of the coefficient of variation of normal strains. From the histograms, the mode of the distributions are about 6% for E_{cc}, 8% for E_{ll}, and 7% for E_{tt}. The spatial distribution and the histogram of the coefficient of variation for shear strains is shown in Fig. 3. The norms of the average shear strains were $\|E_{cl}^{avg}\|_{\infty} = 0.14$, $\|E_{ct}^{avg}\|_{\infty} = 0.37$, and $\|E_{lt}^{avg}\|_{\infty} = 0.36$. For the circumferential-longitudinal, the circumferential-transmural and the longitudinal-transmural shear strains the coefficient of variation ranges are 2.1%–20%, 1.9%–32%, and 2.0%–27%. The maximum values are assumed by E_{cl} in the equatorial region of the endocardium and on the base, by E_{ct} on the endocardium, and by E_{lt} in the basal plane region, while the minimum values occur on the epicardial surface for E_{lt} and in the apical region for E_{cl} and E_{ct}. From the histograms, the mode of the distributions are about 8% for E_{cl}, 10% for E_{ct}, and 9% for E_{lt}.

Figure 4 shows the three sensitivity indices, μ, μ^*, and σ, for the six end systolic strains. In particular, the mean value of the spatial distribution of the indices has been normalized using the absolute value of the mean of the respective average strain and reported as a bar plot. The value of μ is largest for shape

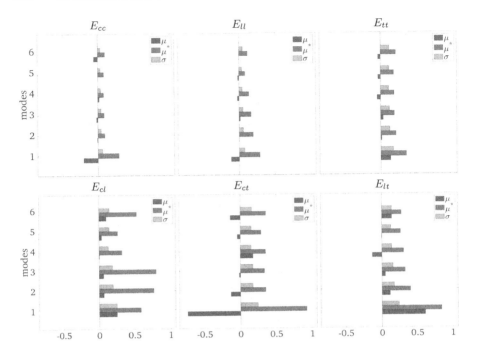

Fig. 4. Mean value of the spatial distribution of the sensitivity indices normalized by the mean value of the respective strain of the population's average.

mode 1, indicating a strong influence on the strains. Almost the same holds for μ^* with the exception of E_{cl} where the third and the second mode have the largest values. μ and μ^* are comparable in modulus only for the first mode of E_{cc}, E_{ct}, and E_{lt}. The distribution of σ between the modes is rather homogeneous. The value is about half of the smallest μ^* for every strain. The largest difference between μ and σ occurs for the first mode of E_{cc}, E_{ct}, and E_{lt}.

4   Discussion

Our results seem to indicate that end systolic shear strains are more sensitive to geometry than normal strains. The histograms in Figs. 2 and 3 show that the distributions of coefficient of variations of shear strains present a larger mode than normal strains and also their spread is wider. Also the bar plots in Fig. 4 support the same fact showing that μ^* assumes larger values for shear strains than for normal strains.

From Fig. 4 also emerges that all strains have the maximum μ for the first shape mode. This suggests that the first shape mode, related to the variation of the size of the left ventricle within the population, plays a major role in determining those strains. This is most pronounced for E_{cc}, E_{ct}, and E_{lt} where μ_1 is almost twofold the value of μ^* of the other modes, suggesting for those strains that the influence of mode 1 is at least two times higher than the others. Moreover, for these strains μ_1 and μ_1^* have similar magnitude, meaning that the first shape mode affects those strains in a specific way. This is due to the fact that the elementary effects are mainly of the same sign. That is, an increase in size of the left ventricle (described by the first shape mode) determines a decrease of both circumferential and circumferential-transmural strains (negative μ_1), and an increase of the longitudinal-transmural shear strain (positive μ_1). For E_{cl} an important role may be played by shape mode 2 and 3 (basically describing the basal plane orientation variation within the population along the anterior-posterior axis and the septal-lateral axis, respectively), since μ_2^* and μ_3^* have the largest values. The corresponding μ are instead small due to the cancelling effect between elementary effects of opposite sign.

Our results show that, for all strains, shape modes 2 to 6 cannot be considered non influential because, even though their respective values of μ^* are typically smaller than μ_1^*, they are still in the same order of magnitude of the predominant shape mode. Therefore, they should be present in a model that aims to predict strains.

Some limited interactions are also present between the shape modes. For those strains with a μ_1 comparable with μ_1^*, such as E_{cc}, E_{ct}, and E_{lt}, μ_1 dominates σ_1. This means that matching LV size in the personalisation process may actually reduce the uncertainty for those strains. For the other strains, σ_1 is comparable with μ_1 and it is quite homogeneously distributed among the other modes. For this reason, for E_{ll}, E_{tt}, and E_{cl} it is not worthwhile to personalise mode 1 only, since, due to interactions, the reduction of uncertainty of this operation would be compromised by the lack of knowledge of the other modes.

5 Conclusions

In this work we tried to assess the sensitivity of LV strains to LV geometry, varied according to six principal modes identified by Zhang et al. [15]. We concluded that mode 1, representing LV size, is most influential but also that none of the shape modes can be considered non influential within the context of this analysis. We also noticed that shear strains are more sensitive to geometry than normal strains. To the best of our knowledge this is the first attempt to quantify the influence of geometry on local quantities such as strains [7,10] using detailed geometrical model of the LV [5].

References

1. Balaban, G., et al.: In vivo estimation of elastic heterogeneity in an infarcted human heart. Biomech. Model. Mechanobiol. **17**(5), 1317–1329 (2018)
2. Bovendeerd, P.H., Kroon, W., Delhaas, T.: Determinants of left ventricular shear strain. Am. J. Physiol. Heart Circulatory Physiol. **297**(3), H1058–H1068 (2009)
3. Campolongo, F., Cariboni, J., Saltelli, A.: An effective screening design for sensitivity analysis of large models. Environ. Model. Softw. **22**(10), 1509–1518 (2007)

4. Chabiniok, R., Moireau, P., Lesault, P.F., Rahmouni, A., Deux, J.F., Chapelle, D.: Estimation of tissue contractility from cardiac cine-MRI using a biomechanical heart model. Biomech. Model. Mechanobiol. **11**(5), 609–630 (2012)
5. Choi, H.F., D'hooge, J., Rademakers, F., Claus, P.: Influence of left-ventricular shape on passive filling properties and end-diastolic fiber stress and strain. J. Biomech. **43**(9), 1745–1753 (2010)
6. Finsberg, H., et al.: Estimating cardiac contraction through high resolution data assimilation of a personalized mechanical model. J. Comput. Sci. **24**, 85–90 (2018)
7. Geerts, L., Kerckhoffs, R., Bovendeerd, P., Arts, T.: Towards patient specific models of cardiac mechanics: a sensitivity study. In: Magnin, I.E., Montagnat, J., Clarysse, P., Nenonen, J., Katila, T. (eds.) FIMH 2003. LNCS, vol. 2674, pp. 81–90. Springer, Heidelberg (2003). https://doi.org/10.1007/3-540-44883-7_9
8. Govindjee, S., Mihalic, P.A.: Computational methods for inverse deformations in quasi-incompressible finite elasticity. Int. J. Numer. Meth. Eng. **43**(5), 821–838 (1998)
9. Marchesseau, S., et al.: Personalization of a cardiac electromechanical model using reduced order unscented kalman filtering from regional volumes. Med. Image Anal. **17**(7), 816–829 (2013)
10. Pluijmert, M., Delhaas, T., de la Parra, A.F., Kroon, W., Prinzen, F.W., Bovendeerd, P.H.: Determinants of biventricular cardiac function: a mathematical model study on geometry and myofiber orientation. Biomech. Model. Mechanobiol. **16**(2), 721–729 (2017)
11. Simo, J.C., Taylor, R.L.: Quasi-incompressible finite elasticity in principal stretches. Continuum basis and numerical algorithms. Comput. Meth. Appl. Mech. Eng. **85**(3), 273–310 (1991)
12. Wang, V.Y., Lam, H., Ennis, D.B., Cowan, B.R., Young, A.A., Nash, M.P.: Modelling passive diastolic mechanics with quantitative MRI of cardiac structure and function. Med. Image Anal. **13**(5), 773–784 (2009)
13. Xi, J., Lamata, P., Lee, J., Moireau, P., Chapelle, D., Smith, N.: Myocardial transversely isotropic material parameter estimation from in-silico measurements based on a reduced-order unscented kalman filter. J. Mech. Behav. Biomed. Mater. **4**(7), 1090–1102 (2011)
14. Xi, J., et al.: The estimation of patient-specific cardiac diastolic functions from clinical measurements. Med. Image Anal. **17**(2), 133–146 (2013)
15. Zhang, X.: Atlas Based Analysis of Heart Shape and Motion in Cardiovascular Disease. Ph.D. thesis, ResearchSpace@ Auckland (2016)

A Simple Multi-scale Model to Evaluate Left Ventricular Growth Laws

Emanuele Rondanina$^{(\boxtimes)}$ and Peter Bovendeerd

Department of Biomedical Engineering, Eindhoven University of Technology,
Eindhoven, The Netherlands
e.rondanina@tue.nl

Abstract. Cardiac growth is the natural capability of the heart of adapting to changes in blood flow demands. Cardiac diseases can trigger the same process leading to an abnormal type of growth. Although several models have been published, details on this process remain still unclear. This study offers an analysis on the driving force of cardiac growth along with an evaluation on the final grown state. Through a zero dimensional model of the left ventricle we evaluate cardiac growth in response to three valve diseases, aortic and mitral regurgitation along with aortic stenosis. We investigate how different combinations of stress and strain based stimuli affect growth in terms of cavity volume and wall volume. All of our simulations are able to reach a converged state without any growth constraint. The simulated grown state corresponded to the experimentally observed state for all valve disease cases, except for aortic regurgitation simulated with a mix of stress and strain stimuli. Thus we demonstrate how a simple model of left ventricular mechanics can be used to have a first evaluation of a designed growth law.

Keywords: Left ventricle · Concentric growth · Eccentric growth

1 Introduction

Cardiac growth is a natural process through which cardiac wall and chamber volume is adapted to deal with a change in hemodynamic load, as for example induced by exercise or chronic disease. It has been demonstrated how pressure overload promotes the thickening of the heart, defined as concentric type of growth, while volume overload generates a dilated heart with a thinning of the tissue, defined as eccentric growth [3]. Valve diseases alter the cardiac pressure-volume balance, leading to an abnormal cardiac growth which is considered one of causes of heart failure [5]. So far several models have been published describing cardiac growth, however the related driving force, namely the growth stimulus, is still under debate [15]. A second point of debate is whether or not to constrain maximum growth.

In this manuscript we evaluate the capability of several growth stimuli to reproduce a clinically realistic and stable response to changes in LV hemodynamic load, as induced by valvular stenosis or regurgitation. The stimuli differ

© Springer Nature Switzerland AG 2019
Y. Coudière et al. (Eds.): FIMH 2019, LNCS 11504, pp. 249–257, 2019.
https://doi.org/10.1007/978-3-030-21949-9_27

in the way they combine changes in tissue load, i.e. stress and strain, to changes in LV geometry, i.e. LV cavity volume and wall volume.

2 Methods

Our analysis of cardiac growth is based on the interaction among a model coupling the mechanics of the Left Ventricle (LV) at tissue level and organ level, a model for the hemodynamics in the systemic circulation, and a growth model in which LV wall (V_{wall}) and cavity (V_{cav}) volumes respond to deviations of actual tissue load from the corresponding homeostatic value.

2.1 LV Mechanics Model

We use the one-fiber model of cardiac function [1,2] to relate mechanics at organ level, expressed in terms of left ventricular pressure p_{cav} and volume V_{cav}, to mechanics at tissue level, expressed with myofiber stress σ_f and sarcomere length l_s. The main equations for LV mechanics are:

$$p_{cav} = \frac{1}{3}\,\sigma_f \ln\left(1 + \frac{V_{wall}}{V_{cav}}\right) \;;\quad \lambda_f = \frac{l_s}{l_{s,0}} = \left(\frac{V_{cav} + \frac{1}{3}V_{wall}}{V_{cav,0} + \frac{1}{3}V_{wall}}\right)^{\frac{1}{3}} \quad (1)$$

Here $V_{cav,0}$ is the cavity volume in the unloaded state. Myofiber stress σ_f is composed of an active component σ_a and two passive components, generated by the collagen matrix along-fiber direction $\sigma_{m,f}$ and radial direction $\sigma_{m,r}$. As consequence σ_f is defined as follows:

$$\sigma_f = \sigma_a(l_s,\, t_a,\, v_s) + \sigma_{m,f}(\lambda_f) - 2\sigma_{m,r}(\lambda_r) \quad (2)$$

where v_s is the sarcomere shortening velocity, t_a is the time elapsed from activation, λ_f is the fiber stretch ratio and λ_r is the resulting tissue stretch in radial direction under the assumption of incompressibility. The functional form for $\sigma_{m,f}$ and $\sigma_{m,r}$ were taken from [2], while σ_a was adopted from [7].

2.2 Systemic Circulation Model

At the organ level, p_{cav} and V_{cav} are determined from the interaction between the LV and the systemic circulation. Arteries, veins and peripheral vessels are modeled by capacitances C and resistances R, as shown in Fig. 1. The flow through aortic valve (q_A) and mitral valve (q_V) is determined by the corresponding R, combined with a dimensionless resistance parameter k which is set to k_f for forward flow ($\Delta p_{(A;V)} > 0$) and k_b for backward flow ($\Delta p_{(A;V)} \leq 0$).

$$q_{(A;V)} = \frac{\Delta p_{(A;V)}}{kR_{(A;V)}} \quad \text{with} \quad \begin{cases} k = k_f & \text{for} \quad \Delta p_{(A;V)} > 0 \\ k = k_b & \text{for} \quad \Delta p_{(A;V)} \leq 0 \end{cases} \quad (3)$$

A healthy valve is represented by k_f equal to 1 and k_b equal to 10^6.

The LV mechanics and systemic circulation model are tuned with the parameters adapted from [7].

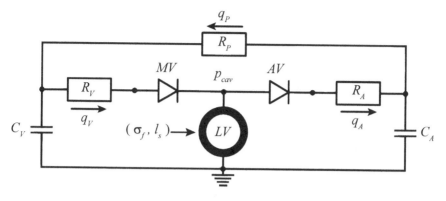

Fig. 1. Lumped parameter model of the circulation. With mitral valve (MV), aortic valve (AV), venous and arterial resistance (R_V and R_A) and capacitance (C_V and C_A), peripheral resistance (R_P) and venous, arterial and peripheral flows (q_V, q_A, q_P). This model is coupled with the one-fiber model of LV mechanics.

2.3   Growth Model

To control cardiac growth, we apply stress or strain related measures. As stress related measure we consider the mean of σ_f (Eq. 2) over a cycle of length T_{cyc}:

$$L_\sigma \;=\; \frac{1}{T_{cyc}} \int_0^{T_{cyc}} \sigma_f(t)\,dt \tag{4}$$

As strain related measure we use the maximum sarcomere strain during the cardiac cycle λ_{max}:

$$L_\epsilon \;=\; \ln(\lambda_{max}) \quad \text{with} \quad \lambda_{max} \;=\; \frac{\max(l_s)}{\min(l_s)} \tag{5}$$

Here $\min(l_s)$ and $\max(l_s)$ indicate the minimum and maximum sarcomere length during one cardiac cycle. Consequently, we define stress-based and strain-based stimuli for growth as follows:

$$S_\sigma \;=\; \frac{L_\sigma - L_{\sigma,\mathrm{hom}}}{L_{\sigma,\mathrm{hom}}} \;;\qquad S_\epsilon \;=\; \frac{L_\epsilon - L_{\epsilon,\mathrm{hom}}}{L_{\epsilon,\mathrm{hom}}} \tag{6}$$

in which L_{hom} represents the homeostatic tissue load for L_σ and L_ϵ.

At tissue level, growth can be either in the direction of the myofibers or perpendicular to their orientation. Considering the organization of the myofibers in the cardiac wall, along-fiber and cross-fiber growth at tissue level correspond to an increase of $V_{\mathrm{cav},0}$ and V_{wall} at organ level respectively. In view of the governing equations for the one-fiber model (Eq. 1) we model the response to a stress-based stimulus as:

$$\frac{1}{V_{\mathrm{wall}}} \frac{dV_{\mathrm{wall}}}{dt} \;=\; +\,\frac{S_\sigma}{\tau_{\mathrm{wall}}} \;;\qquad \frac{1}{V_{\mathrm{cav},0}} \frac{dV_{\mathrm{cav},0}}{dt} \;=\; -\,\frac{S_\sigma}{\tau_{\mathrm{wall}}} \tag{7}$$

Similarly, the response to a strain-based stimulus is modeled as:

$$\frac{1}{V_{\text{wall}}} \frac{dV_{\text{wall}}}{dt} = + \frac{S_\epsilon}{\tau_{\text{cav}}} \; ; \qquad \frac{1}{V_{\text{cav},0}} \frac{dV_{\text{cav},0}}{dt} = + \frac{S_\epsilon}{\tau_{\text{cav}}} \qquad (8)$$

Here the growth rate constant is defined by τ_{cav} and τ_{wall}.

2.4   Model Implementation

The coupling between the models of LV mechanics and cardiac growth works as follows: once the one-fiber model has reached a stable hemodynamic state, one step of Cardiac Growth is introduced through Eqs. 7 and 8. With the new V_{wall}, $V_{\text{cav},0}$ the mechanics model is updated toward a new state in order to evaluate a next growth step. This process is repeated until the growth stimulus has decreased to zero and a new geometry is found.

To characterize the growth obtained, we use a Growth Index (GI) defined as:

$$\text{GI} = \left(\frac{D_{\text{cav}}}{2H_{\text{wall}}} \right) \Big/ \left(\frac{D_{\text{cav}}}{2H_{\text{wall}}} \right)_{\text{hom}} \qquad (9)$$

where D_{cav} is the LV cavity diameter at end diastole while H_{wall} is the LV wall thickness. In order to define D_{cav} and H_{wall} we approximate the LV shape as a thick-walled sphere. As a consequence a GI higher than one identifies a concentric type of growth while a value lower than one indicates an eccentric type of growth.

2.5   Simulations Performed

First we simulate the healthy state (Hom), from which we collect values for $L_{\sigma,\text{hom}}$ and $L_{\epsilon,\text{hom}}$. Second, we simulated three types of valve disease. To simulate aortic valve stenosis (AS) we consider a threefold increase of k_f, based on experimental data [12]. To simulate aortic valve regurgitation (AR) and mitral valve regurgitation (MR) we lowered k_b in order to obtain a regurgitant fraction close to 0.5 for both MR [10,11] and AR [14]. Here the regurgitant fraction is defined as the ratio between backward and forward volume. Each perturbation alters the load at organ level in terms of pressure and volume, causing tissue stress and strain to deviate from their homeostatic values (Fig. 2). Third, we simulate LV growth, resulting in changes of $V_{\text{cav},0}$ and V_{wall} according to Eqs. 7 and 8. We evaluate all possible combinations of these equations. We labeled our simulations σ for a stress based (Eq. 7) and ϵ for a strain based stimulus (Eq. 8). In the resulting label the first letter indicates the growth stimulus used for V_{wall}, while the second refers to $V_{\text{cav},0}$ stimulus. These simulations are labeled AS-G, AR-G and MR-G. The constant τ_{cav} (Eq. 8) is set to $32T_{\text{cyc}}$ while τ_{wall} (Eq. 7) to the a value of $64T_{\text{cyc}}$.

3   Results

The healthy LV has a Stroke Volume of approximately 68 ml, with end diastolic volume (EDV) of 130 ml and end systolic volume (ESV) of about 62 ml (Fig. 2). Cardiac Output is about 5 l/min, maximum systolic pressure ($p_{\text{cav}}^{\text{max}}$) is 18 kPa and Mean Arterial Pressure is 13 kPa. From this healthy heart simulation, we derive the homeostatic tissue load values for $L_{\sigma,\text{hom}}$ (12 kPa) and $L_{\epsilon,\text{hom}}$ (0.14). Figure 2 also shows the effects of changes in pump and tissue functions for the pathological cases. For AS-0, AR-0 and MR-0, $p_{\text{cav}}^{\text{max}}$ changes by +20%, −7% and −19%, respectively, while SV range changes by −19%, +49% and +37%. Characteristic tissue stress L_{σ} changes by +32% and −28% for AS-0 and MR-0, but only +5% for AR-0. Characteristic tissue strain L_{ϵ} changes by +41%, +49%, and −24% for AR-0, MR-0 and AS-0.

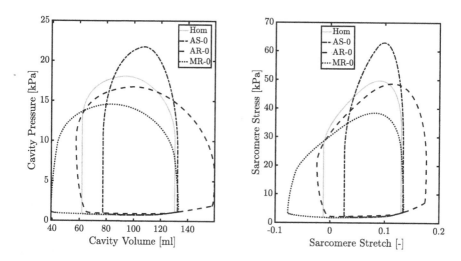

Fig. 2. Pressure-volume (left) loop and sarcomere stress-length (right) loop for the normal heart (Hom) and hearts with Aortic Stenosis (AS-0), Aortic Regurgitation (AR-0) and Mitral Regurgitation (MR-0) without growth.

For all combinations of stimuli, the deviations of L_{σ} and L_{ϵ} from their homeostatic values resulted in stable growth towards a new configuration. Thus for $\epsilon\epsilon$ and $\sigma\sigma$ either strain of stress regained the homeostatic value while $\epsilon\sigma$ and $\sigma\epsilon$ both strain and stress were recovered. In the top row of Fig. 3 we see the unloaded (V_{wall} and $V_{\text{cav},0}$) and the loaded (ESV to EDV range) volumes at the final stage of growth. It is interesting to notice how $\epsilon\sigma$ and $\sigma\epsilon$ lead to the same ending volumes. For AS-G all the simulations have a decrease of $V_{\text{cav},0}$ of about 34% while V_{wall} remains approximately the same for $\epsilon\sigma$ and $\sigma\epsilon$. Simulations $\epsilon\epsilon$ and $\sigma\sigma$ have an opposite type of adaptation, −16% and +16% respectively. For AR-G and MR-G we observe a general increase in $V_{\text{cav},0}$ and V_{wall}, besides $\sigma\sigma$. These changes lead to a reduction in maximum volumes during the cycle for

AS-G (-27% for EDV and -37% for ESV), while we see a general increase in volumes for AR-G and MR-G ($+97\%$ for EDV and $+66\%$ for ESV), with the exception of the ESV for $\sigma\sigma$. In the bottom part of the same figure we compare the resulting GI (left) and $p_{\mathrm{cav}}^{\max}$ (right) with experimental data as collected by [4,8,13,15]. For AS-G and MR-G, both GI and $p_{\mathrm{cav}}^{\max}$ in model and experiment shows similar trends. For AR-G trends match for $\epsilon\epsilon$ and $\sigma\sigma$, but not for $\epsilon\sigma$ and $\sigma\epsilon$, which are characterized by the highest increase in V_{wall}.

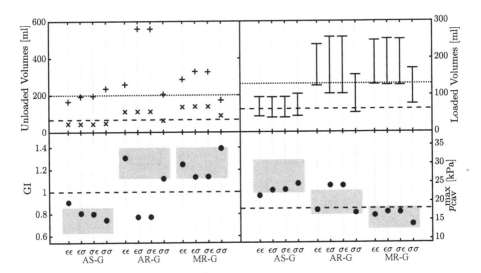

Fig. 3. *Top left*: Unloaded cavity ($\times$) and wall ($+$) volumes after growth. The dotted and dashed line identify the unloaded wall volume V_{wall} and cavity volume $V_{\mathrm{cav,0}}$ before growth. *Top right*: Volume excursion during the cycle. The dotted and dashed line identify the equivalent range before growth. *Bottom left*: Growth Index GI in model ($\bullet$) and experiment (gray boxes) [15]. *Bottom right*: maximum systolic pressure $p_{\mathrm{cav}}^{\max}$ in model ($\bullet$) compared with clinical data (gray boxes) [4,8,13]. The dashed lines in the *bottom* rows are the homeostatic GI and $p_{\mathrm{cav}}^{\max}$. Results are shown for aortic stenosis (AS-G), aortic regurgitation (AR-G) and mitral regurgitation (MR-G) for strain (ϵ) and stress (σ) based stimuli acting on wall volume (first index) and cavity volume (second index).

4   Discussion

In literature several cardiac growth models have been published, based on stress or strain measures [15], however it is still not clear which are the most appropriate loading measures to consider. This work aims to clarify how the chosen stimulus can influence the final grown state. To achieve this goal we test our growth law with a simplified version of the LV [2]. We stimulate concentric and eccentric growth by taking in consideration AS along with AR and MR, which are well known to induce abnormal cardiac growth. AS is known to cause concentric growth [6], while MR [10,11] and AR [14] lead to eccentric growth. We think it is valuable to study both MR and AR because, although they are both interested by volume overload, AR has a lower outflow impedance with higher afterload. Our goal is to describe cardiac growth with only one global law, capable of working for concentric and eccentric growth. This challenge has been already addressed by [9] in which an a priori stopping criterion was used to stabilize the new grown configuration. Differently from [9] we prefer not to limit the growth process. In our analysis we decided to focus on the measures defined in Eq. 4 and 5. Choosing a peak systolic stress over the mean of σ_f, or $\max(l_s)$ over λ_{max}, did lead toward slightly different volumes, but the type of hypertrophy did not change. We designed the growth law such that an hypothetical increase of σ_f (Eq. 1) can be balanced by increasing V_{wall} or decreasing $V_{cav,0}$, hence the negative sign in Eq. 7. An increase of λ_f can be counteracted by increasing both V_{wall} and $V_{cav,0}$ (Eq. 8). As a consequence simulation $\epsilon\epsilon$ has the same change in V_{wall} and $V_{cav,0}$, leading to a GI close to 1, while $\sigma\sigma$ has the same change in volumes but in different directions.

Our study shows how the influence of strain and stress measures on $V_{cav,0}$ and V_{wall} is related to the type of overload considered. It is interesting to notice that although simulations $\epsilon\sigma$ and $\sigma\epsilon$ are defined by opposite type of stimuli, their growth evolution ends on similar volumes. Even if our study is not oriented toward a patient specific analysis we think it is still valuable to compare the results with experimental [15] and clinical [4,8,13] data from literature. In contrast to findings in literature [6], we do not find a significant increase of V_{wall} during AS-G. The corresponding decrease in $V_{cav,0}$ indicates a concentric type of growth. For AR-G and MR-G the increase in $V_{cav,0}$ dominates the increase in V_{wall}, which indicates an eccentric type of growth. The combination of both stress and strain stimuli ($\epsilon\sigma$ and $\sigma\epsilon$) leads to the highest increase of V_{wall}, resulting in a concentric type of growth, which is not in line with the findings of [15]. It is valuable to notice that the collected data [15] is referred to induced pressure and volume overload, without being disease specific. Regarding p_{cav}^{max} we see how all the simulations reproduce the clinical finding of an increased pressure for AS and AR and a decreased pressure for MR.

In all simulations the final configuration represents the condition in which the chosen loading measure (L_σ or L_ϵ) is at the homeostatic state ($L_{\sigma,hom}$ or $L_{\epsilon,hom}$). In contrast with the analysis done by [15], in which only 2 out of 8 laws were not characterized by run away growth, we achieved a stable homeostatic grown state for every simulation. In [16], the most promising law from [15] was used to predict response to valve disease and ischemia. Besides valve parameters, also circulatory parameters like systemic resistance and the stressed blood volume were adapted. Tuning our model to the individual patient as well, would allow for a more strict test of the model. It would be interesting to see how adaptation of circulatory parameters would affect the outcome of our study, in particular whether not only GI would be according to literature, but also the magnitude of changes in cavity and wall volume.

A major limitation of this study is the use of the one fiber LV model which lacks dimensions and spatial variability. The simple transition of anisotropic growth at tissue level to changes in cavity and wall volume can be improved by considering a 3D LV model. However with the introduced spatial variability it might be more complex to reach the model stability.

In conclusion we propose a model coupling organ measures with tissue properties. This allows us to relate hemodynamics perturbations with a myofiber response. We are able to demonstrate that a different choice in loading measures, used for the growth law, might lead toward a different ending state, having however same type of hypertrophy. We observed different results while studying the influence of the chosen stimuli on cavity and wall volumes according to four possible stress and strain combinations. Every simulation converged to a stable ending state which was in accordance with experimental findings for aortic stenosis and mitral regurgitation. Regarding aortic regurgitation, only a strain-based or a stress-based stimulus was in line with the experimental trend.

References

1. Arts, T., Bovendeerd, P.H.M., Prinzen, F.W., Reneman, R.S.: Relation between left ventricular cavity pressure and volume and systolic fiber stress and strain in the wall. Biophys. J. **59**(1), 93–102 (1991)
2. Bovendeerd, P.H.M., Borsje, P., Arts, T., van De Vosse, F.N.: Dependence of intramyocardial pressure and coronary flow on ventricular loading and contractility: a model study. Ann. Biomed. Eng. **34**(12), 1833–1845 (2006)
3. Cantor, E.J.F., Babick, A.P., Vasanji, Z., Dhalla, N.S., Netticadan, T.: A comparative serial echocardiographic analysis of cardiac structure and function in rats subjected to pressure or volume overload. J. Mol. Cell. Cardiol. **38**(5), 777–786 (2005)
4. Carroll, J.D., et al.: Sex-associated differences in left ventricular function in aortic stenosis of the elderly. Circulation **86**(4), 1099–1107 (1992)
5. Cohn, J.N., Ferrari, R., Sharpe, N.: Cardiac remodeling-concepts and clinical implications: a consensus paper from an international forum on cardiac remodeling. J. Am. Coll. Cardiol. **35**(3), 569–582 (2000)
6. Guzzetti, E., et al.: Impact of metabolic syndrome and/or diabetes mellitus on left ventricular mass and remodeling in patients with aortic stenosis before and after aortic valve replacement. Am. J. Cardiol. **123**(1), 123–131 (2019)
7. van der Hout-van, M.B., Oei, S.G., Bovendeerd, P.H.M.: A mathematical model for simulation of early decelerations in the cardiotocogram during labor. Med. Eng. Phys. **34**(5), 579–589 (2012)
8. Kainuma, S., et al.: Pulmonary hypertension predicts adverse cardiac events after restrictive mitral annuloplasty for severe functional mitral regurgitation. J. Thorac. Cardiovasc. Surg. **142**(4), 783–792 (2011)
9. Kerckhoffs, R.C.P., Omens, J.H., McCulloch, A.D.: A single strain-based growth law predicts concentric and eccentric cardiac growth during pressure and volume overload. Mech. Res. Commun. **42**, 40–50 (2012)
10. Kleaveland, J.P., Kussmaul, W.G., Vinciguerra, T., Diters, R., Carabello, B.A.: Volume overload hypertrophy in a closed-chest model of mitral regurgitation. Am. J. Physiol. Heart Circulatory Physiol. **254**(6), H1034–H1041 (1988)

11. Nakano, K., et al.: Depressed contractile function due to canine mitral regurgitation improves after correction of the volume overload. J. Clin. Investig. **87**(6), 2077–2086 (1991)

12. Roger, V.L., Seward, J.B., Bailey, K.R., Oh, J.K., Mullany, C.J.: Aortic valve resistance in aortic stenosis: doppler echocardiographic study and surgical correlation. Am. Heart J. **134**(5), 924–929 (1997)

13. Villari, B., Hess, O.M., Kaufmann, P., Krogmann, O.N., Grimm, J., Krayenbuehl, H.P.: Effect of aortic valve stenosis (pressure overload) and regurgitation (volume overload) on left ventricular systolic and diastolic function. Am. J. Cardiol. **69**(9), 927–934 (1992)

14. Wisenbaugh, T., Spann, J.F., Carabello, B.A.: Differences in myocardial performance and load between patients with similar amounts of chronic aortic versus chronic mitral regurgitation. J. Am. Coll. Cardiol. **3**(4), 916–923 (1984)

15. Witzenburg, C.M., Holmes, J.W.: A comparison of phenomenologic growth laws for myocardial hypertrophy. J. Elast. **129**(1–2), 257–281 (2017)

16. Witzenburg, C.M., Holmes, J.W.: Predicting the time course of ventricular dilation and thickening using a rapid compartmental model. J. Cardiovasc. Trans. Res. **11**(2), 109–122 (2018)

Modeling Cardiac Growth: An Alternative Approach

Nick van Osta, Loes van der Donk, Emanuele Rondanina,
and Peter Bovendeerd$^{(\boxtimes)}$

Department of Biomedical Engineering, Eindhoven University of Technology,
Eindhoven, The Netherlands
p.h.m.bovendeerd@tue.nl

Abstract. Models of cardiac growth might assist in clinical decision making, in particular for long-term prognosis of the effect of interventions. Most growth models strictly enforce the amount and direction of volume change and prevent runaway growth by limiting maximum growth. These assumptions have been questioned. We propose an alternative model for cardiac growth, in which the actual volume change of a tissue element is determined by the desired volume change in that element and the degree to which this change is resisted by the surrounding tissue. The model was evaluated on its ability to reproduce a stable healthy left ventricular configuration under normal hemodynamic load. A homeostatic equilibrium state could not be obtained, which might be due to limitations in the mechanics model or an inadequate stimulus-effect relation in the growth model. Still, the basic idea underlying the model could be an interesting alternative to current growth models.

Keywords: Growth law · Finite element model · Homeostatic state

1 Introduction

Growth, a change in mass, is one of the mechanisms through which the heart can respond to long-term changes in the environment. Changes in hemodynamic load, typically caused by stenotic or leaking valves, tend to cause a spatially homogeneous tissue response. For example, left ventricular (LV) pressure overload results in concentric hypertrophy, characterised by a decrease in the ratio of LV radius-to-wall thickness, while LV volume overload results in eccentric hypertrophy, characterised by an increase in this ratio [12]. More localised changes in tissue load, for example initiated by cardiac pacing, have been demonstrated to cause a spatially inhomogeneous response, characterised by wall thinning at the pacing site and wall thickening at remote sites [8]. Clinical decision making depends on expectations on progression of cardiac growth: would valve replacement lead to reversal of pathologic growth? Will the pacing site that offers the largest benefit in the acute stage still be optimal once the heart has responded through growth?

© Springer Nature Switzerland AG 2019
Y. Coudière et al. (Eds.): FIMH 2019, LNCS 11504, pp. 258–265, 2019.
https://doi.org/10.1007/978-3-030-21949-9_28

Models of cardiac growth might assist in clinical decision making. Such models combine a growth law with a growth evolution model. The growth law specifies the amount and type of growth (e.g. along or perpendicular to the fiber direction) as a function of the deviation of a certain aspect of tissue load (e.g. stress or strain) from its homeostatic value. Recently, several growth laws have been evaluated in a simple test environment [12]. The growth evolution model specifies how the amount of tissue change is actually imposed onto the existing tissue. Most growth models adopt the approach proposed by Skalak et al. [11] and Rodriguez et al. [10], in which the amount and direction of volume change are strictly enforced through a growth deformation tensor F_g. While promising models have been published, they suffer from limitations related to an a priori defined type of growth (eccentric or concentric) [2], or a spatially inhomogeneous homeostatic load value, derived from a simulation of normal LV mechanics [5]. Moreover, throughout the geometry, growth for cases of pressure and volume overload, as predicted in these models, was stopped when it reached a predefined limit. Thus the models could not be evaluated with respect to stability of the growth process. Finally, the growth evolution model used [10,11] has been questioned. The growth tensor F_g has to be accomodated by an elastic deformation tensor F_e in order to maintain an intact, continuous tissue. This elastic deformation is related to a single fixed stress-free reference state and assumed to be the cause of residual tissue stress. Yet, it is more plausible that the stress-free configuration evolves in time as new tissue is deposited, and that residual tissue stress is related to new tissue being deposited under a certain pre-stretch [3].

In this study, we investigate an alternative growth model, based on a model proposed by Kroon et al. [6]. In this model, a growth stimulus leads to a desired amount of growth, that would be realised if tissue elements would not experience mechanical interaction with their neighbouring tissue elements. The actual, realised growth is obtained by including this interaction. Growth-induced residual stresses are neglected. We aim to eventually use the model to describe growth in the adult heart, in response to disease or treatment. However, in this study, we simulate local tissue growth in the LV wall under normal hemodynamic loading conditions, which, theoretically, should lead to a stable final LV geometry.

2 Methods

Cardiac growth is modeled through alternatingly solving a mechanics and a growth problem, see Fig. 1. The mechanics problem provides a stimulus for volumetric growth. The equilibrium configuration obtained from solving the growth problem is adopted as the new stress free configuration for the next mechanics problem.

2.1 Model of Cardiac Mechanics

The model of cardiac mechanics was taken from [1]. Briefly, LV geometry is described as a thickwalled truncated ellipsoid. Myofiber orientation e_f is

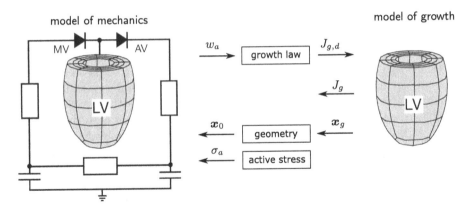

Fig. 1. Model overview: the cardiac mechanics model (left) yields active work density w_a that is converted into a desired growth ratio $J_{g,d}$ that drives the growth model (right) and results in an actual growth ratio J_g; the grown configuration x_g is used as the new unloaded configuration x_0 for the mechanics computation, driven by active stress σ_a.

expressed in terms of a helix angle and a transverse angle, defined as a function of local wall coordinates. Myocardial tissue Cauchy stress σ is composed of a passive component σ_p and an active component σ_a:

$$\sigma = \sigma_p + \sigma_a e_f e_f \tag{1}$$

Active stress σ_a depends on time, sarcomere length and sarcomere shortening velocity. Passive material behavior is assumed nonlinearly elastic, transversely isotropic and nearly incompressible. It is described through a strain energy density function W, composed of a shape part W_s and a volumetric part W_v. Formulated in terms of components of the Green Lagrange strain tensor E with respect to the material bound coordinate system $\{e_f, e_s, e_n\}$, these are given by:

$$W_s = a_0 \left(\exp(Q) - 1\right) \tag{2}$$

$$Q = a_1(E_{ff}^2 + E_{ss}^2 + E_{nn}^2) + a_3 E_{ff}^2$$

$$+ \frac{1}{2} a_2 (E_{fs}^2 + E_{sf}^2 + E_{fn}^2 + E_{nf}^2 + E_{ns}^2 + E_{sn}^2) \tag{3}$$

$$W_v = a_{5,m} \left(\det(F^T \cdot F) - 1\right)^2 \tag{4}$$

with deformation gradient tensor F. Cardiac deformation is computed from solving the equations of conservation of momentum:

$$\nabla \cdot \sigma = 0 \tag{5}$$

Essential boundary conditions are defined at the base: the basal plane is prevented from moving in axial direction, while the rotation of the subendocardial basal ring is suppressed. The epicardial surface is traction free while the endocardial surface is subject to a uniform left ventricular pressure p_{lv}, determined from the interaction of the LV with a lumped parameter circulation model.

2.2 Model of Cardiac Growth

Cardiac growth is also modeled by solving for mechanical equilibrium (5). We neglect active stress σ_a and assume isotropic material properties by setting $a_3 = 0$ in (3). Growth is driven by a modified version of (4):

$$W_{v,g} = a_{5,g} \left(\det(\boldsymbol{F}^T \cdot \boldsymbol{F}) - J_{g,d}^2 \right)^2 \tag{6}$$

Here, $J_{g,d}$ represents a desired relative volume change, which is specified from an evolution equation:

$$\frac{dJ_{g,d}}{dt} = \frac{1}{\tau_g} \left(\frac{w_a - w_{a,hom}}{w_{a,hom}} \right) \tag{7}$$

in which τ_g represents the characteristic time scale for growth. Active work density is obtained from the mechanics model:

$$w_a = \int_{cycle} \sigma_a d\varepsilon_f \quad ; \quad \varepsilon_f = \ln\left(\frac{l_s}{l_{s0}} \right) \tag{8}$$

with fiber strain ε_f derived from the actual (l_s) and reference (l_{s0}) sarcomere length. In solving (5) boundary conditions are the same as those in the mechanics problem, except for setting the endocardial surface traction free. The solution is characterized by the growth deformation tensor $\boldsymbol{F}_g$, which defines the actual relative growth $J_g = \det(\boldsymbol{F}_g)$ that might deviate from the desired relative growth $J_{g,d}$. The grown configuration $\boldsymbol{x}_g$ is used as the new unloaded configuration $\boldsymbol{x}_0$ for the next mechanics computation.

2.3 Simulations

Material parameters in (3) were set to $a_0 = 0.4\,\text{kPa}$, $a_1 = a_3 = 3$ and $a_2 = 6$. In (4) and (6) we set $a_{5,m} = 55\,\text{kPa}$ and $a_{5,g} = 15\,\text{kPa}$. model. In (7) we choose $w_{a,hom} = 6\,\text{kPa}$, which is close to the spatially averaged work density of $6.4\,\text{kPa}$, as obtained in the initial mechanics computation. We used a simple Euler forward scheme to evaluate (7) over a time step Δt, and set $\Delta t/\tau_g$ to $4.2 \cdot 10^{-4}$. Cardiac cycle time was set to $800\,\text{ms}$.

3 Results

Before Growth. Hemodynamic function before growth is characterised by end diastolic and end ejection volumes of $111\,\text{ml}$ and $48\,\text{ml}$, respectively. LV peak pressure is $17\,\text{kPa}$. Fiber stress-strain loops are shown in Fig. 2. The loops are similar, although from endocardium to epicardium fiber strain decreases, while stress increases slightly. In the equatorial part of the LV, active work density ranges from $7\,\text{kPa}$ to $9\,\text{kPa}$, see Fig. 3. Near the apex, w_a drops to $0\,\text{kPa}$. In the midwall basal region, it increases to $12\,\text{kPa}$.

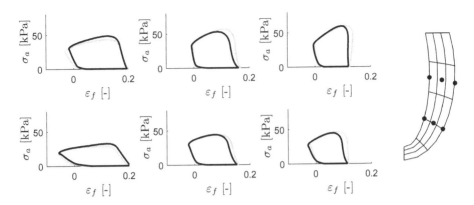

Fig. 2. Active stress - fiber strain loops at six locations in the LV wall, before growth (gray) and after 20 growth cycles (black).

During Growth. During growth, the spatial distribution of w_a, observed in the mechanics before growth, is translated into volume increase near the mid-wall base, and volume decrease near the apex, see Fig. 3. As a consequence, w_a decreases near the base and increases near the apex. After the first 20 growth cycles, changes near the equator are minor, but transmural gradients in fiber strain and stress decrease slightly, see Fig. 2. Over the whole LV wall, spatially averaged work density evolves towards the homeostatic value of 6 kPa, from (6.4 ± 1.6) kPa to (6.2 ± 1.3) kPa at 20 growth cycles. However, thereafter, it keeps decreasing to (5.8 ± 1.5) kPa at 50 cycles and (5.5 ± 1.7) kPa at 59 growth cycles. From cycle 20 onwards, the grown geometry becomes increasingly unphysiological, see Fig. 3. After 59 growth cycles, the mechanics simulation did not converge anymore. Overall wall volume changes from an initial 150 ml, to 157 ml, 161 ml and 160 ml after 20, 40 and 59 growth cycles, respectively. Corresponding unloaded cavity volumes are 43 ml, 47 ml, 50 ml, and 54 ml.

4   Discussion

We present a growth model, in which the actual volume change J_g of a tissue element is determined by the desired volume change $J_{g,d}$ in that element and the degree to which this change is resisted by the surrounding tissue. This resistance to growth is determined by the setting of parameter $a_{5,g}$ in (6). This approach differs from currently used models, is which desired volume change is strictly enforced [2,5,10,11]. The latter models use a fixed reference approach, whereas an updated reference approach might better reflect the idea of evolving natural configurations [3,6]. Whereas current models can be regarded as models tracking an infinite number of configurations, our model represents the other extreme: neglecting growth induced stress might be considered as accounting for the most recent configuration only. In our previous model, we simulated growth in the passive ventricle, driven by sarcomere length at end diastole [6]. In the current

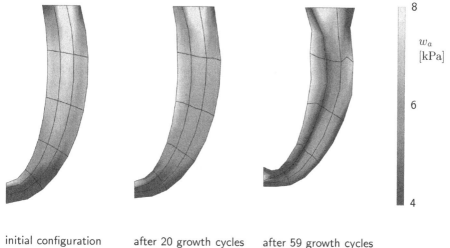

initial configuration   after 20 growth cycles   after 59 growth cycles

Fig. 3. Spatial distribution of active work density w_a displayed in a long axis cross section of the grown geometry in the unloaded configuration.

model, we simulated mechanics at organ and tissue level during the complete cardiac cycle. Consequently, we could apply a stimulus, active work density w_{act}, that also includes tissue load during the complete cycle.

Whereas most models of cardiac growth are tested on their ability to reproduce concentric and eccentric growth as observed in response to valve pathologies, we evaluated our model for the normal healthy case. We reasoned that the healthy case would constitute a dynamic homeostatic equilibrium. Consequently, the ability to reproduce this homeostatic case would be a first test for any growth model.

Although model results evolve towards the homeostatic state during the initial growth steps, they diverge as growth continues. There are several potential causes for this divergence. Firstly, divergence occurs in the basal and apical regions first. In these regions the mechanics model is expected to give less realistic resuls, either to inaccurate boundary conditions (at the base), or to an inaccurate local geometry and fiber field (at the apex). In the model of cardiac mechanics only, we may consider the basal and apical region as suitable boundary conditions for the central region of the LV, and focus on computed mechanics in that region. However, in the growth model the inaccurate predictions of local tissue load near base and apex lead to an inaccurate, often more extreme stimulus for growth, resulting into non-realistic growth in the model. Gradually, this non-physiologic change in geometry progresses towards the central region of the LV, eventually leading to divergence. Obviously, these problems can, at least partly, be avoided by setting the spatial map of the homeostatic tissue load in the growth law identical to the spatial map of local tissue load in the initial mechanics simulation [5]. Alternatively, stability problems may be avoided by imposing limits on maximum growth [2,5]. However, it is doubtful whether these solutions are physiologically realistic.

Secondly, we apply a simple growth law, using only tissue work density as input. This stimulus is plausible as the distribution of tissue work density, as computed in pacing studies [4], seems to match the pattern of local cardiac growth, observed in experiments [8]. However, it has been suggested that growth laws need multiple inputs in order to produce evolution towards a homeostatic state. These inputs should be independent, in terms of spatial directions in the tissue from which the load is retreived, and/or in terms of the moment during the cardiac cycle at which the load is evaluated [12].

Thirdly, whereas the spatial discretization of our model in 108 quadratic elements containing 1071 nodes was found to be sufficiently fine for computation of mechanics, we might need more spatial resolution for the growth model and also need to apply remeshing when growth leads to deterioration of element quality.

Finally, homeostatic equilibrium in the real heart may be expected to be the result of many growth and remodeling processes acting and interacting simultaneously. For example, a model of remodeling of myocardial fiber orientation has been proposed [7,9]. Eventually, a combination of growth and remodeling models might be required to produce a realistic homeostatic cardiac state.

5   Conclusion

We proposed an alternative model for cardiac growth, in which the actual volume change of a tissue element is determined by the desired volume change in that element and the degree to which this change is resisted by the surrounding tissue. The model was evaluated on its ability to reproduce a normal healthy left ventricular configuration. Even though a homeostatic equilibrium state could not be obtained, the model could be an interesting alternative to current growth models, that strictly enforce volumetric growth and prevent runaway growth by imposing limits to growth.

References

1. Bovendeerd, P.H.M., Kroon, J.W., Delhaas, T.: Determinants of left ventricular shear strain. Am. J. Physiol. **297**, H1058–H1068 (2009)
2. Göktepe, S., Abilez, O.J., Parker, K.K., Kuhl, E.: A multiscale model for eccentric and concentric cardiac growth through sarcomerogenesis. J. Theor. Biol. **265**, 433–442 (2010)
3. Humphrey, J.D., Rajagopal, K.R.: A constrained mixture model for growth and remodeling of soft tissues. Math. Models Meth. Appl. Sci. **12**, 407–430 (2002)
4. Kerckhoffs, R.C.P., Bovendeerd, P.H.M., Prinzen, F.W., Smits, K., Arts, T.: Intra- and interventricular asynchrony of electromechanics in the ventricularly paced heart. Eng. Math. **47**, 201–216 (2003)

5. Kerckhoffs, R.C.P., Omens, J.H., McCulloch, A.D.: A single strain-based growth law predicts concentric and eccentric cardiac growth during pressure and volume overload. Mech. Res. Commun. **42**, 40–50 (2012)
6. Kroon, W., Delhaas, T., Arts, T., Bovendeerd, P.H.M.: Constitutive modeling of cardiac tissue growth. In: Sachse, F.B., Seemann, G. (eds.) FIMH 2007. LNCS, vol. 4466, pp. 340–349. Springer, Heidelberg (2007). https://doi.org/10.1007/978-3-540-72907-5_35
7. Kroon, W., Delhaas, T., Bovendeerd, P.H.M., Arts, T.: Computational analysis of the myocardial structure: adaptation of cardiac myofiber orientations through deformation. Med. Image Anal. **13**, 346–353 (2009)
8. van Oosterhout, M.F.M., et al.: Asynchronous electrical activation induces asymmetrical hypertrophy of the left ventricular wall. Circulation **98**, 588–595 (1998)
9. Pluijmert, M.H., Delhaas, T., Flores de la Parra, A., Kroon, W., Prinzen, F.W., Bovendeerd, P.H.M.: Determinants of biventricular cardiac function: a mathematical model study on geometry and myofiber orientation. Biomech. Mod. Mechanobiol. **16**, 721–729 (2017)
10. Rodriguez, E.K., Hoger, A., McCulloch, A.D.: Stress-dependent finite growth in soft elastic tissues. J. Biomech. **27**, 455–467 (1994)
11. Skalak, R., Dasgupta, G., Moss, M., Otten, E., Dullumeijer, P., Vilmann, H.: Analytical description of growth. J. Theor. Biol. **94**, 555–577 (1982)
12. Witzenburg, C.M., Holmes, J.W.: A comparison of phenomenologic growth laws for myocardial hypertrophy. J. Elast. **129**, 257–281 (2017)

Minimally-Invasive Estimation of Patient-Specific End-Systolic Elastance Using a Biomechanical Heart Model

Arthur Le Gall[1,2,3]($\boxtimes$) (iD), Fabrice Vallée[1,2,3], Dominique Chapelle[1,2] (iD), and Radomír Chabiniok[1,2,4] (iD)

[1] Inria, Paris-Saclay University, Palaiseau, France
arthur.le-gall@aphp.fr
[2] LMS, École Polytechnique, CNRS, Paris-Saclay University, Palaiseau, France
[3] Anaesthesia and Intensive Care department, Lariboisière hospital, Paris, France
[4] School of Biomedical Engineering and Imaging Sciences (BMEIS),
St Thomas Hospital, King's College London, London, UK

Abstract. The end-systolic elastance (E_{es}) – the slope of the end-systolic pressure-volume relationship (ESPVR) at the end of ejection phase – has become a reliable indicator of myocardial functional state. The estimation of E_{es} by the original multiple-beat method is invasive, which limits its routine usage. By contrast, non-invasive single-beat estimation methods, based on the assumption of the linearity of ESPVR and the uniqueness of the normalised time-varying elastance curve $E^N(t)$ across subjects and physiology states, have been applied in a number of clinical studies. It is however known that these two assumptions have a limited validity, as ESPVR can be approximated by a linear function only locally, and $E^N(t)$ obtained from a multi-subject experiment includes a confidence interval around the mean function. Using datasets of 3 patients undergoing general anaesthesia (each containing aortic flow and pressure measurements at baseline and after introducing a vasopressor noradrenaline), we first study the sensitivity of two single-beat methods—by Sensaki et al. and by Chen et al.—to the uncertainty of $E^N(t)$. Then, we propose a minimally-invasive method based on a patient-specific biophysical modelling to estimate the whole time-varying elastance curve $E^{model}(t)$. We compare E_{es}^{model} with the two single-beat estimation methods, and the normalised varying elastance curve $E^{N,model}(t)$ with $E^N(t)$ from published physiological experiments.

Keywords: Time-varying elastance ·
End-systolic elastance estimation ·
Patient-specific biophysical modelling

1 Introduction

The relation between ventricular pressure (P) and volume (V) at the end of ejection is described by the preload and afterload independent end-systolic pressure-volume relationship (ESPVR). The slope of ESPVR – the so-called end-systolic

© Springer Nature Switzerland AG 2019
Y. Coudière et al. (Eds.): FIMH 2019, LNCS 11504, pp. 266–275, 2019.
https://doi.org/10.1007/978-3-030-21949-9_29

elastance, E_{es} – and its volume intercept (V_0) allow to derive the time-varying elastance $E(t) = P(t)/(V(t)-V_0)$, see Fig. 1. Under physiological loading ranges, the E_{es} is known to be closely related to the active properties of the myocardium [1,13], and is assumed to be itself preload and afterload independent, and so the ESPVR to be linear. Even though the load dependency of E_{es} has been experimentally shown [1], the linear approximation of ESPVR and the subsequent analysis of its derived indicators have been proven to be clinically useful for performance assessment and monitoring of failing hearts, and for studying the interaction between the heart and vasculature (e.g. by assessing the so-called ventricular-arterial coupling, $V_{va} = E_{es}/E_a$, with E_a being the arterial elastance [13]).

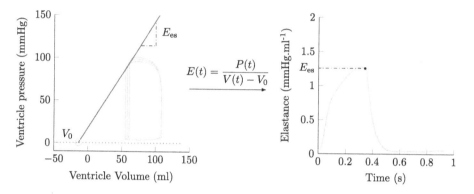

Fig. 1. Example of the model-based determination of E_{es}, V_0 and $E(t)$

Originally, E_{es} and ESPVR were obtained using a multiple-beat measurement technique. A linear regression was fitted on the end-systolic P-V points measured at different loading conditions (*e.g.* by inferior vena cava occlusion, or administration of vasopressors), during cardiac catheterisation. The associated technical issues led to a development of single-beat methods [14,15], which allow to estimate E_{es} and ESPVR non-invasively [5,7]. We hypothesised that the estimation of E_{es} using such methods is, however, too sensitive to their parameters to obtain a reliable patient-specific result.

By calibrating a biomechanical model of heart and vasculature [2,4] using aortic pressure and flow data, we can simulate the entire P-V loop and reproduce minimally-invasively the original multiple-beat measurement method. We aimed at comparing the E_{es} obtained by the single-beat methods or obtained by a method based on patient-specific biophysical modelling. Furthermore, we evaluated the properties of the derived time-varying elastance after spatial-temporal normalisation [16].

2   Methods

2.1   Data and Models

Patients and Procedures. Three patients undergoing general anaesthesia (GA) for neuroradiological intervention, for whom a continuous arterial pressure and cardiac output monitoring were indicated, were included in the presented observational study, approved by the ethical committee of the Société de Réanimation de Langue Française (CE-SRLF 14-356). The data collection is described in detail in [10]. Briefly, after GA induction, a transthoracic echocardiography (TTE) was performed to obtain cardiac geometry information. A transoesophageal Doppler probe (Deltex Medical, Chichester, UK) was inserted into the oesophagus in order to continuously measure the aortic flow. During the procedure, the anaesthetist could need to raise blood pressure using intravenous administration of $5\,\mu g$ of noradrenaline (NOR). The neuroradiologist cannulated aorta through femoral puncture, and inserted a guidewire. A fluid-filled mechanotransducer was connected to obtain the aortic pressure waveform.

Biomechanical Model of Heart and Vasculature. The model used in this study was a combination of a biomechanical heart and Windkessel circulation models connected together to represent the cardiovascular system. The heart model was derived from a previously validated complete three-dimensional (3D) model [4] by model reduction [2]. While the entire geometry was reduced to a sphere, all the passive and active properties were kept as in the 3D model. The passive part of myocardium was modelled according to Holzapfel and Ogden [8], and adjusted using experimental data [11]. The active contraction was based on Huxley's sliding filament theory [9]. The circulation was represented by a 2-stage Windkessel model connected in series (proximal and distal capacitances and resistances). In turning the model into patient-specific regime, see also [3,12], first the Windkessel model parameters were calibrated by imposing the measured aortic flow and tuning the resistance and capacitance parameters, in order to fit the simulated and measured aortic pressure. Then, geometry and passive myocardial properties were calibrated using TTE data. The timing of the electrical activation was adjusted using ECG timings (in particular, the action potential duration in line with the ST interval). Finally, the myocardial contractility was tuned to minimise the difference between the simulated and measured aortic flow and pressure. The model calibration as described above was performed in two different conditions – at baseline and at maximal effect of NOR – to explore comparatively the cardiovascular effect of NOR. Data processing and signal analysis were performed in Matlab, (Natick, Massachussets, USA) in which the model [2] was implemented into a library named CardiacLab.

2.2   Single-Beat Estimation of E_{es}

Method by Senzaki et al. [14]. The method is based on the characteristics of the time-varying elastance $E(t)$ described in detail by Suga et al. [16].

In brief, the normalised $E(t)$ – with respect to the time at end-systole (t_{max}) and maximal elastance value (E_{es}), i.e. $E^N(t) = \frac{E(t/t_{max})}{E_{es}}$ – was found to be consistent across subjects and across varying loading conditions [16]. This principle therefore allows to identify a particular time-point on the subject and physiology independent "universal" $E^N(t)$, if the ratio $\frac{t}{t_{max}}$ is known.

To estimate E_{es}, the following values need to be obtained: (1) the end of isovolumic contraction (t_d) assessed by TTE; (2) ventricular pressure at the opening of aortic valve (measured by aortic catheter); (3) ejection time (from end-diastole to end-systole) by TTE; and (4) ventricular volumes (end-diastole and end-systole), accessed by TTE. We can then apply the formula by Senzaki et al. [14]:

$$E_{es}^{senzaki} = \left(\frac{P_{ed}}{E_d^N} - P_{es} \right) / SV, \qquad (1)$$

where $E_d^N = E^N(t_d)$, P_{ed} being the aortic end-diastolic pressure, P_{es} the aortic end-systolic pressure, SV the stroke volume.

Method by Chen et al. [5]. This method is derived from the original method of Senzaki by optimising the following linear regression to estimate E_d^N:

$$E_d^{N,modified} = 0.0275 - 0.165 \cdot EF + 0.3656 \cdot \frac{P_{ed}}{P_{es}} + 0.515 \cdot E_d^N, \qquad (2)$$

EF being the ejection fraction measured by TTE as the ratio between SV and the end-diastolic volume. Finally, $E_d^{N,modified}$ is used in Eq. (1) to estimate E_{es}^{chen}. We used the methods of Senzaki and Chen to predict E_{es}.

2.3 Estimation of E_{es} Using Biomechanical Heart Model

Model-Based E_{es} and Time-Varying Elastance Curve Estimation. The original multiple-beat technique involves first a construction of ESPVR, which is given by linear regression performed on the consecutive End-Systolic Pressure-Volume points obtained in PV loops measured in different loading conditions. Then the slope and the intercept of the ESPVR with the volume axis represent E_{es} and V_0, respectively. To reproduce this procedure in silico using the calibrated patient-specific model described in Sect. 2.1, we modified sequentially the afterload by varying Windkessel model parameters. We obtained 5 P-V loops with varying loading conditions and identified the end-systolic pressure-volume point of every P-V loop (corresponding to the aortic valve closing). We performed a linear regression to obtain E_{es}^{model} and V_0, and computed the simulated varying elastance $E^{model}(t) = P(t)/(V(t) - V_0)$ (see Fig. 1).

Study of the Simulated Time-Varying Elastance. The simulated and normalised time-varying elastances $E^{N,model}(t)$ were compared with the $E^N(t)$ obtained experimentally by Suga et al. [16].

3   Results

Table 1 shows the main characteristics of patients and data indicators at baseline obtained from the three patients included in the study.

Table 1. Patients characteristics at baseline. LVEDV: left ventricular end-diastolic volume; SV: stroke volume; P_{ed}: end-diastolic aortic pressure; P_{es}: end-systolic aortic pressure; t_d: pre-ejection time; t_{max}: ejection time.

Patient	Age (yo)	Weight (kg)	Height (cm)	LVEDV (ml)	SV (ml)	P_{ed} (mmHg)	P_{es} (mmHg)	t_d (ms)	t_{max} (ms)
Patient 16	40	58	160	109	71	61	77	107	329
Patient 21	15	58	158	140	91	56	79	84	283
Patient 69	58	88	178	81	45	55	72	77	321

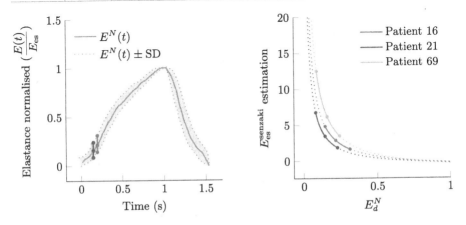

Fig. 2. Normalised elastance curve and effect of the standard deviation (SD) of E_d^N on $E_{es}^{senzaki}$ estimation [14]. Left: Normalised elastance curve (reproduced from Suga et al. [16]) and E_d^N prediction (colored dots) using $\frac{t_d}{t_{max}}$ as obtained by TTE. Right: $E_{es}^{senzaki}$ estimation as function of E_d^N, for the data obtained in the 3 patients. The plain lines represent the $E_{es}^{senzaki}$ for E_d^N inside the ranges given by Suga et al. The dashed line represent the extrapolation of the $E_{es}^{senzaki}$ for E_d^N outside these ranges. (color figure online)

Sensitivity Analysis of the Existing Methods. To assess the sensitivity of the Senzaki method, we used the $E^N(t)$ curve from the study of Suga et al. [16]. We generated an interpolation of all the outliers of the curve, to be able to evaluate the effect of the error in measuring E_d^N on the $E_{es}^{senzaki}$ estimation (see Fig. 2, left). We can see in the right panel of Fig. 2 that the standard deviation in estimating E_d^N had a significant impact on the predicted $E_{es}^{senzaki}$. When considering the standard deviation in the $E^N(t)$ data, the value of E_{es} is ranging between half and twice times the predicted value, for all 3 patients.

In order to appreciate the consistency between the method of Senzaki and the method of Chen, we compared the values of E_{es} given by the two methods at baseline and at maximum effect of NOR. The results are presented in Table 2.

Table 2. Results of the end-systolic elastance E_{es} (in mmHg.ml^{-1}) estimation using the method of Senzaki et al. [14], Chen et al. [5], and the biophysical model [2].

Method	Challenge	Patient 16	Patient 21	Patient 69
$E_{es}^{senzaki}$	Baseline	2.88 [1.65 − 4.83]	3.45 [1.85 − 6.76]	6.17 [3.5 − 12.5]
	Noradrenaline	4.68 [2.66 − 8.3]	3.39 [1.86 − 6.17]	7.95 [4.28 − 15.02]
E_{es}^{chen}	Baseline	1.59 [1.23 − 1.94]	1.58 [1.22 − 1.93]	2.51 [2 − 3.08]
	Noradrenaline	2.12 [1.68 − 2.55]	1.83 [1.39 − 2.27]	3.71 [2.83 − 4.57]
E_{es}^{model}	Baseline	2.42	3.35	3.64
	Noradrenaline	5.82	3.05	5.51

Method from Biomechanical Model. For the 3 subjects, we were able to calibrate the biomechanical model and obtain the P-V loop at baseline and after the administration of NOR (see example of calibration in Fig. 3).

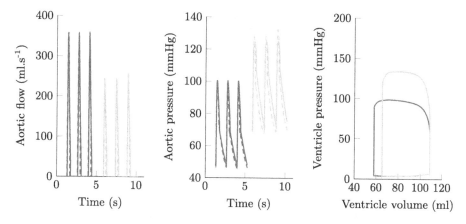

Fig. 3. Example of a calibrated model. The left and the middle panels display respectively the aortic blood flow and pressure at baseline (blue) and at maximal effect of noradrenaline(orange). The simulations (bold) are calibrated using patient's measured data (dashed). The right panel displays pressure-volume loops generated using the results of the aforementioned calibrated simulations at baseline (blue) and at maximal effect of noradrenaline (orange). (Color figure online)

Table 2 shows the patient-specific E_{es} prediction from the $E_{es}^{senzaki}$, E_{es}^{chen} and E_{es}^{model}. We can see that the E_{es}^{model} is close to $E_{es}^{senzaki}$ for Patients 16 and 21, and located within the ranges given in Senzaki et al. [14] for all 3 subjects. We can also see that the usage of NOR was associated with an increase of $E_{es}^{senzaki}$ and in E_{es}^{model} except for Patient 21, in whom all methods suggested no change in E_{es}.

Out of uncertainties in the estimated E_{es}^{model}, we performed a sensitivity analysis with respect to the error in measuring the input parameters for the

272   A. Le Gall et al.

Table 3. Sensitivity analysis for E_{es} estimation to relative change in wall thickness, for the Patient 16. The measured wall thickness is in bold.

Wall thickness	(% of measured value)	60	80	**100**	120	150	avg (SD)
E_{es}^{model}	(ml.mmHg^{-1})	2.56	2.38	**2.42**	2.73	3.09	2.64 ± 0.29

model, specifically the error in the wall thickness. The wall thickness was varied by ±50% from the measured value, and the passive and active properties of the model were calibrated accordingly. Table 3 displays the results in E_{es}^{model} estimation for each wall thickness.

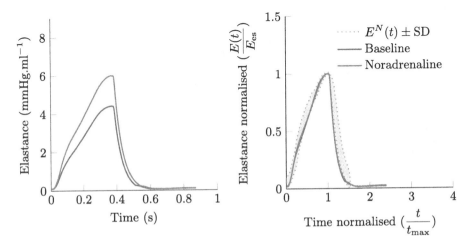

Fig. 4. Example of a patient-specific model-derived time-varying elastance curve. Left: Time-varying elastance at baseline and after NOR administration. Right: Normalised time-varying elastance curve at baseline and after NOR administration, plotted against experimental data (reproduced from [16]).

Figure 4 demonstrates that the biomechanical model was able not only to estimate E_{es}, but did provide the overall time-varying elastance curve. This example shows that the $E^{model}(t)$ was higher when using noradrenaline (Fig. 4 left panel), according to the expected effect of noradrenaline (enhancement of contractility). Furthermore, when $E^{model}(t)$ curves were normalised as described in Sec. 2.2 (see Fig. 4 right panel), both the time-varying elastance curves (baseline and NOR) were within the physiological ranges described by Suga et al. [16].

4 Discussion

In this paper, we described a minimally invasive multi-beat method to estimate patient-specific time-varying and end-systolic elastance by using biomechanical modelling. As originally described, the E_{es} estimation involves a multiple-beat measurement of P-V loop. For technical concerns, single-beat estimation methods were developed. These methods assume the ESPVR being linear with constant slope of E_{es}, which is extrapolated from the end-diastolic measurement point. However, the slope of the real ESPVR is decreasing when approaching the end-diastolic point [1], deviating the estimated E_{es}. Our method allows to modify loading conditions in order to estimate E_{es} around the measured end-systolic P-V point, where the linearity of ESPVR can be assumed. Furthermore, we demonstrated a very high sensitivity of the single-beat estimation method by Senzaki et al. [14] to E_d^N parameter. Then, we demonstrated a limited accuracy of the method by Chen et al. [5], which introduced some phenomenological terms to the equation (1). Indeed, no concordance with the method of Senzaki et al. [14] could have been observed, the mean values and confidence intervals for E_{es}^{chen} estimation falling outside the ranges of $E_{es}^{senzaki}$. The reproducibility of the results obtained in the validation studies [5,14] is therefore questioned. To address these issues, we would have to compare our results against invasive P-V loop measurements, which were not available in our study. Despite the aforementioned limitations, we used the $E_{es}^{senzaki}$ estimation as a comparator for an indirect validation of our method. We verified that our simulated E_{es}^{model} at least fell within the ranges of the outliers of the $E_{es}^{senzaki}$, and that the normalised time-varying elastance $E^{N,model}(t)$ was consistent with the $E^N(t)$ from Suga et al. [16], in all three subjects even when varying physiology (baseline vs. administration of NOR). We remark in addition that neglecting the standard deviation of the experimentally obtained $E^N(t)$ in the single-beat estimations of E_{es} effectively means decreasing the individuality of the considered subject. We showed, however, that the specificity of patients had a great impact on the E_{es} estimation by these methods. Our framework – based on patient-specific biomechanical modelling – allows a more detailed personalisation. The output of the model is the actual P-V loop and the entire time-varying elastance curve – both being important when considering management of individual patients. Additionally to study the sensitivity of the $E^N(t)$ given by the range of values in the experiments [16], we could have also explored the sensitivity of the single-beat estimation methods to the accuracy of time measurement. Clearly, during the steepest part of the $E^N(t)$ curve, a small error on the t_d measurement will have a significant impact on the E_d^N, and therefore on the estimation of E_{es}. Finally, the model-derived elastance would be as well a subject of analysis of sensitivity to the parameters of the model. In this paper, we considered only an example of the uncertainty in ventricular wall thickness, see Table 3. A thorough sensitivity analysis including other input parameters remains to be done in the future.

Our study suffers from several limitations. Indeed, while the Chen's method, allows a non-invasive estimation of E_{es}, our presented framework involves aortic pressure measurement. This preliminary setup will be improved by the methods of transferring the peripheral arterial pressure – practically always available during GA – into the central aortic pressure [6]. Also, the $E^N(t)$ curve was reproduced manually from the study of Suga et al. [16], involving experimental setup from dogs. In the study of Senzaki et al. [14], the authors presented an $E^N(t)$ curve from human data. They showed an absence of variability during the pre-ejection period supporting their final results regarding the reproducibility of their method. This lack of variability is however questioned by Shishido et al. [15]. In a preliminary *in silico* study (data not shown), we also observed a great variability in $E^N(t)$ during the pre-ejection period. For this reason, we did not use the human data made available by Senzaki et al. [14]. We could have compared our method with the method by Shishido et al. [15] aiming at considering the patient's variability of E_d^N estimation, by using a bilinear interpolation of the $E^N(t)$ curve. The comparison with our modelling framework will be explored in the future.

5   Conclusion

By using a patient-specific modelling framework, we proposed a method to estimate E_{es} and the entire time-varying elastance curve, considering individual normalised time-varying elastance variability, at the expense of minimally invasive data measurements. This method provides patient-specific time-varying curve and estimates the value of maximum elastance. Our proposed method could be used both clinically – to assess the patients' heart function – and in cardiac modelling community to provide patient-specific input for simplified models of the heart contraction.

Acknowledgment. We acknowledge Prof. Alexandre Mebazaa, and Prof. Etienne Gayat (Anaesthesiology and Intensive Care department, Lariboisière hospital, Paris, France) for their support in conducting the study. In addition, we would like to acknowledge Dr. Philippe Moireau, Inria research team MΞDISIM, for the development of the cardiac simulation software CardiacLab used in this work.

References

1. Burkhoff, D.: Assessment of systolic and diastolic ventricular properties via pressure-volume analysis: a guide for clinical, translational, and basic researchers. AJP: Heart Circulatory Physiol. **289**(2), H501–H512 (2005)
2. Caruel, M., Chabiniok, R., Moireau, P., Lecarpentier, Y., Chapelle, D.: Dimensional reductions of a cardiac model for effective validation and calibration. Biomech. Model. Mechanobiol. **13**(4), 897–914 (2014)
3. Chabiniok, R., Moireau, P., Kiesewetter, C., Hussain, T., Razavi, R., Chapelle, D.: Assessment of atrioventricular valve regurgitation using biomechanical cardiac modeling. In: Pop, M., Wright, G.A. (eds.) FIMH 2017. LNCS, vol. 10263, pp. 401–411. Springer, Cham (2017). https://doi.org/10.1007/978-3-319-59448-4_38
4. Chapelle, D., Le Tallec, P., Moireau, P., Sorine, M.: Energy-preserving muscle tissue model: formulation and compatible discretizations. Int. J. Multiscale Comput. Eng. **10**(2), 189–211 (2012)

5. Chen, C.H., et al.: Noninvasive single-beat determination of LV end-systolic elastance in humans. JACC **38**(7), 2028–2034 (2001)
6. Gaddum, N., Alastruey, J., Chowienczyk, P., Rutten, M.C., Segers, P., Schaeffter, T.: Relative contributions from the ventricle and arterial tree to arterial pressure and its amplification: an experimental study. Am. J. Physiol. Heart Circulatory Physiol. **313**(3), H558–H567 (2017)
7. Gayat, E., Mor-Avi, V., Weinert, L., Yodwut, C., Lang, R.M.: Noninvasive quantification of LV elastance and ventricular-arterial coupling using 3D echo and arterial tonometry. AJP: Heart Circulatory Physiol. **301**(5), H1916–H1923 (2011)
8. Holzapfel, G.A., Ogden, R.W.: Constitutive modelling of passive myocardium: a structurally based framework for material characterization. Phil. Trans. R. Soc. A: Math. Phys. Eng. Sci. **367**(1902), 3445–3475 (2009)
9. Huxley, A.F.: Muscular contraction. J. Physiol. **243**(1), 1–43 (1974)
10. Joachim, J., et al.: Velocity-pressure loops for continuous assessment of ventricular afterload: influence of pressure measurement site. J. Clin. Monit. Comput. **32**(5), 833–840 (2017)
11. Klotz, S., et al.: Single-beat estimation of end-diastolic pressure-volume relationship: a novel method with potential for noninvasive application. AJP: Heart Circulatory Physiol. **291**(1), H403–H412 (2006)
12. Ruijsink, B., Zugaj, K., Pushparajah, K., Chabiniok, R.: Model-based indices of early-stage cardiovascular failure and its therapeutic management in Fontan patients. In: Coudière, Y., et al. (eds.) FIMH 2019. LNCS, vol. 11504, pp. 379–387. Springer, Cham (2019)
13. Sagawa, K., Maughan, L., Suga, H., Sunagawa, K.: Cardiac Contraction and the Pressure Volume Relationship. Oxford University Press, New York (1988)
14. Senzaki, H., Chen, C.H., Kass, D.A.: Single-beat estimation of end-systolic pressure-volume relation in humans: a new method with the potential for noninvasive application. Circulation **94**(10), 2497–2506 (1996)
15. Shishido, T., Hayashi, K., Shigemi, K., Sato, T., Sugimachi, M., Sunagawa, K.: Single-beat estimation of end-systolic elastance using bilinearly approximated time-varying elastance curve. Circulation **102**(16), 1983–1989 (2000)
16. Suga, H., Sagawa, K., Shoukas, A.A.: Load independence of the instantaneous pressure-volume ratio of the canine left ventricle and effects of epinephrine and heart rate on the ratio. Circulation Res. **32**(3), 314–322 (1973)

Domain Adaptation via Dimensionality Reduction for the Comparison of Cardiac Simulation Models

Nicolas Duchateau[✉], Kenny Rumindo, and Patrick Clarysse

Creatis, CNRS UMR5220, INSERM U1206, Université Lyon 1, INSA Lyon,
Lyon, France
nicolas.duchateau@creatis.insa-lyon.fr

Abstract. We tackle the determination of a relevant data space to quantify differences between two databases coming from different sources. In the present paper, we propose to quantify differences between cardiac simulations from two different biomechanical models, assessed through myocardial deformation patterns. At stake is the evaluation of a given model with respect to another one, and the potential correction of bias necessary to merge two databases. We address this from a domain adaptation perspective. We first represent the data using non-linear dimensionality reduction on each database. Then, we formulate the mapping between databases using cases that are shared between the two databases: either as a linear change of basis derived from the learnt eigenvectors, or as a non-linear regression based on the low-dimensional coordinates from each database. We demonstrate these concepts by examining the principal variations in deformation patterns obtained from two cardiac biomechanical models personalized to 20 and 15 real healthy cases, respectively, from which 11 cases were simulated with both models.

1 Introduction

Simulations stand as a powerful support to understand complex physiological phenomena, and as a potential way to enrich real databases with large amounts of realistic data [1,2]. However, strong differences may exist between the simulations obtained from distinct models, and between simulated and real data. For cardiac biomechanical models, personalization often focuses on matching global parameters such as volumes or pressures [3], although some works went further by matching finer traits such as myocardial deformation patterns [4,5].

Nonetheless, matching the data from different but comparable sources can be performed a-posteriori, and falls under the umbrella of domain adaptation techniques [6,7]. Simple examples in medical imaging include the global adaptation of data distributions, illustrated on real cardiac meshes from different magnetic resonance protocols [8] or real vs. simulated fluoroscopy images to train a transducer localization algorithm [1]. The merging of databases can be

© Springer Nature Switzerland AG 2019
Y. Coudière et al. (Eds.): FIMH 2019, LNCS 11504, pp. 276–284, 2019.
https://doi.org/10.1007/978-3-030-21949-9_30

made more sample-specific after reducing the dimensionality of the data. Recent works investigated the estimation of joint spaces that merge heterogeneous data features [9,10]. We are more interested in adapting the data from one database to another one, as pursued globally in [1,8]. A mathematically sound framework for this was developed in [11] with diffusion maps, a spectral embedding technique relevant for clustered data samples.

Here, we demonstrate the interest of this strategy to quantify differences between cardiac simulations from two different biomechanical models, assessed through myocardial deformation patterns. Such simulations build upon different input segmentations, modeling strategies, and personalization. We specifically explore linear and non-linear ways to formulate correspondences between the two populations and therefore to perform domain adaptation, and benchmark them against simple data mapping using mesh registration.

2 Methods

2.1 Data and Pre-processing

We processed healthy left ventricular deformation patterns generated using two distinct cardiac biomechanical models personalized to two different populations, as summarized in Table 1. Population #2 consists of all the cases from the cMAC-STACOM 2011 challenge [12]. Population #1 consists of 11 of these cases, as the 4 remaining cMAC-STACOM cases could not be loaded to the segmentation software CVI42, and 9 other cases from our clinical collaborators at CHU Saint Etienne, France. The 11 cases simulated with both models were used to define the mapping between the two databases as described in Sect. 2.2. We therefore focused the analysis on myocardial deformation parameters that are common to the output of the two models (radial, circumferential and longitudinal strain).

Table 1. Summary of populations and models used, with 11 cases shared between the two populations and simulated with both models.

	Population	Model
#1	20 healthy cases processed as in [13]: - 11 cases from cMAC-STACOM 2011 [12], - 9 cases from our clinical collaborators. Left ventricular segmentations from CVI42 software[a]	- hexahedral elements - finite elements model - ABAQUS software[b]
#2	15 healthy cases processed as in [2], All from cMAC-STACOM 2011 [12]. Biventricular segmentations from a statistical atlas	- tetrahedral elements - finite elements model - SOFA framework[c]

[a] https://www.circlecvi.com/cardiac-mri/
[b] https://www.3ds.com/products-services/simulia/products/abaqus/
[c] https://www.sofa-framework.org/

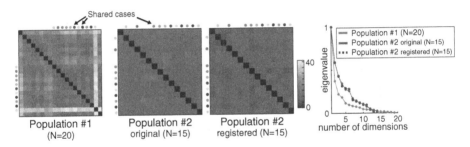

Fig. 1. *Left*: pairwise comparison of cases in each population, using the Euclidean distance between deformation patterns—treated as column vectors, after concatenating the radial, circumferential and longitudinal components. Colored dots point out the 11 cases shared between the two databases. *Right*: eigenvalues normalized by the first eigenvalue. (Color figure online)

These data are originally defined on subject-specific meshes, and were matched to a given template for each database before the population analysis (here, an arbitrarily chosen mesh in each population). For population #1, correspondences between meshes were obtained by mesh parameterization along the radial, circumferential, and longitudinal directions taking advantage of the hexahedral elements. For population #2, such parameterization is more challenging due to the tetrahedral elements and the biventricular segmentations. Correspondences between meshes were therefore obtained by registering left ventricular surface meshes using the currents representation and the large deformation diffeomorphic metric mapping framework, with the open-source software Deformetrica [14], and finally propagating the transformation to the left ventricular volumetric meshes. For both populations, data from each individual were finally interpolated at the cells and vertices of the template mesh using kernel ridge regression and the estimated mesh correspondences as in [13].

A similar pipeline was used to match the template meshes from the two populations, and therefore transport the data from one population to the other without domain adaptation, for benchmarking purposes.

2.2   Dimensionality Reduction and Domain Adaptation

Let's denote $\mathbf{y}_{m,k} \in \mathcal{Y}_m$ the high-dimensional input data associated to subject $k \in [1, K_m]$ for the dataset $m \in \{1, 2\}$ (in our case, the deformation patterns defined over the left ventricular meshes for each individual—treated as column vectors, after concatenating the radial, circumferential and longitudinal components). For each dataset, standard dimensionality reduction is applied to the K_m samples, and provides the low-dimensional coordinates $\mathbf{x}_{m,k} \in \mathcal{X}_m$. Here, we tested our methods with the Isomap algorithm [15], a non-linear dimensionality reduction algorithm that builds a neighborhood graph between the high-dimensional input samples, and looks for a low-dimensional space in which the Euclidean distance approximates the geodesic distance in the high-dimensional space.

The small population size may limit the generalization ability of the representation, but may be less critical compared to recently popular techniques such as auto-encoders, which may require larger populations. Also, domain adaptation via dimensionality reduction was thoroughly investigated for the diffusion maps algorithm [11], but this algorithm is more relevant for clustered data samples. In any case, our approach is generic and can be applied to any manifold learning algorithm.

Domain adaptation from $\mathcal{X}_1$ to $\mathcal{X}_2$ (similar formulations for $\mathcal{X}_2$ to $\mathcal{X}_1$) is performed as a linear change of basis derived from the eigenvectors learnt for each database, or as a non-linear regression based on the low-dimensional coordinates from each database. For both options, the mapping was defined using the 11 cases that are present in both populations, denoted as the "shared cases" in Figs. 1 and 2. Without lack of generalizability, we can assume that these cases correspond to the first cases in each database, and denote Ω the subset of indices associated to these cases.

The *linear mapping* from $\mathcal{X}_1$ to $\mathcal{X}_2$ can be formulated through the following change of basis [11]:

$$\forall\, \mathbf{x}_1 \in \mathcal{X}_1, \quad f_{1\to 2}(\mathbf{x}_1) = (\mathbf{\Psi}_1)^t \mathbf{\Psi}_2 \mathbf{x}_1 \quad \in \mathcal{X}_2, \tag{1}$$

where $\mathbf{\Psi}_1$ and $\mathbf{\Psi}_2$ are the matrices of eigenvectors of $\mathcal{X}_1$ and $\mathcal{X}_2$ associated to the shared cases of indices Ω, and $.^t$ is the transposition operator.

The *non-linear mapping* between these two spaces can be estimated through non-linear regression. We chose kernel ridge regression, which is formulated as:

$$\forall\, \mathbf{x}_1 \in \mathcal{X}_1, \quad g_{1\to 2}(\mathbf{x}_1) = \sum_{i\in\Omega} k(\mathbf{x}_1, \mathbf{x}_{1,i})\mathbf{c}_i \quad \in \mathcal{X}_2, \tag{2}$$

and in practice computed using the matrix formulation $\mathbf{C} = (\mathbf{K} + \mathbf{I}/\gamma)^{-1} \mathbf{X}_2$, where $\mathbf{C} = (\mathbf{c}_i)_{i\in\Omega}$, $\mathbf{K} = (k(\mathbf{x}_{1,i}, \mathbf{x}_{1,j}))_{(i,j)\in\Omega^2} = \exp(-\|\mathbf{x}_{1,i} - \mathbf{x}_{1,j}\|^2/\sigma^2)$ and σ controls the scale of the regression, $\mathbf{X}_2 = (\mathbf{x}_{2,i})_{i\in\Omega}$, $\mathbf{I}$ is the identity matrix, and γ balances the contributions of the data fidelity and regularization terms.

We used a multi-scale version of this algorithm[1] [16], which is suitable for non-uniformly distributed samples, as in our case and in most real-life applications. This multi-scale algorithm starts with a scale σ equal to the largest distance between samples, and reduces it by a factor 2 at each iteration s to estimate the residual of the regression, until $\sigma/2^s$ is smaller than a fraction α of the average distance between each sample and its nearest neighbors (the samples density).

---

[1] Source code and demo available at https://nicolasduchateau.wordpress.com/downloads/.

3 Experiments and Results

3.1 Original Pattern Comparisons

The distances between the deformation patterns in each population are displayed in the left part of Fig. 1. They already point out differences between the two populations, which are lower for population #2. Such changes come from differences in the mesh segmentation process used before the simulations (3D interpolation and remeshing from segmented 2D slices for population #1, against statistical atlas for population #2), different biomechanical models and therefore different modeling errors, and different personalization strategies—and potential errors in the personalization.

3.2 Parameters Setting

We used 5 neighbors to construct the neighborhood graph of the Isomap dimensionality reduction algorithm. This value was chosen to minimize the relative importance of the second eigenvalue and the sum of eigenvalues (the compactness of the representation, illustrated in the right part of Fig. 1), within a reasonable amount of possible neighbors (up to half the database).

The maximum amount of dimensions for the low-dimensional space was set to 14 (one unit less than the size of the smallest studied population—population #2). Although the intrinsic dimensionality of the data may be lower (right part of Fig. 1), some artifacts were observed in our implementation of the linear mapping when computing the scalar product $(\mathbf{\Psi}_1)^t \mathbf{\Psi}_2$ with cut dimensions, which deserves further investigation.

Leave-one-out cross correlation was used to determine the regression parameters (the weight γ and the bandwidth at which to stop the multi-scale scheme, determined as a fraction α of the samples density). The retained values were $\gamma = 10$ and $\alpha = 0.5$, respectively.

3.3 Transported Coordinates

As observed in Fig. 2, the cases that are shared between the two populations and that are used to compute the mapping were exactly matched (linear mapping) or almost exactly matched (non-linear mapping) in the target population space. The main modes of variation substantially changed after the mapping (a given mode in the original population is expressed as a combination of several modes of the target population after mapping).

We also observed the limits of the linear mapping, through substantial shrinkage of the variations around zero, due to the non-guaranteed correlation between the coordinates of the shared cases in the two populations, and therefore the potentially opposing contribution of some cases to the mapping. This shrinkage effect is still present for the non-linear mapping, but much less pronounced. Besides, the curvature of the modes of variation after the non-linear mapping supports the relevance of this mapping strategy against the linear one.

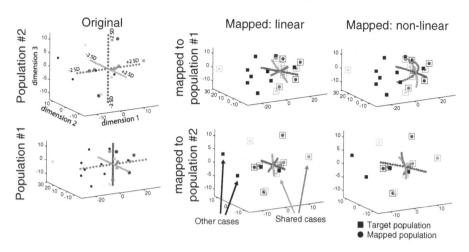

Fig. 2. Low-dimensional coordinates before and after domain adaptation. Dots correspond to the samples that are mapped to the target population, whose samples are represented by squares. Colored dots and squares stand for the 11 cases shared between the two populations: after mapping the colored dots lay near the center of the colored squares. The black dots and squares correspond to the remaining cases that also served to obtain the low-dimensional embedding, but were not used in the domain adaptation process. The red-green-blue curves represent the main directions of variation in the original population (from -2 SD to $+2$ SD), before and after mapping. *Top*: population #2 mapped to population #1, using the linear (*middle*) or non-linear *right* mapping. *Bottom*: similar plots for population #1 mapped to population #2. (Color figure online)

3.4   Domain-Adapted Patterns

Figure 3 illustrates the deformation patterns encoded in the first mode of variations, before and after the domain adaptation. Patterns were reconstructed from low-dimensional coordinates evolving from -2 SD to $+2$ SD along the first dimension. Reconstruction was achieved through the multi-scale regression presented above, with $\gamma = 1$ and $\alpha = 0.5$.

As visible in Fig. 3, the deformation patterns from populations #1 and #2 before domain adaptation substantially differ in terms of amplitude of the deformation, and spatial distribution of the patterns. This is even more visible when the patterns in the original population are transported to the target population by registration (as described in Sect. 2.1). After domain adaptation, the mapped patterns are much closer from the target population both regarding amplitude and spatial distribution of the patterns. Subtle differences remain, which should correspond to the intrinsic differences between these two populations.

Figure 4 complements this information by quantifying differences between the original deformation patterns and the deformation patterns after domain adaptation, across all the modes of variation. With the linear mapping, patterns near the average are unchanged, due to the definition of this mapping (linear

282   N. Duchateau et al.

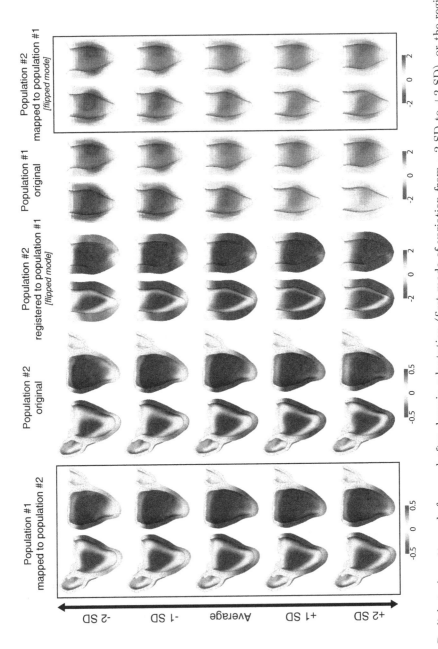

Fig. 3. Radial strain patterns before and after domain adaptation (first mode of variation from −2 SD to +2 SD), or the registration process described in Sect. 2.1.

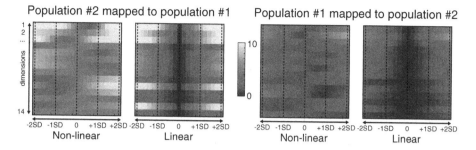

Fig. 4. Differences between the original deformation patterns and the deformation patterns reconstructed from the low-dimensional coordinates after domain adaptation, across all the modes of variation (Euclidean distance on the radial-circumferential-longitudinal deformation patterns treated as column vectors—as in Fig. 1).

change of basis, centered around the average). Differences are not guaranteed to be concentrated on the first dimensions, as visible for dimensions 10 and 13. With the non-linear mapping, patterns near the average are also transported and therefore exhibit differences. Also, the modes of variation become curved (Fig. 2) after domain adaptation, and differences between patterns are therefore not guaranteed to be symmetric on each side of the average.

4 Conclusion

We have demonstrated the potential of domain adaptation to compare databases from distinct sources, upon the constraint that some cases are shared between the databases to define the mappings. In our tested populations, non-linear mapping stands as a relevant solution, given that the distribution of the shared cases is not guaranteed to be the same across the two populations.

In terms of application, the approach is promising to better understand differences between populations from different sources (here, from different biomechanical models), and may pave the ground for enriching a real database with simulated data or the comparison of different clinical cohorts. These perspectives will require further exploring up to which limit two different databases are actually compatible, and scaling the proposed methods to larger populations—including a substantial amount of cases shared between the two populations.

Acknowledgements. GK Rumindo was supported by the European Commission H2020 Marie Sklodowska-Curie Training Network (VPH-CaSE-642612). The authors also acknowledge the partial support from the French ANR (LABEX PRIMES of Université de Lyon [ANR-11-LABX-0063], within the program "Investissements d'Avenir" [ANR-11-IDEX-0007]). Finally, the authors thank M Sermesant (INRIA Sophia-Antipolis, France) and J Ohayon (TIMC-IMAG Grenoble, France), who contributed to the development of the cardiac simulations, as well as P Croisille and M Viallon (CREATIS, CHU Saint Etienne, France), who provided the 9 cases used in population #1.

References

1. Heimann, T., Mountney, P., John, M., et al.: Real-time ultrasound transducer localization in fluoroscopy images by transfer learning from synthetic training data. Med. Image Anal. **18**, 1320–1328 (2014)
2. Duchateau, N., Sermesant, M., Delingette, H., et al.: Model-based generation of large databases of cardiac images: synthesis of pathological cine MR sequences from real healthy cases. IEEE Trans. Med. Imaging **37**, 755–766 (2018)
3. Molléro, R., Pennec, X., Delingette, H., et al.: Multifidelity-CMA: a multifidelity approach for efficient personalisation of 3D cardiac electromechanical models. Biomech. Model. Mechanobiol. **17**, 285–300 (2018)
4. Wang, V.Y., Lam, H.I., Ennis, D.B., et al.: Modelling passive diastolic mechanics with quantitative MRI of cardiac structure and function. Med. Image Anal. **13**, 773–84 (2009)
5. Chabiniok, R., Moireau, P., Lesault, P.F., et al.: Estimation of tissue contractility from cardiac cine-MRI using a biomechanical heart model. Biomech. Model. Mechanobiol. **11**, 609–30 (2012)
6. Csurka, G. (ed.): Domain Adaptation in Computer Vision Applications. ACVPR. Springer, Cham (2017). https://doi.org/10.1007/978-3-319-58347-1
7. Wang, M., Deng, W.: Deep visual domain adaptation: a survey. Neurocomputing **312**, 135–53 (2018)
8. Medrano-Gracia, P., Cowan, B.R., Bluemke, D.A., et al.: Atlas-based analysis of cardiac shape and function: correction of regional shape bias due to imaging protocol for population studies. J. Cardiovasc. Magn. Reson. **15**, 80 (2013)
9. Sanchez-Martinez, S., Duchateau, N., Erdei, T., et al.: Characterization of myocardial motion patterns by unsupervised multiple kernel learning. Med. Image Anal. **35**, 70–82 (2017)
10. Puyol-Antón, E., Sinclair, M., Gerber, B., et al.: A multimodal spatiotemporal cardiac motion atlas from MR and ultrasound data. Med. Image Anal. **40**, 96–110 (2017)
11. Coifman, R.R., Hirn, M.J.: Diffusion maps for changing data. Appl. Comp. Harm. Anal. **36**, 79–107 (2014)
12. Tobon-Gomez, C., De Craene, M., McLeod, K., et al.: Benchmarking framework for myocardial tracking and deformation algorithms: an open access database. Med. Image Anal. **17**, 632–648 (2013)
13. Rumindo, G.K., Duchateau, N., Croisille, P., Ohayon, J., Clarysse, P.: Strain-based parameters for infarct localization: evaluation via a learning algorithm on a synthetic database of pathological hearts. In: Pop, M., Wright, G.A. (eds.) FIMH 2017. LNCS, vol. 10263, pp. 106–114. Springer, Cham (2017). https://doi.org/10.1007/978-3-319-59448-4_11
14. Bône, A., Louis, M., Martin, B., Durrleman, S.: Deformetrica 4: an open-source software for statistical shape analysis. In: Reuter, M., Wachinger, C., Lombaert, H., Paniagua, B., Lüthi, M., Egger, B. (eds.) ShapeMI 2018. LNCS, vol. 11167, pp. 3–13. Springer, Cham (2018). https://doi.org/10.1007/978-3-030-04747-4_1
15. Tenenbaum, J., De Silva, V., Langford, J.: A global geometric framework for non-linear dimensionality reduction. Science **290**, 2319–23 (2000)
16. Bermanis, A., Averbuch, A., Coifman, R.R.: Multiscale data sampling and function extension. Appl. Comp. Harm. Anal. **34**, 15–29 (2013)

Large Scale Cardiovascular Model Personalisation for Mechanistic Analysis of Heart and Brain Interactions

Jaume Banus[1(✉)], Marco Lorenzi[1], Oscar Camara[2], and Maxime Sermesant[1]

[1] Inria, Epione team, Université Côte d'Azur, Sophia Antipolis, France
jaume.banus-cobo@inria.fr
[2] PhySense, Department of Information and Communication Technologies,
Universitat Pompeu Fabra, Barcelona, Spain

Abstract. Cerebrovascular diseases have been associated with a variety of heart diseases like heart failure or atrial fibrillation, however the mechanistic relationship between these pathologies is largely unknown. Until now, the study of the underlying heart-brain link has been challenging due to the lack of databases containing data from both organs. Current large data collection initiatives such as the UK Biobank provide us with joint cardiac and brain imaging information for thousands of individuals, and represent a unique opportunity to gain insights about the heart and brain pathophysiology from a systems medicine point of view. Research has focused on standard statistical studies finding correlations in a phenomenological way. We propose a mechanistic analysis of the heart and brain interactions through the personalisation of the parameters of a lumped cardiovascular model under constraints provided by brain-volumetric parameters extracted from imaging, i.e: ventricles or white matter hyperintensities volumes, and clinical information such as age or body surface area. We applied this framework in a cohort of more than 3000 subjects and in a pathological subgroup of 53 subjects diagnosed with atrial fibrillation. Our results show that the use of brain feature constraints helps in improving the parameter estimation in order to identify significant differences associated to specific clinical conditions.

Keywords: 0D model · Cardiovascular modelling · Personalisation · White matter damage · Brain damage · Atrial fibrillation

1 Introduction

Cerebrovascular diseases are related to a variety of heart diseases such as heart failure [1] or atrial fibrillation (AF) [2], sharing several risk factors such as cholesterol, diabetes or high blood pressure. In parallel, it has been shown that stroke doubles the risk of dementia [3]. All these connections suggest a common underlying pathological process that links cardiac function with brain atrophy. Large scale analysis on databases combining cardiovascular and brain data from the

© Springer Nature Switzerland AG 2019
Y. Coudière et al. (Eds.): FIMH 2019, LNCS 11504, pp. 285–293, 2019.
https://doi.org/10.1007/978-3-030-21949-9_31

same individuals are thus required to demonstrate and better understand the interaction between brain and heart. To this end, studies such as the UK Biobank aim at the acquisition of multi-modal databases containing both heart and brain imaging information [4]. Thanks to these databases ongoing studies have focused on the study of the relationships between cardiovascular risk factors and image-derived features, such as subcortical volumes [5]. However, a number of fundamental descriptors of the cardiac function are not possible to obtain in-vivo, i.e: heart contractility or fibers stiffness. Personalised modelling approaches allow us to estimate these descriptors and gain insight of the cardiac function, allowing us to obtain more reliable results and relate them to brain damage information.

Personalizing a cardiovascular model for a given subject is an ill-posed problem that implies estimating the model parameters so that the simulation behaves as close as possible to the available clinical data. In this work we will focus on a 0D model of the whole cardiovascular system. Previous studies have used multi-scale models to describe the whole-body circulation and study the venous blood flow in the brain [6]. However, their modeling of the heart chambers does not take into account the contractile and elastic properties of the heart. While other models of the whole-body circulation are available [7,8], to our knowledge, no explicit modelling study relating neurodegeneration and cardiovascular parameters has been done.

In this paper we aim to study the relationship between cardiovascular indicators and brain volumetric features extracted from the imaging data available in UK Biobank, through the personalisation of a cardiovascular lumped model using the approach presented in [9]. The use of this approach allows us to tackle the ill-posedness nature of the personalisation and identify plausible and coherent solutions across the population. To achieve that, we define a regularisation term that can be extended to take into account features not present in the lumped model, allowing to explore the effect of including brain features as additional constraint. We apply this framework to a large cohort composed by more than 3 000 subjects for which cardiac and brain information was jointly available in the UK Biobank. To illustrate how to exploit the framework to identify meaningful clinical relationships, we applied it in a subset of subjects diagnosed with AF, which is considered as an independent risk factor for stroke and dementia [2,10]. We identified statistically significant associations between the personalised model parameters and brain volumetric features that match findings reported in previous clinical studies.

The paper is structured as follows: in Sect. 2.1 we detail the data preprocessing and inclusion criterion for the whole-population analysis. Following, in Sect. 2.2 we present the lumped model and how to take into account the subject's information to constrain the solution space in the personalisation. Next, in Sect. 2.3 we assess the impact of our model in determining significant relationships between the estimated cardiac parameters and brain damage using the AF subset. Finally, in Sect. 3 we present the obtained results.

2   Methods

2.1   Data Pre-processing and Inclusion Criterion

Our analysis includes data from UK Biobank participants for which all brain image modalities and all cardiac-image derived indicators were available, for a total of 3783 subjects. In the available cardiac images it was possible to quantify the cardiac function using indicators such as stroke volume (SV), cardiac output (CO) or ejection fraction (EF). Multi-modal brain MRI images allowed the extraction of image-derived features such as brain tissue volumes and white matter hyperintensities (WMHs), one of the most common indicators used to assess neurological damage.

Using FLAIR MR images, WMHs were segmented by the lesion prediction algorithm (LPA), available in the lesion segmentation toolbox (LST) [11] for SPM[1]. FLAIR MR images were pre-processed following the protocol described in [12], in which gradient distortion correction and defacing were performed. After discard subjects for which pre-processing (449) or segmentation of WMHs (250) failed, the final number of available subjects was 3 084. From the segmentations we extracted the total volume of WMHs and the number of lesions. All brain-related volumes were normalized by head size.

2.2   Cardiovascular Lumped Model

The cardiovascular personalisation of the subjects was performed by using the 0D model shown in Fig. 1 which is a simplification of a 3D cardiac electromechanical model [13] derived in [14]. In the 0D version, which assumes spherical symmetry, the myocardial forces and motion can be described by the inner radius (R_0) of the ventricle. Deformation and stress tensors are also reduced to 0D forms, which allow us to characterise the heart contractile (σ_0) and elastic (C_1) properties of the heart.

The model M consists in a set of ordinary differential equations with P_M parameters, e.g. maximum contraction of the heart fibers or its stiffness. The state variables of the model are denoted by O_M, e.g. arterial or venous pressures, and they describe the state of the system. During the personalisation we are interested in a subset of n state variables, such that $O = (O_1, O_2, ..., O_n)$, and we vary a subset θ of the P_M model parameters. We consider $O(\theta)$ the set of state variables generated by the model for a given set of θ. The goal is to find $\theta*$ such that $O(\theta*)$ best approximates the target features $\hat{O}$.

Due to the high dimensionality and non-convexity of this inverse problem, we solve it with the CMA-ES optimization algorithm based on evolution strategies [15]. CMA-ES minimizes a given error function by combining maximum likelihood principles with natural gradient descent on the ranks of the point scores (i.e: the score of each individual at every generation). The error function $S(\theta, \hat{O})$ is defined as the L_2 distance between $O(\theta)$ and $\hat{O}$. Since each target feature has different

---

[1] https://www.fil.ion.ucl.ac.uk/spm.

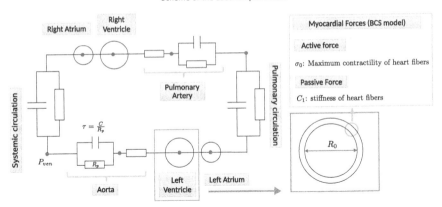

Fig. 1. Simplified schematic representation of the lumped model showing the parameters used in the personalisation. The 0D representation of the myocardial forces has been omitted for the sake of clarification. τ characterizes the contractility of the aorta, R_p the peripheral resistance and P_{ven} the venous pressure.

range of values we defined a tolerance interval, Tol, for each feature i to be able to compare the different outputs. This can be formalized as shown in Eq. 1:

$$S(\theta, \hat{O}) = \sum_{i=1}^{n} \frac{(O_i(\theta) - \hat{O}_i)^2}{Tol_i} \tag{1}$$

Based on the available clinical data, we selected the following target features for the personalisation; stroke volume (SV), ejection fraction (EF), diastolic blood pressure (DBP), mean blood pressure (MBP) and end-diastolic volume (EDV). Considering the uncertainty of the measured data, the tolerance interval for each feature was set to 10 ml for the SV and the EDV, 200 Pa for the DBP and the MP, and 5% for the EF. Finally, the personalized parameters of the cardiovascular model were maximum contraction of the heart fibers σ_0, stiffness of the heart fibers C_1, peripheral resistance R_p, venous pressure P_{ven}, and the characteristic time τ of the aorta, which defines the time that takes for blood pressure to decrease from systolic to the systemic, or "asymptotic" value. We selected these parameters based on a sensitivity analysis in which we assessed the influence of each parameter over the selected target features.

Since the solution of Eq. 1 is non-unique, there is an observability difficulty in this personalisation problem. To tackle this issue, we used the iterative-update prior (IUP) approach presented in [9] to introduce constraints in the fitting process. In the IUP method a regularization term, $R(\theta, \mu, \Sigma)$, is used to reduce the variability in the estimation of the parameters. The regularization constrains the directions in which we explore the parameter-space by using the relationships among the model parameters. Formally, the regularization term is parameterized by an expected value μ and by a covariance matrix Σ encoding the relationships across parameters.

$$R(\theta, \mu, \Sigma) = (\theta - \mu)^T \Sigma^{-1} (\theta - \mu). \tag{2}$$

Therefore, the fitting score becomes:

$$S(\theta, \hat{O}, \mu, \Sigma) = S(\theta, \hat{O}) + \gamma R(\theta, \mu, \Sigma), \tag{3}$$

where γ defines the relative importance of the regularization term. This term is updated at each IUP iteration, using the obtained mean value of the fitted parameters and the estimated covariance in the previous iteration.

Accounting for Brain Information in the 0D Model. $R(\theta, \mu, \Sigma)$ can be extended to incorporate relationships with features not present in the cardiovascular model. In our setting, we included in the regularization term the extended feature space corresponding to the concatenation of the model parameters, θ, with the brain and clinical information, here denoted by ϕ. We used the total brain volume, the ventricles volume, the obtained WMHs features, age, sex and body surface area (BSA). Therefore, the problem in Eq. 3 becomes:

$$S(\theta, \hat{O}, \mu, \Sigma, \phi) = S(\theta, \hat{O}) + \gamma R(\theta, \phi, \mu, \Sigma). \tag{4}$$

Equation 4 now accounts for a covariance term constraining the parameters according to the extended set of information. We have used 10 IUP iterations and assessed the results at different γ levels (0.1, 0.5, 2 and 10). The optimisation is performed over the logarithm of the parameter values.

2.3 Atrial Fibrillation Analysis

Considering the dataset obtained after the pre-processing described in Sect. 2.1 we had access to 53 subjects diagnosed with AF. Using bootstrapping we sampled 100 control groups of the same sample size of the AF group and without any significant difference in age, sex and BSA. The sampled controls came from the subset of subjects without any diagnosed cardiovascular disease ($n = 2022$). We applied the framework described in the previous section to each bootstrap subset composed by the AF group and sampled control group, to obtain the bootstrap distributions of the correlations between cardiac and external parameters. This approach allowed us to exploit the dataset variability for assessing the difference between cardiac and brain associations.

3 Results

Whole-Population Analysis. As expected, we observed that as the value of γ is increased (i.e. more regularisation), the fitting error increases and at the same time the number of outliers is reduced and the estimated distributions have lower variability, as can be seen in Fig. 2a. We can observe that strong

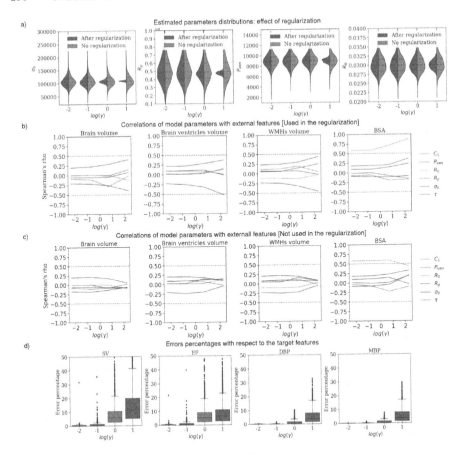

Fig. 2. (a) Estimated density distributions of the fitted parameters at different regularization levels. Initial and final distributions after 10 iterations in brown and blue respectively. The variability among the initial distributions is due to the variability in the sampling that CMA-ES performs during the optimization. (b) Evolution of Spearman's rank correlation coefficient between the model parameters and the external parameters as the regularization level increases when external features are considered in the regularization and (c) when external features are not considered. Model parameters being: maximum heart fibers contractility σ_0, heart fibers stiffness, C_1, left ventricle size R_0, peripheral resistance R_p, aorta characteristic time τ, and venous pressure P_{ven}. (d) Error percentages with respect to the target features. Stroke volume (SV), ejection fraction (EF), diastolic blood pressure (DBP) and mean blood pressure (MBP). While end-diastolic volume (EDV) is not shown due to space issues, its error pattern was similar to the one observed in the SV (Color figure online)

regularisation even shrinks some parameters close to a constant value, implying that those parameters cannot be observed from the available data. Looking at the correlation of the model parameters with the external features we note the strong correlation between the left ventricle size, R_0, and the BSA, even for low

γ values. Moreover, there is a positive correlation between peripheral resistance, the WMHs volume, and brain ventricles volume, which are at the same time negatively correlated to the aorta characteristic time, τ. The number of WMHs lesions and age followed the same correlations pattern, but due to space issues they have been omitted. On the other hand, brain volume is positively correlated with τ and peripheral resistance. An increase in peripheral resistance can be associated to higher DBP, while a decrease in contractility, τ, can be interpreted as an increase in arterial stiffness leading to high SBP. Both, DBP and SBP, have been previously associated to WMHs [16]. In Fig. 2b we note that the significant correlations present when no regularization is applied become stronger as regularization increases, while the non-significant correlations stay close to zero. This behavior is expected since regularization is constraining the space of feasible solutions. Therefore, as we increase γ we further limit the feasible parameter-space towards the set of solutions that satisfy the existing relationships between the parameters. In Fig. 2c we observe the obtained correlations when the external features are not taken into account. In this case, the solutions are constrained into a different parameter-space in which the relationships between the model parameters and the external features are lost. Limiting the interpretability when assessing the parameters estimations with respect external factors not present in the mechanistic model.

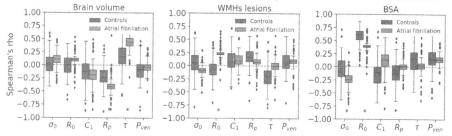

Bootstrap distribution of correlation between cardiac parameters and brain information

Fig. 3. Comparison of the Spearman's rank correlation coefficient bootstrap distributions obtained at $\gamma = 0.5$ between the personalised model parameters and the external features. Blue boxplots correspond to control groups and brown to AF subjects. * denotes that the correlations are significantly different according to the Wilcoxon ranksum test, and that in the AF group correlations are significantly greater or smaller than 0 (5% significance level). (Color figure online)

Atrial Fibrillation. In Fig. 3 we observe a statistical description of the empirical distribution of correlations obtained from the bootstrap analysis done in the AF subset. The results are obtained with trade-off $\gamma = 0.5$, which in the whole-population analysis provided a good a balance between data-fit and regularization. We assessed the difference between the controls and AF groups correlations distributions using the Wilcoxon rank-sum test with a significance level of $\alpha = 0.05$. Moreover, to consider the obtained results as significant we assessed if in the AF group the obtained correlations were statistically greater or smaller than 0 with a 5% significance level.

In the brain volume we observe the same correlations found in the whole-population analysis, but it can be seen that for the AF group these correlations are stronger, which suggests that in the AF subjects brain is more susceptible to cardiovascular factors. For the BSA we found the positive correlation with the left ventricle size observed in Sect. 3 and a negative correlation with the maximum heart contractility σ_0. In the number of WMHs lesions we can observe a positive correlation with the left ventricle size. The associations of BSA with σ_0 and WMHs with R_0 could be related to cardiac dilation due to an increased impairment of the functioning heart in AF. Interestingly, in the AF control subjects the correlation between WMHs and left ventricle size is negative. These findings suggest an association between AF and WMHs. Moreover, they agree with previous studies reporting an association between left ventricle remodelling and AF [17]. No significant associations were found for brain ventricles volume, WMHs volume and age.

4 Conclusions

We have modeled 3 084 subjects with a 0D cardiovascular model and we constrained the available parameter-space during personalisation by incorporating external features in the regularization term, allowing us to study their influence in the estimated model parameters. The use of this approach gives access to a generative model that allows to analyze the relationships between external features and non-observable parameters such as the characteristic time of aorta, τ, which we found to be related with brain-volumetric features. Using the same framework we assessed a clinical subgroup in which we have found meaningful clinical relationships, linking AF with WMHs and heart remodelling. Our model does not currently simulate the cerebral blood flow, while previous studies [6] suggest that WMHs are due to more localize vascular impairments. This highlights the need to obtain a local flow characterization to estimate more relevant parameters. Moreover, the presented approach can be seen as a parameter selection approach. It allows to identify which parameters cannot be estimated from the available data and find a parameter subspace of solutions in which the non-observable parameters get close to constant values. The identification of the non-observable parameters coupled with human modelling expertise can help in the selection of a reduced subset of observable cardiovascular parameters for personalisation. Future work will go towards the local blood flow characterization in the brain, as well as towards the assessment of its spatial patterns, and the modelling of more brain atrophy indicators.

Acknowledgements. This work was supported by the Inria Sophia Antipolis - Méditerranée, "NEF" computation cluster. This research has been conducted using the UK Biobank Resource under Application Number 20576 (PI Nicholas Ayache). Additional information can be found at: https://www.ukbiobank.ac.uk.

References

1. Ois, A., et al.: Heart failure in acute ischemic stroke. J. Neurol. **255**(3), 385–389 (2008)
2. Benjamin, E.J., et al.: Heart disease and stroke statistics-2018 update: a report from the American heart association. Circulation **137**(12), e67 (2018)
3. Azarpazhooh, M.R., et al.: Concomitant vascular and neurodegenerative pathologies double the risk of dementia. Alzheimer's Dement. **14**(2), 148–156 (2018)
4. Sudlow, C., et al.: UK Biobank an open access resource for identifying the causes of a wide range of complex diseases of middle and old age. PLoS Med. **12**, 1–10 (2015)
5. Cox, S.R., et al.: Associations between vascular risk factors and brain MRI indices in UK Biobank. bioRxiv (2019)
6. Müller, L.O., Toro, E.F.: Enhanced global mathematical model for studying cerebral venous blood flow. J. Biomech. **47**(13), 3361–3372 (2014)
7. Safaei, S., et al.: Bond graph model of cerebral circulation: toward clinically feasible systemic blood flow simulations. Front. Physiol. **9**, 1–15 (2018)
8. Blanco, P., et al.: An anatomically detailed arterial network model for one-dimensional computational hemodynamics. IEEE Trans. Biomed. Eng. **62**(2), 736–753 (2015)
9. Molléro, R., Pennec, X., Delingette, H., Ayache, N., Sermesant, M.: Population-based priors in cardiac model personalisation for consistent parameter estimation in heterogeneous databases. Int. J. Numer. Methods Biomed. Eng. **35**, e3158 (2018)
10. Alonso, A., de Larriva, A.P.A.: Atrial fibrillation, cognitive decline and dementia. Eur. Cardiol. Rev. **11**(1), 49 (2016)
11. Schmidt, P.: Bayesian inference for structured additive regression models for large-scale problems with applications to medical imaging. Ph.D. thesis, Ludwig-Maximilians-Universität München (2017)
12. Alfaro-Almagro, F., et al.: Image processing and quality control for the first 10,000 brain imaging datasets from UK Biobank. NeuroImage **166**, 400–424 (2018)
13. Chapelle, D., Le Tallec, P., Moireau, P., Sorine, M.: Energy-preserving muscle tissue model: formulation and compatible discretizations. Int. J. Multiscale Comput. Eng. **10**(2), 189–211 (2012)
14. Caruel, M., Chabiniok, R., Moireau, P., Lecarpentier, Y., Chapelle, D.: Dimensional reductions of a cardiac model for effective validation and calibration. Biomech. Model. Mechanobiol. **13**, 897–914 (2014)
15. Hansen, N.: The CMA evolution strategy: a comparing review. In: Lozano, J.A., Larrañaga, P., Inza, I., Bengoetxea, E. (eds.) Towards a New Evolutionary Computation. Studies in Fuzziness and Soft Computing, vol. 102, pp. 75–102. Springer, Berlin (2016). https://doi.org/10.1007/3-540-32494-1_4
16. Modir, R., Gardener, H., Wright, C.B.: Stroke blood pressure and white matter hyperintensity volume—a review of the relationship and implications for stroke prediction and prevention. US Neurol. **8**(1), 33–36 (2012)
17. Seko, Y., et al.: Association between atrial fibrillation, atrial enlargement, and left ventricular geometric remodeling. Sci. Rep. **8**(1), 1–8 (2018)

Model of Left Ventricular Contraction: Validation Criteria and Boundary Conditions

Aditya V. S. Ponnaluri[1], Ilya A. Verzhbinsky[2] (ID), Jeff D. Eldredge[1] (ID),
Alan Garfinkel[3], Daniel B. Ennis[2] (ID), and Luigi E. Perotti[4(✉)] (ID)

[1] Department of Mechanical and Aerospace Engineering,
University of California Los Angeles, Los Angeles, CA 90095, USA
[2] Department of Radiology, Stanford University, Stanford, CA 94305, USA
[3] Departments of Medicine (Cardiology) and Integrative Biology and Physiology,
University of California Los Angeles, Los Angeles, CA 90095, USA
[4] Department of Mechanical and Aerospace Engineering,
University of Central Florida, Orlando, FL 32816, USA
Luigi.Perotti@ucf.edu

Abstract. Computational models of cardiac contraction can provide critical insight into cardiac function and dysfunction. A necessary step before employing these computational models is their validation. Here we propose a series of validation criteria based on left ventricular (LV) global (ejection fraction and twist) and local (strains in a cylindrical coordinate system, aggregate cardiomyocyte shortening, and low myocardial compressibility) MRI measures to characterize LV motion and deformation during contraction. These validation criteria are used to evaluate an LV finite element model built from subject-specific anatomy and aggregate cardiomyocyte orientations reconstructed from diffusion tensor MRI. We emphasize the key role of the simulation boundary conditions in approaching the physiologically correct motion and strains during contraction. We conclude by comparing the global and local validation criteria measures obtained using two different boundary conditions: the first constraining the LV base and the second taking into account the presence of the pericardium, which leads to greatly improved motion and deformation.

Keywords: Cardiac contraction · Validation criteria ·
Boundary conditions · MRI

1 Introduction

Heart Failure (HF) remains a widespread health problem worldwide. In order to improve our understanding of cardiac contraction in health and disease,

The research reported in this publication was supported by NIH/NHLBI K25-HL135408 and R01-HL131823 grants, and UCLA URSP. The content is solely the responsibility of the authors and does not necessarily represent the official views of the National Institutes of Health.

© Springer Nature Switzerland AG 2019
Y. Coudière et al. (Eds.): FIMH 2019, LNCS 11504, pp. 294–303, 2019.
https://doi.org/10.1007/978-3-030-21949-9_32

on-going research seeks to develop electromechanical computational models based on medical imaging data. Computational models can help uncover the mechanisms underlying HF and, based on this improved mechanistic understanding, more effective therapies and treatment plans can be designed and proposed (see, e.g., [1,4]). Progress toward designing new therapies for patients with HF, however, is hindered by an incomplete understanding of the normal functioning of the heart in healthy subjects. Current models of cardiac mechanics are often based on idealized assumptions (e.g., idealized ellipsoidal geometries, rule-based aggregate cardiomyocyte –"myofiber"– orientations, over-simplified boundary conditions) that limit their applicability in patient-specific simulations and/or fail to reproduce key characteristics of cardiac contraction. Recent studies [2,11,12] have shown the importance of including a pericardial boundary condition (BC) in reproducing realistic left ventricle (LV) motion. In this work we construct a series of boundary conditions, including a pericardial boundary condition, to replicate the *in vivo* constraints and motion [12].

Our objectives were: (1) to construct a subject-specific LV model that integrates experimental MRI data for cardiac anatomy and microstructure without using rule-based approaches; (2) to establish a set of criteria based on cardiac MRI data to validate the computed cardiac motion; and (3) to vet the simulation results obtained using the subject-specific LV model and compare against the proposed validation criteria.

2   Validation Criteria

LV systolic motion is characterized by global and local measures that a model of cardiac contraction should reproduce before it can be used to evaluate pathological conditions or plan therapeutic interventions.

2.1   Global Measures

Ejection Fraction (EF) and twist angle are global measures of cardiac function that serve as clinical markers to detect the onset and monitor the progression of cardiac diseases.

- EF: The average left ventricular EF is $\approx$56.9% $\pm$ 4.6 [18] in healthy human subjects and an EF below 50% is considered a symptom of heart failure with reduced ejection fraction (HFrEF) [8]. However, normal healthy subject EF values may vary in other species.
- Twist angle: Peak LV twist equal to 11.5° $\pm$ 3.3° has been reported in [14] for healthy human subjects. Peak twist is measured as the difference in rotation at the LV base (-3.9° $\pm$ 1.3°) and LV apex (7.5° $\pm$ 3.6°).

2.2   Local Measures

Strain measures are clinical biomarkers of regional cardiac function. Strains are usually reported along the longitudinal, circumferential, and radial directions, although this reference system depends on an arbitrary geometrical definition and cannot be uniquely defined. As local measures we use both the widely reported strains in a cylindrical coordinate system and strain in the direction of aggregate cardiomyocytes, which is directly related to the average cardiomyocyte shortening driving cardiac contraction. The direction of aggregate cardiomyocytes does not depend on an arbitrary geometric definition but directly reflects the microstructure of the myocardium and can be measured using diffusion tensor imaging (DTI).

Average Green-Lagrange strain values at peak systole for the mid-ventricular LV are:

- Longitudinal strain E_{ll}: -0.16 ± 0.02 [20], -0.15 ± 0.02 [9]. Longitudinal strain corresponds to the LV base-to-apex shortening observed during contraction.
- Circumferential strain E_{cc}: -0.17 ± 0.02 [20], -0.19 ± 0.02 [9], with a gradient across the myocardial wall ($E_{cc} = -0.16 \pm 0.02$ at the epicardium and $E_{cc} = -0.20 \pm 0.02$ at the endocardium [20]).
- Radial strain E_{rr}: a larger range is present in the literature for E_{rr} with respect to other strain measures. Zhong et al. [20] report average peak systolic mid-wall $E_{rr} = 0.33 \pm 0.10$ while Moore et al. [9] report $E_{rr} = 0.42 \pm 0.11$. E_{rr} also exhibits a transmural gradient: $E_{rr} = 0.29 \pm 0.11$ at the epicardial wall and $E_{rr} = 0.38 \pm 0.10$ at the endocardium [20]. Radial strains correspond to the transmural wall thickening observed during contraction.
- Aggregate cardiomyocyte strain E_{ff}: it ranges from -0.13 [10] and -0.12 ± 0.01 [17] to -0.18 [18].
- Incompressibility: in vivo MRI-based study [15] reported 1–2% myocardial volume change during contraction. The low compressibility of the myocardium has also been confirmed by ex vivo tissue studies [5,19] that reported a volume change of approximately 2–4% during contraction. This limited compressibility is attributed to blood outflow during contraction.

3   Methods

The finite element model is based on: (1) MRI-based subject-specific LV anatomy and microstructure acquired in a healthy swine; (2) a sliding boundary condition to model the interaction between visceral (epicardium) and the deformable parietal pericardium; (3) a boundary condition to model the basal surface constraint; (4) an auxiliary electrophysiology model to compute the activation times throughout the LV [6]; and (5) active and passive material laws to model the myocardial mechanical behavior during filling and ventricular systole adapted from [13]. Here, the force velocity curve in [13] governing cardiac contraction is calibrated to achieve EF $\approx 50\%$.

In the following sections, we describe the construction of the finite element model, the corresponding boundary conditions, and the calculation of quantitative measures of cardiac function at peak systole to evaluate the fulfillment of the listed validation criteria.

3.1 Subject-Specific LV Anatomy and Microstructure

The anatomy and microstructure (cardiomyocyte aggregate orientations) data of the FE model were computed from DTI of an *ex vivo* swine heart. The animal experiments were conducted in accordance to research protocol # 2015-124 approved by the UCLA Chancellor's Animal Research Committee. After euthanasia, the heart was extracted and the four chambers were filled with a silicone rubber compound to approximate the heart configuration corresponding to the lowest intra-ventricular pressure. The heart was then submersed in perfluoropolyether solution with no MR signal and scanned overnight for eight hours (readout-segmented, diffusion weighted spin echo sequence, 30 directions, b-value = $1000 \, \text{s/mm}^2$, Echo Time/Repetition Time = $62 \, \text{ms}/18100 \, \text{ms}$, 5 signal averages, and a spatial resolution of $1.0 \times 1.0 \times 1.0 \, \text{mm}^3$).

The DTI data acquired *ex vivo* was segmented to determine the epicardial and endocardial contours that were subsequently edited to smooth surfaces (3-matic, Materialise). The myocardial volume enclosed by the endocardial and epicardial surfaces was meshed with 6109 quadratic tetrahedral elements (Fig. 1A-B) (average element edge length = $5.3 \, \text{mm}$ [95% CI: $4.8 \, \text{mm}$, $6.5 \, \text{mm}$], average element Jacobian = 0.97 [95% CI: 0.96, 1.0]). Diffusion tensors were reconstructed at each DTI voxel and interpolated at each mesh quadrature point using an in house Matlab code and linear tensor invariant interpolation [3]. Aggregate cardiomyocytes orientations were computed as the primary eigenvector of the diffusion tensor at each quadrature point (Fig. 1C).

The pericardial surface was constructed by projecting outward the LV epicardial surface by a distance δ in the normal direction (Fig. 1D). The pericardial surface was then meshed using linear triangular elements.

Since shape and volume are important contributors to LV function, we verify that our *ex vivo* based geometry represents the *in vivo* LV reference configuration at the lowest intraventricular pressure, which was measured during MR exams using a fiber optic pressure transducer. The *ex vivo* myocardium was within 2.4% of the *in vivo* tissue volume measured from CINE images at diastasis, end diastole, and peak systole. The *ex vivo* LV cavity volume was within 4.2% of the *in vivo* cavity volume in the reference configuration. The computed Dice Similarity Coefficient between the *ex vivo* and *in vivo* configurations was 0.80, indicating that the differences between the two configurations are similar to differences due to intra-observer segmentation [21].

3.2 Boundary Condition: Epicardial and Basal Surfaces

The heart is connected to the great vessels and is contained in the parietal pericardium (Fig. 2) wherein it contracts and twists with minimal resistance. These boundary conditions are modeled by including: (1) a flexible surface that represents the pericardium and exerts a reaction force only in the direction normal to itself; and (2) a constraint to limit the warping and out of plane rotation of the LV basal surface due to the presence of the valves' structure and great vessels connected to the heart.

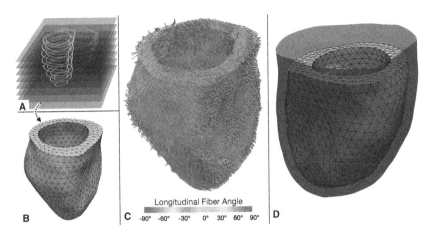

Fig. 1. Model of LV Anatomy and Microstructure. (A) Segmentation of LV anatomy from *ex vivo* DTI data. (B) Generation of smooth LV surface from segmented contours and finite element discretization of LV volume. Mesh is constructed using quadratic tetrahedral elements. (C) Incorporation of aggregate cardiomyocyte orientations into LV volume mesh. Aggregate cardiomyocyte orientations are interpolated from DTI voxels to all mesh quadrature points using linear tensor invariant interpolation. Colorbar represents the aggregate cardiomyocytes elevation angle. (D) Construction of the pericardial surface boundary condition (red) enclosing the LV myocardium (yellow). (Color figure online)

Epicardial Surface. The pericardium is modeled as a flexible elastic membrane with bending and in-plane stretching energies. For simplicity, the stretching and bending energies are modeled using a network of springs between the element nodes and the element normals. Although approximate, this simple approach to model elastic shells has been used in several studies in large deformations, e.g., [7]. The total pericardium elastic energy is obtained by summing the element contributions W_e^s (in-plane stretching energy) and W_e^b (bending energy). $W_e^s = \frac{1}{2}k_p \sum_{i=1}^{3}(l_e^i - L_e^i)^2$, where k_p is the spring elastic constant, $i = 1 \ldots 3$ refers to the edges of a linear triangular element e, l_e^i is the current length of edge i, and L_e^i is the corresponding length in the reference configuration. $W_e^b = \frac{1}{2}k_b \sum_{i=1}^{3} \left(\|\mathbf{n}_e - \mathbf{n}_i\| - \bar{\theta}_{ei} \right)^2$, where k_b is the angular spring elastic constant, $i = 1 \ldots 3$ refers to the elements sharing an edge with element e, $\mathbf{n}_e$ and $\mathbf{n}_i$ are the unit normals to element e and i, and $\bar{\theta}_{ei}$ is the angle between $\mathbf{n}_e$ and $\mathbf{n}_i$ in the reference configuration. Consistent with the small angle approximation, $\bar{\theta}_{ei} \approx \|\bar{\mathbf{n}}_e - \bar{\mathbf{n}}_i\|$, where $\bar{\mathbf{n}}_e$ and $\bar{\mathbf{n}}_i$ are the unit normals in the reference configuration.

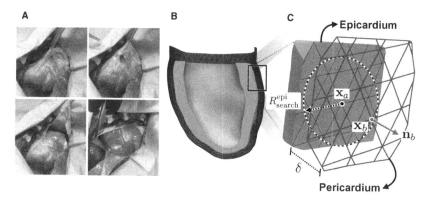

Fig. 2. (A) Parietal pericardium during post euthanasia heart extraction. (B) Long axis view of finite element mesh of the myocardium (yellow) with surrounding pericardial mesh (red). (C) Zoomed in view of epicardial and pericardial surfaces with components used to model the pericardium boundary condition (see Eq. 1). (Color figure online)

In order to minimize any artefactual constraints on cardiac twist and longitudinal motion, the pericardium only exerts forces in the direction normal to its surface according to the following interaction energy W^{int}:

$$W^{int} = \frac{1}{2} k_{int} \sum_{a=1}^{N^{epi}} \frac{1}{N^{search}} \sum_{b=1}^{N^{search}} \left((\mathbf{x}_b - \mathbf{x}_a) \cdot \mathbf{n}_b - \delta \right)^4, \tag{1}$$

where k_{int} scales the interaction forces between the parietal pericardium and the LV, a and b represent a node on the epicardium and parietal pericardium, respectively, N^{search} is the number of nodes within a search R^{epi}_{search} from node a, $\mathbf{x}_a$ and $\mathbf{x}_b$ are the current nodal positions of nodes a and b, and $\mathbf{n}_b$ is the unit normal to the parietal pericardium at node b (Fig. 2). In our simulations we used: $\delta = 0.5$ cm and $R^{epi}_{search} = 1.5$ cm, but limit N^{search} to three.

The mesh of the parietal pericardium was extended above the LV basal plane (see Fig. 2) and anchored by constraining the nodes in its most basal region. Since the constrained nodes are above the LV basal plane, this boundary condition does not limit the in-plane and longitudinal motion of the LV. Similarly, the nodes in the apical region of the parietal pericardium mesh were constrained in order to prevent motion of the apex in the short-axis plane without affecting physiological LV twist.

Basal Surface. The LV basal surface out of plane rotation and warping is constrained by the valves' structure and the great vessels connected to the heart. This constraint is represented by the following energy that penalizes deviation from the basal surface reference configuration while it allows its free translation along the longitudinal axis.

$$W^{base\text{-}n} = \frac{1}{2} k_{base\text{-}n} \int_{\Gamma_{base}} \|\mathbf{n} - \bar{\mathbf{n}}\|^2 \, d\Gamma, \tag{2}$$

where $k_{base\text{-}n}$ is the bending stiffness of the LV basal surface Γ_{base} due to the valves' structure and great vessels while $\bar{\mathbf{n}}$ and $\mathbf{n}$ are the local unit normals on the basal surface in the reference and current configurations, respectively. This energy is adapted from [16].

The great vessels and the valves' structure also limit the rigid rotation of the LV around its longitudinal axis. In order to incorporate this effect, we apply torsional springs at every node on the LV base by including the following energy:

$$W^{base\text{-}t} = \frac{1}{2} k_{base\text{-}t} \sum_{a=1}^{N^{base}} [(\mathbf{x}_a - \mathbf{X}_a) \cdot \mathbf{c}]^2, \tag{3}$$

where $k_{base\text{-}t}$ is the torsional stiffness, N^{base} is the total number of nodes a on the basal surface, $\mathbf{X}_a$ and $\mathbf{x}_a$ are the reference and current positions of basal node a, respectively, and $\mathbf{c}$ is the in-plane (perpendicular to the longitudinal axis) circumferential unit vector.

3.3 Computing Measures of Cardiac Contraction

Based on the LV configuration at peak systole, the following output measures were computed to be compared with the reference values in the list of validation criteria:

- LV Ejection Fraction. EF is the ratio between the stroke volume (SV) during systole and the end diastolic cavity volume (EDV). SV is equal to the difference between EDV and ESV, where ESV is the end systolic cavity volume. All cavity volumes are computed using the divergence theorem.
- LV Twist Angle. Since the LV longitudinal axis is aligned here with the Z-axis and the chosen basal boundary condition constrains rotations at the LV base, LV twist is computed as the average rotation of an apical slice around Z.
- Characteristic Strains. Based on the deformation gradient tensor $\mathbf{F}$, we compute the Green-Lagrange strain tensor $\mathbf{E} = \frac{1}{2}\left(\mathbf{F}^T\mathbf{F} - \mathbf{I}\right)$, where $\mathbf{I}$ is the identity tensor. By projecting $\mathbf{E}$ along different directions, we compute strains along the cylindrical axes $\mathbf{r}$, $\mathbf{c}$, $\mathbf{l}$, and aggregate cardiomyocyte direction $\mathbf{f}$. LV strain measures are then divided in epicardial, mid, end endocardial regions based on a continuous scalar field φ: $\Omega_{epi} := \{X : 0 \le \varphi(X) < \frac{1}{3}\}$, $\Omega_{mid} := \{X : \frac{1}{3} \le \varphi(X) \le \frac{2}{3}\}$, and $\Omega_{endo} := \{X : \frac{2}{3} < \varphi(X) \le 1\}$. φ is the solution of the Laplace equation solved in the LV domain Ω with boundary conditions $\varphi = 0$ on the epicardial wall and $\varphi = 1$ on the endocardial wall.
- Incompressibility. Tissue compressibility was evaluated based on $J = \det(\mathbf{F})$.

4   Results

During filling the pericardial boundary condition (Fig. 3) supports the apex while leaving the base free to move upward. The LV epicardial surface moves outward only slightly, which is qualitatively consistent with observed *in vivo* motion patterns. During contraction, the LV with the pericardial boundary condition shows significant twist (12.5°), longitudinal shortening, and wall thickening (Fig. 3). The corresponding ejection fraction at peak systole is 51.2%. On the contrary, the LV with pinned base presents very limited twist (0.1°) and longitudinal shortening. The corresponding ejection fraction at peak systole is 53.1%. Strain values corresponding to wall thickening (E_{rr}) and longitudinal shortening (E_{ll}) are shown in Fig. 4 together with circumferential and aggregate cardiomyocytes strains. Both radial and circumferential strains present a transmural gradient that agrees with trends presented in the literature. The average Jacobian for both boundary conditions indicates very limited tissue volume change (pericardial boundary condition: 3.3%; pinned boundary condition: 1.6%).

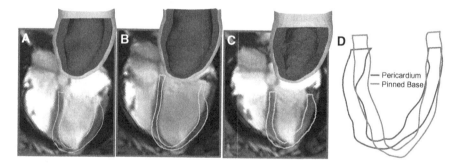

Fig. 3. LV motion from diastasis (A), through late filling (B), to peak systole (C). Section of the LV FE model with pericardial boundary condition (top) and LV outline superimposed to long-axis MR images (bottom). (D) Comparison of LV cross-sectional deformation obtained at peak systole with pinned and pericardial boundary conditions.

	Region	E_{ff}	E_{rr}	E_{cc}	E_{ll}
Pericardium BC	Endo	-0.24	0.40	-0.21	-0.18
	Mid	-0.21	0.34	-0.18	-0.18
	Epi	-0.19	0.32	-0.12	-0.14
Pinned Base BC	Endo	-0.23	0.36	-0.32	-0.06
	Mid	-0.25	0.29	-0.28	-0.06
	Epi	-0.22	0.28	-0.20	-0.06

Fig. 4. Peak systolic strains obtained with pericardial and pinned base boundary conditions.

5 Discussion

We have presented a set of validation criteria based on MRI measures that can help develop and validate cardiac models of LV contraction. The simulation using the pericardial boundary condition shows far more physiologically accurate cardiac motion and deformation when compared to the simulation using the pinned based boundary condition. This is despite the fact that both simulations lead to an EF close to the *in vivo* value of 48.7%. Therefore, EF alone is not sufficient to evaluate a model. The need for a correct pericardial boundary condition agrees with other studies, e.g., [2,11]. The pericardial boundary condition allows eliminating LV rigid body motions without imposing unphysiological constraints on cardiac motion. This is possible also because the pericardial surface is not rigid but flexible, allowing the heart (which does not have an axially-symmetric geometry) to better slide while twisting during contraction. A flexible pericardial boundary condition better represents the *in vivo* conditions surrounding the heart.

The model presented here was built using subject-specific data and contained the essential components to meet the presented validation criteria. However several improvements are possible, including a direct coupling with the electrophysiology model, the inclusion of the right ventricle, and possibly the atria and great vessels to anchor the heart. Furthermore, the current pericardial boundary condition does not take into account the stiffness of the surrounding organs, such as the lungs and the liver. Here, the stiffness of the pericardial boundary condition was calibrated to minimally constrain the heart motion while maintaining convergence of the finite element model.

The presented list of validation criteria is not meant to be exhaustive and several other criteria may be considered depending also on the simulation goals. For example, additional validation criteria may focus on the passive filling, active relaxation phase, and/or may include measures of aggregate cardiomyocytes kinematics. In addition, the validation criteria listed here have been derived from literature mostly focusing on human data. Inter-species variability should also be taken into account and species-specific reference values for the listed validation criteria should be used as they become available.

References

1. Chabiniok, R., et al.: Multiphysics and multiscale modelling, data-model fusion and integration of organ physiology in the clinic: ventricular cardiac mechanics. Interface focus 6(2), 20150083 (2016)
2. Fritz, T., Wieners, C., Seemann, G., Steen, H., Dössel, O.: Simulation of the contraction of the ventricles in a human heart model including atria and pericardium. Biomech. Model. Mechanobiol. 13(3), 627–641 (2014)
3. Gahm, J.K., Ennis, D.B.: Dyadic tensor-based interpolation of tensor orientation: application to cardiac DT-MRI. In: Camara, O., Mansi, T., Pop, M., Rhode, K., Sermesant, M., Young, A. (eds.) STACOM 2013. LNCS, vol. 8330, pp. 135–142. Springer, Heidelberg (2014). https://doi.org/10.1007/978-3-642-54268-8_16

4. Genet, M., et al.: Distribution of normal human left ventricular myofiber stress at end diastole and end systole: a target for in silico design of heart failure treatments. J. Appl. Physiol. **117**(2), 142–152 (2014)

5. Judd, R.M., Levy, B.I.: Effects of barium-induced cardiac contraction on large-and small-vessel intramyocardial blood volume. Circ. Res. **68**(1), 217–225 (1991)

6. Krishnamoorthi, S., et al.: Simulation methods and validation criteria for modeling cardiac ventricular electrophysiology. PloS One **9**(12), e114494 (2014)

7. Lidmar, J., Mirny, L., Nelson, D.R.: Virus shapes and buckling transitions in spherical shells. Phys. Rev. E **68**(5), 051910 (2003)

8. Mahadevan, G., et al.: Left ventricular ejection fraction: are the revised cut-off points for defining systolic dysfunction sufficiently evidence based? Heart **94**(4), 426–428 (2008)

9. Moore, C.C., Lugo-Olivieri, C.H., McVeigh, E.R., Zerhouni, E.A.: Three-dimensional systolic strain patterns in the normal human left ventricle: characterization with tagged MR imaging. Radiology **214**(2), 453–466 (2000)

10. Perotti, L.E., Magrath, P., Verzhbinsky, I.A., Aliotta, E., Moulin, K., Ennis, D.B.: Microstructurally anchored cardiac kinematics by combining in vivo DENSE MRI and cDTI. In: Pop, M., Wright, G.A. (eds.) FIMH 2017. LNCS, vol. 10263, pp. 381–391. Springer, Cham (2017). https://doi.org/10.1007/978-3-319-59448-4_36

11. Pfaller, M.R., et al.: The importance of the pericardium for cardiac biomechanics: from physiology to computational modeling. Biomech. Model. Mechanobiol. **18**, 503–529 (2018)

12. Ponnaluri, A.V.S.: Cardiac Electromechanics Modeling and Validation. Ph.D. thesis, UCLA (2018)

13. Ponnaluri, A., Perotti, L., Ennis, D., Klug, W.: A viscoactive constitutive modeling framework with variational updates for the myocardium. Comput. Methods Appl. Mech. Eng. **314**, 85–101 (2017)

14. Reyhan, M., et al.: Left ventricular twist and shear in patients with primary mitral regurgitation. J. Magn. Reson. Imaging **42**(2), 400–406 (2015)

15. Rodriguez, I., Ennis, D.B., Wen, H.: Noninvasive measurement of myocardial tissue volume change during systolic contraction and diastolic relaxation in the canine left ventricle. Magn. Reson. Med. **55**(3), 484–490 (2006)

16. Szeliski, R., Tonnesen, D.: Surface modeling with oriented particle systems, vol. 26. ACM (1992)

17. Tseng, W.Y.I., Reese, T.G., Weisskoff, R.M., Brady, T.J., Wedeen, V.J.: Myocardial fiber shortening in humans: initial results of MR imaging. Radiology **216**(1), 128–139 (2000)

18. Wang, V.Y., et al.: Image-based investigation of human in vivo myofibre strain. IEEE Trans. Med. imaging **35**(11), 2486–2496 (2016)

19. Yin, F., Chan, C., Judd, R.M.: Compressibility of perfused passive myocardium. Am. J. Physi.-Heart Circulatory Physiol. **271**(5), H1864–H1870 (1996)

20. Zhong, X., Spottiswoode, B.S., Meyer, C.H., Kramer, C.M., Epstein, F.H.: Imaging three-dimensional myocardial mechanics using navigator-gated volumetric spiral cine DENSE MRI. Magn. Reson. Med. **64**(4), 1089–1097 (2010)

21. Zijdenbos, A.P., Dawant, B.M., Margolin, R.A., Palmer, A.C.: Morphometric analysis of white matter lesions in MR images: method and validation. IEEE Trans. Med. Imaging **13**(4), 716–724 (1994)

End-Diastolic and End-Systolic LV Morphology in the Presence of Cardiovascular Risk Factors: A UK Biobank Study

Kathleen Gilbert[1]($\boxtimes$), Avan Suinesiaputra[2], Stefan Neubauer[5],
Stefan Piechnik[5], Nay Aung[3], Steffen E. Petersen[3],
and Alistair Young[4]

[1] Auckland Bioengineering Institute, University of Auckland, Auckland,
New Zealand
kat.gilbert@auckland.ac.nz
[2] Department of Anatomy and Medical Imaging, University of Auckland,
Auckland, New Zealand
[3] William Harvey Research Institute, NIHR Barts Biomedical Research Centre,
Queen Mary University of London, London, UK
[4] Department of Biomedical Engineering, King's College London, London, UK
[5] Oxford NIHR Biomedical Research Centre,
Division of Cardiovascular Medicine, Radcliffe Department of Medicine,
University of Oxford, Oxford, UK

Abstract. Left ventricular function and morphology have been shown to be important factors in clinical and pre-clinical cardiovascular disease. In this paper we used atlas-based techniques to capture the full extent of morphological changes at end-diastole, end-systole and in a coupled functional atlas for 4547 UK Biobank participants. The morphological differences between participants with risk factors for cardiovascular disease were tested using a logistic regression model. The result was compared to a model built from traditional mass and volume measures, and the strength of associations were tested using a Delong's test. Atlas based models had stronger associations with risk factors than mass and volume parameters in most risk factors. The functional atlas showed better performance than the separate end-diastole and end-systole atlases.

Keywords: Cardiac MRI · Cardiac atlases · Cardiovascular risk factors

1 Introduction

Cardiovascular disease is the leading cause of death worldwide, and is responsible for one in every four deaths [1]. Left ventricular (LV) function and morphology are important predictors of cardiac health. The LV changes shape, in a process known as remodeling, in response to a variety of clinical and pre-clinical disease processes [2].

Previous work has demonstrated changes to LV morphology in those with higher blood pressure [3], higher fat mass [4] and smoking [5]. Many measures of changes to morphology ignore large amounts of information, which is available in cardiac MRI. Statistical shape atlases capture the variety of LV morphological differences [6].

© Springer Nature Switzerland AG 2019
Y. Coudière et al. (Eds.): FIMH 2019, LNCS 11504, pp. 304–312, 2019.
https://doi.org/10.1007/978-3-030-21949-9_33

Atlases of the left ventricle have been used to quantify shape differences in participants of longitudinal studies [5, 7]. While some previous studies have defined shape measures end-diastole (ED), e.g. mass to volume ratio and sphericity, others are defined at end-systole (ES), e.g. ES volume index.

UK Biobank is a longitudinal cohort study in the United Kingdom, which has performed cardiac MRI on participants to examine the pre-clinical determinants of cardiovascular disease. The cohort contains a subset of healthy participants who have no cardiovascular risk factors.

Here, we extend previous work which showed that different shape atlases perform similarly in quantifying relationships with cardiovascular risk factors in UK Biobank, and are robust to different methods used in their construction. That work combined the scores from the first 20 modes of the separate ED and ES statistical shape models to quantify the relationships with risk factors. Here, we provide the following novel contributions:

1. We show that separate ED and ES statistical shape models have similar discrimination power for the risk factors in UK Biobank, with ES having higher AUC, suggesting that LV morphology at ES may be more sensitive to the presence of risk factors.
2. We test a functional atlas constructed by concatenating ED and ES shape models, thereby forming a coupled atlas which simultaneously describes coupled ED and ES shape variations. The functional atlas outperformed the separate ED and ES atlases for the diabetes and high cholesterol.

2 Methods

2.1 Imaging Protocol and Analysis

The CMR protocol has been described previously in [8]. Briefly, all images were acquired on a Siemens AERA 1.5T wide-BORE MRI scanner (Siemens Healthineers, Erlangen, Germany) using retrospectively gated cine balanced steady-state free precession breath-hold acquisitions. A short axis stack covering the left and right ventricles, as well as horizontal and vertical long axis slices and the left ventricular outflow track were acquired. The short axis images were manually contoured at ED and ES as reported in [9]. All contouring was completed in accordance with Society of Cardiovascular Magnetic Resonance recommendations as described in [9].

2.2 Atlas Creation

The atlas was constructed using the method described in [5]. Briefly, the manual contours were extracted and fitted to a finite element model of the left ventricle using a least squares minimization. The model was orientated using landmarks from the manual contours to denote the LV apex, mitral valve inserts and right ventricle. The orientated model was then fitted to the manual contours at ED and ES, creating participant specific models of the left ventricle.

The participant specific models were then used to create three atlases (ED, ES and a functional atlas). For the individual ED and ES atlases each model was first aligned to the average model using Procrustes (rotation and translation only). The coupled functional atlas (EDES atlas) aligned at both ED and ES using the Procrustes calculated for the ED model. Principal component analysis (PCA) was then used to create each atlas. The ED and ES atlas describe shape at each individual time point. The functional atlas (EDES atlas) was created from participants coupled shapes where the surfaces were concatenated at ED and ES during the PCA, thus describes the variations in ED and ES that occur together.

2.3   Statistics

Z-scores were extracted for each participant in each mode of each of the three atlases. Associations between cardiovascular risk factors and shape were examined using a fivefold cross-validated logistic regression model in the R package Caret (R version 3.5.1 and Caret version 6.0-81). A separate model was created for each risk factor, which compared the healthy risk-free cohort and those with the particular risk factor. Participants with angina, and those who reported having had a heart attack were excluded from the analysis, except in the case of angina as a risk factor where only those who reported having a heart attack were excluded. Separate models were created for each atlas, as well as a mass and volume model. Area under the curve was recorded. Models were tested using the first 50 modes of variation from each atlas, and the final number of modes was selected where the area under the curve stabilized. Differences between the result of the mass volume model and the atlas models were tested using a Delong's test and a Bonferroni corrected p-value of 0.0125 was considered significant. Risk factors of self-reported hypertension, high cholesterol, angina, smoking, diabetes as well as those who reported taking blood pressure medication or cholesterol medication at the time of imaging were selected for analysis.

2.4   Visualization of the Statistical Models

Let $X \in R^{N \times P}$ be the centered data shape matrix with N cases and P shape dimensions. Principal component analysis of X can be estimated by singular value decomposition of covariance of X, i.e. $C = X^T X/(N-1)$, as follows

$$C = \Phi L \Phi^T \qquad (1)$$

where $\Phi \in R^{P \times W}$ is an orthonormal matrix of the principal components, $L \in R^{P \times P}$ is a diagonal matrix of eigenvalues (variances of each mode), and $W \leq P$ is the number of principal components or modes of variation. Let x represent an individual column shape vector of X with P shape dimensions. The normalized PCA z-scores for each participant in each mode can be defined as:

$$T_z = x\Phi\sqrt{L}^{-1} \qquad (2)$$

The statistical model defines a regression score as:

$$Y = T_z\beta = T\sqrt{L}^{-1}\beta \tag{3}$$

Where Y is the score (log-odds in the logistic regression), $\beta \in R^{1\times Q}$ are the coefficients of the statistical model where Q is the number of variables included in the regression model. We can define the following morphometric scores:

$$Y_1 = T\beta_1 = X\Phi\beta_1 = Xm \tag{4}$$

Where β_1 is the normalized scaled coefficients $\left(\sqrt{L}^{-1}\beta/\left|\sqrt{L}^{-1}\beta\right|\right)$, and m is a unit length morphometric shape vector. Thus we can visualize shape from the statistical model as

$$x_M = xmm^T = Y_1 m^T \tag{5}$$

The resulting logistic regression models were evaluated for all participants included in the model and visualized at the 5th and 95th percentiles.

3 Results

Table 1 shows the LV volumes and mass of participants with each risk factor, and the risk free cohort, as well the number in each group. Figure 1 shows the first five PCA modes in the independent ED and ES atlases. The atlases were visually inspected to categorize mode shapes. Modes 1 of both atlases show that size accounts for 46% of the variation in LV morphology at ED and 43% at ES. Mode 2 shows variations in sphericity of the ventricle accounting for 15% and 10% at ED and ES respectively. The ED atlas shows variation in the mitral valve tilt in modes 3, 4 and apical sphericity in mode 5. The ES atlas shows variation in global wall thickness in mode 3 and mitral valve tilt in modes 4 and 5.

Table 1. Cardiovascular function and number of participants in each risk factor or cohort included in analysis. LV volumes and mass are reported as mean ± standard deviation

	n	Male	LV EDV (ml)	LV ESV (ml)	LV mass (g)
Risk free cohort	751	344	143.74 ± 34.32	58.85 ± 18.26	85.42 ± 24.00
Blood pressure medication	885	498	145.86 ± 33.02	59.29 ± 19.84	96.06 ± 24.86
Cholesterol medication	822	539	146.08 ± 32.88	59.86 ± 18.86	96.06 ± 24.62
High blood pressure	1100	621	146.48 ± 31.49	59.85 ± 19.64	97.02 ± 25.67
High cholesterol	276	170	143.48 ± 31.49	58.85 ± 18.31	94.06 ± 24.09
Angina	83	51	150.55 ± 31.04	60.98 ± 18.75	98.94 ± 24.77
Diabetes	207	121	144.82 ± 33.68	60.83 ± 20.32	97.29 ± 24.76
Smoking	279	151	145.43 ± 33.20	61.20 ± 18.94	93.04 ± 24.17

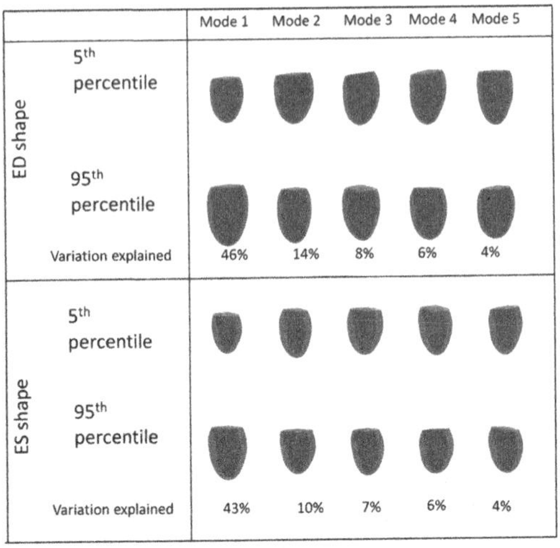

Fig. 1. The first five modes of the ED and ES atlas at the 5th and 95th percentile. The variance explained by each mode is also displayed. Septum is on the left.

Figure 2 shows the first five modes of variation in the functional atlas. Mode 1 shows variation in size at both ED and ES, accounting for 39% of the variation in the LV morphology of the participants. Mode 2 shows changes in sphericity at both ED and ES which explains 11% of the variation in LV morphology across the participants. Mode 3 shows a variation in mitral valve tilt at ED similar to that seen in mode 4 of the ED atlas, and variation at wall thickness at ES. Mode 4 shows variation in mitral valve tilt at both ED and ES. Mode 5 can be characterized as variations in apical sphericity at ED and mitral valve tilt at ES.

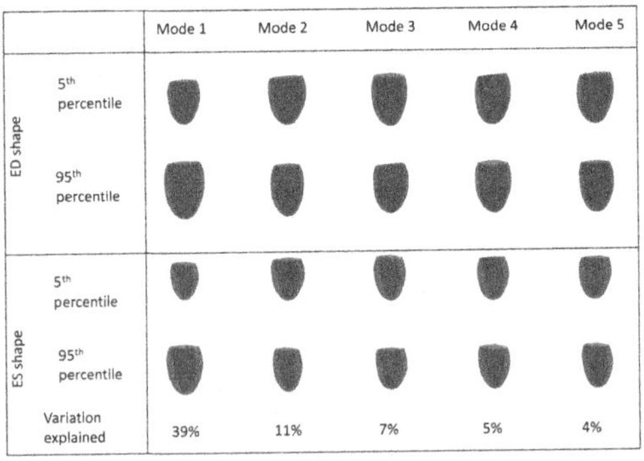

Fig. 2. The first five modes of the function atlas at the 5th and 95th percentile. The variance explained by each mode is also displayed. Septum is on the left.

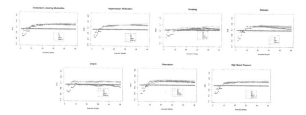

Fig. 3. Cumulative area under the curve plots in a logistic regression model with 1 to 50 modes included. Black is the area under the curve for a mass volume model, blue is ED, red is ES and green is the functional atlas EDES. (Color figure online)

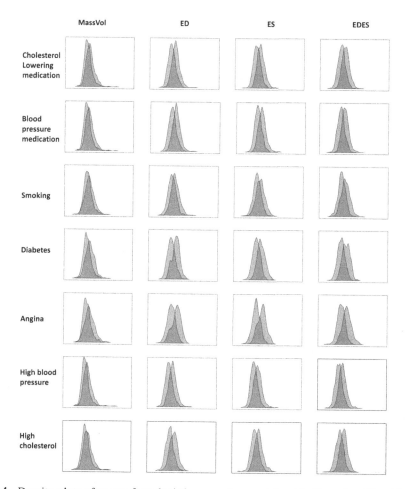

Fig. 4. Density plots of scores from logistic regression models with 40 modes. The risk free cohort are shown in blue and the risk factor positive cases are shown in orange. (Color figure online)

Figure 3 shows the area under the curve for each risk factor, with each of the four logistic regression models, varying the number of mode scores included. 40 modes were chosen as an appropriate cut off for inclusion in the statistical model as it allowed for the AUC to stabilize in all variables considered.

The results of the logistic regression models are shown in Table 2 and in Fig. 4. Table 2 shows the area under the curve for each model and the result of the Delong's test between the mass volume model and the atlas. The atlases had a significantly higher association with cholesterol lowering medication, blood pressure medication, diabetes, high blood pressure and high cholesterol than LV mass and volume (Fig. 5).

Table 2. Area under the curve for fivefold cross validation logistic regression models for each risk factor. Mass volume model (MV) included LV EDV, ESV and mass, each model built from an atlas used the first 40 modes. * p < 0.0125 for Delong's test

	MV	ED	ES	EDES
Cholesterol lowering medication	0.67	0.74*	0.75*	0.76*
Blood pressure medication	0.67	0.74*	0.75*	0.76*
Smoking	0.61	0.65	0.63	0.68
Diabetes	0.69	0.73	0.76*	0.77*
Angina	0.67	0.66	0.71	0.73
High blood pressure	0.68	0.74*	0.74*	0.76*
High cholesterol	0.65	0.72*	0.70	0.73*

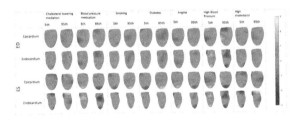

Fig. 5. Morphometric risk factor shapes from the logistic regression model of the functional atlas at the 5th and 95th percentiles. The average shapes were drawn with the color scale shown. Yellow denotes an outward displacement of the surface, and blue an inward displacement. View point is from the anterior with the septum on the left. (Color figure online)

4   Discussion

The results of this study indicate a strong association between cardiovascular risk factors and LV morphology. Three separate atlases were tested and all showed strong association cholesterol lowering medication, blood pressure medication and high blood pressure. The ED model also showed a strong association with high cholesterol, and the functional atlas (EDES) showed a strong association with diabetes, high blood pressure and high cholesterol. The results suggest that LV morphology changes at both ED and ES in the presence of risk factors. Previously [7] used a five-fold cross-validated logistic regression to test the association of risk factors to two atlases created using different methods. The method used in this paper was similar to the surface atlas in the previous work; which built a logistic regression model with the first 20 mode scores at ED and ES combined from independent atlases. Our current study showed similar area under the curves for the risks factors studied, however the ES model and EDES model described in this paper result in a larger area under the curve than ED in most cases. The functional atlas performed similarly to the previous combined score analysis.

The functional atlases was created from the participants' specific shapes at ED and ES, thus having the advantage of describing variation in LV morphology that occur at ED and ES in the same participant. Thus building a logistic regression from this atlas has the advantage of making sure morphological features selected by the logistic regression model at ED and at ES can occur in the same participant.

The study is limited by the cross-sectional nature of UK biobank. However in the future, the shape atlases could be compared with follow-up information, such as cardiac events to better understand remodeling in pre-clinical cardiovascular disease.

5 Conclusions

Shape atlases of LV morphology have a stronger relationship with the cardiovascular risk factors of high blood pressure, high cholesterol, and diabetes, than standard measures of mass and volume. The functional atlas has the largest area under the curve, and provides accurate simultaneous deformations at ED and ES.

References

1. Finegold, J.A., et al.: Mortality from ischaemic heart disease by country, region, and age: statistics from World Health Organisation and United Nations. Int. J. Cardiol. **168**, 934–945 (2013)
2. Bluemke, D.A., et al.: The relationship of left ventricular mass and geometry to incident cardiovascular events: the MESA (Multi-Ethnic Study of Atherosclerosis) study. J. Am. Coll. Cardiol. **52**, 2148–2155 (2008)
3. de Marvao, A., et al.: Precursors of hypertensive heart phenotype develop in healthy adults: a high-resolution 3D MRI study. JACC Cardiovasc. Imaging **8**, 1260–1269 (2015)
4. Corden, B., et al.: Relationship between body composition and left ventricular geometry using three dimensional cardiovascular magnetic resonance. J. Cardiovasc. Magn Reson. **18**, 32 (2016)
5. Medrano-Gracia, P., et al.: Left ventricular shape variation in asymptomatic populations: the multi-ethnic study of atherosclerosis. J. Cardiovasc. Magn Reson. **16**, 56 (2014)
6. Abdi, H., Williams, L.J.: Principal component analysis. Wiley Interdiscip. Rev. Comput. Stat. **2**, 433–459 (2010)

7. Gilbert, K., et al.: Independent left ventricular morphometric Atlases show consistent relationships with cardiovascular risk factors: a UK Biobank study scientific reports **9**, 1130 (2019). https://doi.org/10.1038/s41598-018-37916-6

8. Petersen, S.E., et al.: Reference ranges for cardiac structure and function using cardiovascular magnetic resonance (CMR) in Caucasians from the UK Biobank population cohort. J. Cardiovasc. Magn. Reson. **19**, 18 (2017)

9. Petersen, S.E., et al.: UK Biobank's cardiovascular magnetic resonance protocol. J. Cardiovasc. Magn. Reson. **18**, 8 (2015)

Solution to the Unknown Boundary Tractions in Myocardial Material Parameter Estimations

Anastasia Nasopoulou[1(✉)], David A. Nordsletten[1,2], Steven A. Niederer[1], and Pablo Lamata[1]

[1] Department of Biomedical Engineering, Division of Imaging Sciences and Biomedical Engineering, King's College London, London, UK
anastasia.nasopoulou@kcl.ac.uk
[2] Department of Biomedical Engineering and Cardiac Surgery, University of Michigan, Ann Arbor, MI, USA

Abstract. Passive material parameter estimation can facilitate the in vivo assessment of myocardial stiffness, an important biomarker for heart failure stratification and screening. Parameter estimation strategies employing biomechanical models of various degrees of complexity have been proposed, usually involving a significant number of cardiac mechanics simulations. The clinical translation of these strategies however is limited by the associated computational cost and the model simplifications. A simpler and arguably more robust alternative is the use of data-based approaches, which do not involve mechanical simulations and can be based for example on the formulation of the energy balance in the myocardium from imaging and pressure data. This approach however requires the estimation of the mechanical work at the myocardial boundaries and the strain energy stored, tasks that are challenging when external loads are unknown - especially at the base which deforms extensively within the cardiac cycle. In this work we employ the principle of virtual work in a strictly data-based approach to uniquely identify myocardial material parameters by eliminating the effect of the unknown boundary tractions at the base. The feasibility of the method is demonstrated on a synthetic data set using a popular transversely isotropic material model followed by a sensitivity analysis to modelling assumptions and data noise.

1 Introduction

It is a known hypothesis that myocardial remodelling occurring in cardiac disease may result in changes in myocardial tissue stiffness. Recently stiffness was identified as a powerful biomarker for the diagnosis and monitoring of heart failure (HF) with preserved ejection fraction, a syndrome affecting 50% of HF patients [11]. Stiffness estimation however is not directly available in vivo and for this reason biomechanical models are used to identify the material parameters that best match model predictions to data observations.

© Springer Nature Switzerland AG 2019
Y. Coudière et al. (Eds.): FIMH 2019, LNCS 11504, pp. 313–322, 2019.
https://doi.org/10.1007/978-3-030-21949-9_34

Most of the available parameter estimation pipelines are based on the comparison between data (usually deformations) and model predictions. As such, these methods require the execution of many mechanical simulations and/or nonlinear optimisation routines, which are accompanied by a significant computational burden. Moreover, there is a limited overlap between the space of deformations in actual data and those reproduced by mechanical simulations, caused by the presence of data artefacts and a series of model assumptions [5,9].

There is thus a need to progress towards real-time and accurate myocardial tissue stiffness estimation for translation to the clinic. Data based approaches in parameter estimation are a valid candidate to achieve this goal, since they do not require to run simulations. Instead, they propose a physical interpretation of the deformation and pressure data, based for example on energetics analysis. Arguably, the expression of energy conservation can provide a richer metric in terms of information than the point by point comparison of a distance or strain based metric, especially in the context of material parameter estimation. And it has been shown to solve the lack of identifiability of material parameters of the myocardium [7,8], thus having the potential of rendering a more accurate parameter estimation pipeline for diagnostic and prognostic purposes. One of the limitations of this approach is that it requires the estimation of the work at the myocardial boundaries and the strain energy stored, tasks that are challenging when external loads are unknown, especially at the base which deforms extensively within the cardiac cycle.

In this work we employ the principle of virtual work in a strictly data-based approach (parameter estimation is based on data analysis without the need for mechanical simulations) to eliminate the effect of the unknown boundary tractions while keeping the advantage of uniquely identifying myocardial material parameters. We thus take a closer look into the energetics analysis of the myocardium, quantify the impact associated with the unaccounted boundary tractions at the base and identify a virtual deformation field that can be used to remove their impact from the parameter identification task.

2   Materials and Methods

2.1   Material Model

The myocardium in this study was modelled according to the transversely isotropic model introduced by Guccione et al. [3]. Here we use a modified formulation of this expression proposed by Xi et al. [12], focusing on the scaling parameter C_1 and the bulk exponential parameter α (Eq. (1)). The parameters r_f, r_t, r_{ft} scale the Green-Lagrange strain tensor (E) components along the fiber direction (f), in the transverse plane (t) and in the fiber-transverse shear plane and E is expressed in the local fiber coordinate system where f, s, n denote the fiber, sheet and sheet normal directions.

$$\Psi = \frac{1}{2}C_1(e^Q - 1)$$

$$Q = \alpha[r_f E_{ff}{}^2 + r_{ft}(2E_{fs}{}^2 + 2E_{fn}{}^2) + r_t(E_{ss}{}^2 + E_{nn}{}^2 + 2E_{sn}{}^2)] \tag{1}$$

In this study we focus on C_1-α estimation (as they represent the main direction of coupling in the Guccione model [12]) and assume the anisotropy parameters are fixed to the ground truth values (see Sect. 2.2), following a common approach (see more in [8]). This assumption is revisited in the sensitivity analysis (Sect. 2.6).

2.2 Generation of Synthetic Data Sets

The left ventricle (LV) was modelled using a truncated prolate spheroidal geometry with human dimensions (bottom left corner in Fig. 1). The constructed finite element (FE) mesh consisted of 320 elements (4 circumferential, 4 transmural, 4 longitudinal and 16 in the apical cap) and 9685 nodes.

The three synthetic cases (SC1, SC2, SC3) were generated by prescribing three variations of basal displacements (shown in top left corner of Fig. 1 in green) and passively inflating the reference geometry to different pressure levels (180 Pa for SC1, SC3 and 40 Pa for SC2). The myocyte orientation in the LV was assumed to follow an idealised $-90°/+90°$ distribution from the epi-to the endocardium and the Guccione material model parameters were: $C_1 = 100Pa$, $\alpha = 15$, $r_f = 0.55$, $r_{ft} = 0.25$ and $r_t = 0.2$. The prescribed pressure and the deformed 'LV geometry' (the deformed FE mesh) at each simulation increment represent the pressure-'imaging' data set for each synthetic case.

The deformation was found by solving the linearised total potential energy equations using the *CHeart* nonlinear mechanics solver [6], using a split $\boldsymbol{u}$-p formulation outlined in [4]. The reference ($\boldsymbol{X}$) and deformed geometry ($\boldsymbol{x}$), and local fiber orientation vectors ($\boldsymbol{f}, \boldsymbol{s}, \boldsymbol{n}$) fields were interpolated with cubic-Lagrange shape functions, hydrostatic pressure field (p) interpolation was linear Lagrange. We used *MATLAB* for all pre- and post- processing[1] and *cmGui* for the visualisations[2].

2.3 Estimation of the Exponential Parameter α from the Reformulated Energy-Based CF

Parameter estimation in this work is based on previous research [8], where unique parameter estimation was achieved with the use of an energy-based cost function (CF), f^{EC}, that allowed determination of the α parameter. This CF used the principle of energy conservation (EC), dictating that the work of internal stresses inside the tissue (W_{int}) stored as elastic energy and the external work of external forces (W_{ext}) are equal, where assumptions of quasi-static loading and absence of residual active tension in the diastolic window of relevance apply. Equation (3) expresses the external work in terms of the cavity volume (Ω)-cavity pressure p relationship, assuming negligible contribution of epicardial tractions on W_{ext}, as well as that the contribution of basal work on the total external work

---

[1] https://uk.mathworks.com/products/matlab.html.
[2] http://www.cmiss.org/cmgui.

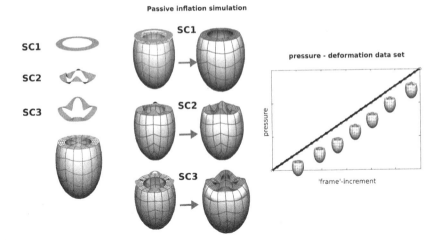

Fig. 1. Generation of the 3 synthetic data sets. A prolate spheroidal mesh representing the left ventricle (LV) was prescribed 3 types of basal displacements and passively inflated to different cavity pressures. SC1 involves a small displacement gradient in the radial direction and none in the circumferential. SC2 is highly varying circumferentially and radially and there is a restriction of zero displacements at the epicardial-basal junction simulating a stiff epicardial rim. SC3 is similar to SC2 but allows for radial expansion. The LV meshes at each simulation increment and the corresponding inflation pressures were treated as LV geometry/pressure data at each 'frame'. (Color figure online)

can be largely accounted for by the increase in cavity volume due to the atrioventricular plane motion. In Eqs. (2), (3) Ψ is given by (1), V represents the myocardial domain of the LV and Ω_0, Ω_{def} denote the cavity volume at the reference (unloaded and unstressed) state and deformed state (corresponding to the data frame in question) respectively. Expressing the energy conservation principle over two diastolic frames (DF), DF_1 and DF_2, yielded f^{EC} shown in Eq. (4).

$$W_{int} = \int_V \Psi dV. \tag{2}$$

$$W_{ext} = \int_{\Omega_0}^{\Omega_{def}} p \, d\Omega \tag{3}$$

$$f^{EC} = \left| \frac{W_{ext}^{DF_1}}{W_{ext}^{DF_2}} - \frac{W_{int}^{DF_1}}{W_{int}^{DF_2}} \right| \tag{4}$$

Instead of energy conservation, the CF employed in this study (f^{VW}) is based on the principle of virtual work (VW), which is a weak expression of the equilibrium. It states the equality of the works of the internal (δW_{int}) and external (δW_{ext}) forces acting on the myocardium along an arbitrary displacement field δu called the virtual field. The internal and external components of the virtual work are given in Eqs. (5) and (6) respectively. Equation (5) expresses the work of the internal 2nd Piola-Kirchhoff stresses ($\partial\Psi/\partial E$) on the linearised Green-Lagrange strains in the direction of δu ($DE[\delta u]$) given in Eqs. (7) and (8). Equation (6) expresses the work of the cavity pressure p acting on the endocardial boundary along δu, where da and F denote the infinitesimal area vector in the deformed configuration and deformation gradient respectively [1,12]. Expressing the principle of virtual work over 2 diastolic frames DF_1 and DF_2, the modified energy-based CF f^{VW} can then be expressed as in Eq. (9).

Cost functions f^{EC} and f^{VW} allow the unique estimation of α as they are independent of C_1. Clearly since both the numerator and denominator of the ratio of W_{int} or δW_{int} in Eqs. (4) and (9) respectively contain the parameter C_1, it cancels out.

$$\delta W_{int} = \int_V \frac{\partial\Psi}{\partial E} : DE[\delta u]dV. \tag{5}$$

$$\delta W_{ext} = \int_a pda \cdot \delta u \tag{6}$$

$$\frac{\partial\Psi}{\partial E} = C_1\alpha e^Q \begin{bmatrix} r_f & r_{ft} & r_{ft} \\ r_{ft} & r_t & r_t \\ r_{ft} & r_t & r_t \end{bmatrix} \circ \begin{bmatrix} E_{ff} & E_{fs} & E_{fn} \\ E_{sf} & E_{ss} & E_{sn} \\ E_{nf} & E_{ns} & E_{nn} \end{bmatrix} \tag{7}$$

$$DE[\delta u] = \frac{1}{2}\left(DF[\delta u]^T F + F^T DF[\delta u]\right) \tag{8}$$

$$f^{VW} = \left| \frac{\delta W_{ext}^{DF_1}}{\delta W_{ext}^{DF_2}} - \frac{\delta W_{int}^{DF_1}}{\delta W_{int}^{DF_2}} \right| \tag{9}$$

2.4   Estimation of the Scaling Parameter C_1

Following determination of α from f^{VW} (Sect. 2.3), parameter C_1 can be estimated from the expression of the principle of virtual work ($\delta W_{ext} = \delta W_{int}$) in one of the two frames DF_1 or DF_2 used in Eq. (9). The estimated value of C_1 using the principle of virtual work and using as input the α parameter from minimised f^{VW} is denoted by C_1^{VW}. The choice of frame (DF_1 or DF_2) was found to be insignificant with differences in C_1 in the order of 1% and here we present the results for DF_2 for consistency (Eq. (10)). $C_1{}^f$ denotes the assumed C_1 value in Eqs. (4), (9) for the estimation of f^{EC}, f^{VW} (in this analysis $C_1{}^f$ was assigned a fixed value of 1 Pa).

For comparison to the previous method based on energy conservation [8] C_1 was also estimated following the α estimation from f^{EC} by expressing the energy balance ($W_{ext} = W_{int}$) in frame DF_2 - see Eq. (11) - and is denoted by C_1^{EC}.

$$C_1{}^{VW} = \frac{\delta W_{ext}^{DF_2}}{\left(\frac{\delta W_{int}^{DF_2}}{C_1{}^f}\right)} \tag{10}$$

$$C_1{}^{EC} = \frac{W_{ext}^{DF_2}}{\left(\frac{W_{int}^{DF_2}}{C_1{}^f}\right)} \tag{11}$$

2.5   Generation of Virtual Field for Tackling Basal Displacements

The virtual field (VF) used in Eqs. (5, 6) can be any displacement field which conforms with the inherent boundary conditions (BCs) of the problem. In this case the additional restrictions involve respect to incompressibility constraints (in order to overcome the indeterminacy of the Lagrange multiplier p from just the deformation data [1]) and zero displacements at the basal boundary so that the unknown boundary tractions due to tissue deformation don't participate in δW_{ext} in Eq. (6). Specifically, the chosen VF was obtained from passively inflating the LV as specified in Sect. 2.2 under a fully fixed base (with Guccione material model parameters: $C_1 = 100$, $\alpha = 15$ and $r_f, r_{ft}, r_t = 1/3$).

2.6   Sensitivity Study

To provide an estimation of the severity of data quality and model-data discrepancies on the estimated parameter values with the VW based approach, a sensitivity study was performed. For estimating effects of miscalibration in pressure measurements, synthetic data sets with modified pressure traces (pressure offset by $\pm10\%$ of mean pressure value) were used. Additionally data sets with modified deformation fields were examined, were white Gaussian noise with 1–10% STD of the mean value was independently applied to the distribution of each of the 9 deformation gradient tensor ($\boldsymbol{F}$) components at the employed Gauss points used for the integration of the energy density function (Eq. (1)). The impact of the assumed fiber field in the model was evaluated by assuming an alternative $-60°/+60°$ fiber angle variation from epi-to endocardium. To investigate the effect of the assumed anisotropy ratio parameter values (r_f, r_{ft}, r_t) alternative assumptions of isotropic ($r_f = 0.34, r_{ft} = 0.33, r_t = 0.33$) or highly anisotropic ($r_f = 0.85, r_{ft} = 0.1, r_t = 0.05$) material assumptions were made in the analysis (see Sect. 2.3). The results of the sensitivity study are shown in Table 1.

2.7   Basal Work Contribution Quantification at the Boundaries

To provide context for the choice of prescribed basal displacements in the 3 synthetic data sets (SC1-3) we estimated the external work at the base (W_{ext}^{base}) with respect to the strain energy (W_{int}) in the myocardium for the ground truth parameters. This ratio was also quantified in 7 clinical data sets from 6 HF patients (PC1, PC3-7, case PC2 was removed as an outlier) and one healthy volunteer (HC) [8]. The estimation of W_{ext}^{base} is given by Eq. (12) [1,6], where J is the determinant of the deformation gradient $\boldsymbol{F}$. The bottleneck in this estimation is the indeterminate Lagrange multiplier p, which here was provided by the simulation outputs, but is not available from deformation data. (W_{ext}^{base} estimation in the clinical cases was based on passive inflation simulations with the

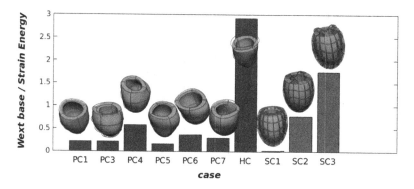

Fig. 2. Comparison of basal work to strain energy ratios at the last 'frame' in the three synthetic data sets (SC1-3) to the ratios of the clinical cases (PC1-7,HC) in [8] at the end diastolic frame (Sect. 2.7). The reference configuration of each case is shown as green surfaces and the red lines denote the deformed configuration. Note that the highest ratio in the clinical data sets occurs for the healthy case (HC) in accordance with the hypothesis that atrio-ventricular plane displacement is compromised in HF. (Color figure online)

identified parameters from [8] with cavity pressure and prescribed basal displacements from end diastolic frame 'DF' data). The non-dimensionalised results are shown in Fig. 2 in terms of ratios of W_{ext}^{base} with respect to W_{int} in the myocardial domain. Please note that these results provide an 'order of magnitude' level of quantification of the possible contributions of the basal boundary tractions (there are possible numerical errors associated with stress recovery from FE meshes, which have not been accounted for in this work [10]).

$$W_{ext}^{base} = \int_{u=0}^{u_{DF}} \int_{\Gamma_b} [(1/J F \frac{\partial \Psi}{\partial E} F^T - pI) \cdot da] \cdot du \qquad (12)$$

3   Results

The results of the parameter estimation study with and without the VW method and the sensitivity analysis for the basal deformation data set SC2 (with the highest offset between identified and ground truth values) are given in Table 1.

4   Discussion

In this work we propose a solution to the presence of unknown basal tractions while still providing a unique myocardial material parameter estimation. The solution is based on an energy-based cost function that overcomes the parameter coupling problem, and the novelty is a modified version of this CF based on the principle of virtual work. Using a virtual field within a weak expression of the equilibrium brings fundamental advantages, since it enables us to render

Table 1. Upper part: Results of parameter estimation with and without use of the modified CF with the virtual fields method in the 3 synthetic data sets (ground truth values: $\alpha = 15$, $C_1 = 100$ Pa). Lower part: Sensitivity analysis results of parameter estimation against data noise and modelling assumptions for the synthetic data set SC2 where the identified parameters varied more from the ground truth (ground truth values: $\alpha = 15$, $C_1 = 100$ Pa). n/a: cost function was unable to identify α parameter (since the estimation of C_1 is derivative, its estimation was also unsuccessful)

Data set/Analysis:	f^{EC}		f^{VW}	
Analysis results	α^{EC}	C_1^{EC} (Pa)	α^{VW}	C_1^{VW} (Pa)
SC1	15	114	16	108
SC2	n/a	n/a	10	175
SC3	n/a	n/a	16	106
Sensitivity analysis				
Data/Model modification	α^{EC}	C_1^{EC} (Pa)	α^{VW}	C_1^{VW} (Pa)
Default SC2	n/a	n/a	10	175
Pressure $+10\%$ $\bar{p}$ offset	n/a	n/a	9	208
-10% $\bar{p}$ offset	n/a	n/a	11	147
$\boldsymbol{F}$ noise STD 1% $\bar{F}_{ij}$	n/a	n/a	9	197
STD 5% $\bar{F}_{ij}$	n/a	n/a	11	152
STD 10% $\bar{F}_{ij}$	2	< 0	4	455
Fibers $-/+$ 60 ^{o}	n/a	n/a	10	168
r_f-r_{ft}-r_t 0.85-0.1-0.05	n/a	n/a	5	580
0.34-0.33-0.33	n/a	n/a	11	115

conspicuous aspects of deformation irrelevant to the analysis (i.e. remove the impact of the unknown basal traction). The choice of a suitable virtual field is the core aspect, and prescribing a fixed base in a passive inflation was a relatively simple way of finding it, with easy implementation in clinical data sets. The only simulation required is thus the one that generates the virtual field (a passive inflation of the reference configuration mesh with a fixed base), and the result does no more depend on the accurate estimation of boundary conditions the prediction of deformations.

The feasibility of the method is demonstrated in the analysis of three synthetic data sets (each including cavity pressure and myocardial deformation measurements) generated from a passive inflation simulation with varying degrees of prescribed basal displacements. The virtual work based parameter estimation (α^{VW}, C_1^{VW}) shows an improved performance in estimating myocardial material parameters in the presence of large basal deformations compared to the original energy based approach (α^{EC}, C_1^{EC}), where there is no valid solution found to the original cost function (f^{EC}) in most of the cases (Table 1). In order to put our results into perspective, it is worth noting that the errors in identified parameters of a 50% are much smaller than the fundamental limitation caused by a non-unique material parameter estimation [7]. Moreover, to provide context for the introduced basal deformations in the synthetic data sets, we first estimated the basal work present in real cases (see Sect. 2.7). Our situations of basal deformation are thus in the range of observed deformations in clinical data (see Fig. 2). As it may be expected, the clear tendency observed is that the bias is more pronounced in cases with larger deformations (i.e. the healthy case HC), and it has a variable range across the heart failure cases.

The robustness of the method is shown in the sensitivity analysis to basic modelling assumptions and data noise (Table 1). Consistent to original results [7,8], the factor most compromising accuracy was shown to be the deformation noise, which is modelled in the worst case for the estimation of strain: white noise that has no spatial correlation. Pressure data quality, which is often compromised in clinical data sets, was also shown to be important. On the modelling assumptions side, the least influential factor is the myofiber orientation in the model, but the anisotropy ratios was shown to be critical. A result which comes in antithesis to previous analysis with a $-60°/+60°$ fiber field in the data generating simulation, highlighting the effect of the fixed anisotropy assumption in the presence of a myocardium with very longitudinally oriented fibers at the boundaries.

One last fundamental advantage is that this approach removes the requirement for a complete cavity pressure trace measurement in diastole as the modified CF f^{VW} focuses on two frames only in the diastolic window. Hence only relative pressure measurements are required at these two frames since f^{VW} utilises a pressure ratio only (Eq. 9), unlike the original approach where the cost function f^{EC} requires knowledge of the cavity pressure throughout most of diastole (Eq. 4).

The main limitation of this approach is that it depends more on the quality of the data, in this case of the deformation field that is used to estimate the strain energy. Cost functions can be defined in terms of much more robust observations, such as volume, but at the cost of lack of identifiability of estimated parameters. Image analysis techniques that extract deformation fields that respect model assumptions [2] can provide a solution to alleviate this dependence. Moreover, the proposed parameter estimation procedure relies on a single 'frame' combination (DF_1, DF_2) for α and on DF_2 only for C_1 based on our results that indicated frame independent in silico. However this step will need to be revised before application of the method to clinical data sets and more frames will need to be incorporated in the pipeline in anticipation of data noise.

Future extensions of this work involve exploration of the application of the virtual fields method to eliminate the effect of external tractions in the remaining boundaries of the ventricle (i.e. the impact of the right ventricle and the kinematic constraints on the epicardium) and its translation to clinical data sets.

Acknowledgements. This work was supported by the Wellcome/EPSRC Centre for Medical Engineering [WT 203148/Z/16/Z] and by the National Institute for Health

Research (NIHR) Cardiovascular MedTech Co-operative. PL holds a Wellcome Trust
Senior Research Fellowship [209450/Z/17/Z].

References

1. Bonet, J., Wood, R.D.: Nonlinear Continuum Mechanics for Finite Element Analysis, 2nd edn. Cambridge University Press, Cambridge (2008)
2. Genet, M., Stoeck, C., von Deuster, C., Lee, L., Kozerke, S.: Equilibrated warping: finite element image registration with finite strain equilibrium gap regularization. Med. Image Anal. 50, 1–22 (2018)
3. Guccione, J.M., McCulloch, A.D., Waldman, L.K.: Passive material properties of intact ventricular myocardium determined from a cylindrical model. J. Biomech. Eng. 113(1), 42–55 (1991)
4. Hadjicharalambous, M., Lee, J., Smith, N.P., Nordsletten, D.A.: A displacement-based finite element formulation for incompressible and nearly-incompressible cardiac mechanics. Comput. Methods Appl. Mech. Eng. 274(100), 213–236 (2014)
5. Lamata, P., et al.: Images as drivers of progress in cardiac computational modelling. Prog. Biophys. Mol. Biol. 115(2–3), 198–212 (2014)
6. Lee, J., et al.: Multiphysics computational modeling in CHeart. SIAM J. Sci. Comput. 38(3), C150–78 (2016)
7. Nasopoulou, A., Nordsletten, D.A., Niederer, S.A., Lamata, P.: Feasibility of the estimation of myocardial stiffness with reduced 2D deformation data. In: Pop, M., Wright, G.A. (eds.) FIMH 2017. LNCS, vol. 10263, pp. 357–368. Springer, Cham (2017). https://doi.org/10.1007/978-3-319-59448-4_34
8. Nasopoulou, A., et al.: Improved identifiability of myocardial material parameters by an energy-based cost function. Biomech. Model. Mechanobiol. 16(3), 971–988 (2017)
9. Palit, A., Franciosa, P., Bhudia, S.K., Arvanitis, T.N., Turley, G.A., Williams, M.A.: Passive diastolic modelling of human ventricles: effects of base movement and geometrical heterogeneity. J. Biomech. 52, 95–105 (2017)
10. Sharma, R., Zhang, J., Langelaar, M., van Keulen, F., Aragón, A.M.: An improved stress recovery technique for low-order 3D finite elements. Int. J. Numer. Methods Eng. 114(1), 88–103 (2018)
11. Westermann, D., et al.: Role of left ventricular stiffness in heart failure with normal ejection fraction. Circulation 117(16), 2051–2060 (2008)
12. Xi, J., et al.: The estimation of patient-specific cardiac diastolic functions from clinical measurements. Med. Image Anal. 17(2), 133–146 (2013)

Advanced Cardiac Image Analysis Tools for Diagnostic and Interventions

Fully Automated Electrophysiological Model Personalisation Framework from CT Imaging

Nicolas Cedilnik[1,2(✉)], Josselin Duchateau[2], Frédéric Sacher[2], Pierre Jaïs[2], Hubert Cochet[2], and Maxime Sermesant[1]

[1] Université Côte d'Azur, Epione Research Project, Inria,
Sophia Antipolis, France
nicolas.cedilnik@inria.fr
[2] Liryc Institute, Bordeaux, France

Abstract. There has been a recent growing interest for cardiac computed tomography (CT) imaging in the electrophysiological community. This imaging modality indeed allows to locate and assess post-infarct scar heterogeneity, allowing to predict zones of abnormal electrical activity and even personalise EP models.

To this end, most of the literature uses manually segmented CT images where one fundamental information is extracted, the myocardial wall thickness. In this paper, we evaluate the impact of using an automated deep learning (DL) methodology to segment the left ventricular wall and extract relevant scar information on the resulting personalised models.

Using CT images from 8 patients that were not used during the DL training, we show that the automated segmentation is very similar to the manual one (median Dice score: 0.9). Thickness information obtained this way is also very close to the manual one (median difference: 0.7 mm). A wavefront propagation model personalisation framework based on this thickness information does not show relevant differences in its output (median difference in local activation time: 2 ms), proving its robustness. Bipolar electrograms, simulated through a novel approach, do not differ significantly between manual and automated segmentations (Pearson's r: 0.99).

Keywords: Imaging · Deep learning · Segmentation ·
Model personalisation

1 Introduction

Catheter radiofrequency ablation of the ischemic arrhythmogenic substrate has been shown to be efficient to prevent sudden cardiac deaths. This efficiency can partially be attributed to the integration of information extracted from cardiac imaging data both prior to and during the intervention [11].

© Springer Nature Switzerland AG 2019
Y. Coudière et al. (Eds.): FIMH 2019, LNCS 11504, pp. 325–333, 2019.
https://doi.org/10.1007/978-3-030-21949-9_35

Fig. 1. The model personalisation pipeline. Orange: image processing steps; Green: modelling steps (Color figure online)

While cardiac magnetic resonance imaging (MRI) is still widely considered the gold standard for ischemic scar assessment and model personalisation [9], computed tomography (CT) has recently gained interest in the electrophysiological (EP) community. Cardiac CT is indeed able to locate the chronic ischemic scar and evaluate its heterogeneity [4,7]. This is possible by identifying zones of myocardial wall thinning and by evaluating the severity of this thinning; this approach is even able to predict abnormal electrical activity [2,10]. Moreover, CT is less affected than MRI by the presence of a implantable cardioverter defibrillator, a device commonly found in patients susceptible to undergo catheter ablation of ventricular tachycardia.

Given these scar characteristics on CT images, it is no surprise that CT has been successfully used as a way to personalise EP models [1]. However, up until now, this personalisation relies on manual or semi-automated segmentation of the left ventricular (LV) wall, despite the availability of efficient three-dimensional medical image automated segmentation methods [5].

In this paper we evaluate the impact of a deep learning automated segmentation approach on CT ischemic scar assessment and the robustness of the related model personalisation framework. This framework (Fig. 1) takes a cardiac CT image as input where the LV is automatically segmented using a neural network, allowing myocardial thickness to be automatically computed [12]. After choosing a virtual pacing point, activation maps can be simulated using the Eikonal model. Finally, electrograms can be generated through a novel approach presented in Subsect. 2.4.

2 Methods

2.1 Deep Learning Segmentation

The deep learning approach (Fig. 2) we applied is based on a previously described methodology to segment the left atrium [6]. It relies on the use of two successive specialised U-nets [5]. The first is used to coarsely segment the full original CT image. Its output is used to compute the bounding box of the region of interest. This allows a cropping of the original image for a higher resolution segmentation of the desired structures.

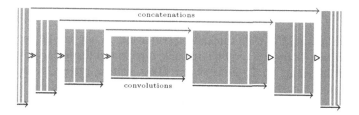

Fig. 2. Architecture of the U-net used to segment cardiac CT images (green blocks: 3D features, ≫: max-pooling, ▷: up-convolution). A first "low-resolution" network is used to determine the left ventricle location on the original CT image. A cropped version of the image is then fed to a "high resolution" net. Adapted from [6].

Database and Training. The network was trained from scratch using a database of 500 cardiac manual segmentations of contrast-enhanced CT images. These segmentations comprise the LV endocardium, the LV epicardium and the right ventricular epicardium. We used 450 cases for training *per se* and 50 cases for validation with a loss function defined as the opposite of a label-wise Dice score. The model was fitted using an nVidia GeForce 1080 Ti provided by the NEF computing platform.

"Low Resolution" U-Net. The first network's input is the original CT image resampled to $128 \times 128 \times 128$ voxels and it outputs 3 ventricular masks: 2 for the left ventricle (epicardium and endocardium), 1 epicardial right ventricle. The training data was augmented twice, by 2 random rotation of the original image along each axis in the $\left[-\frac{\pi}{8}; \frac{\pi}{8}\right]$ interval.

"High Resolution" U-Net. The second network was trained using cropped CT images around the LV, with 5 mm margins and resampled to $144 \times 144 \times 144$ voxels. Each original image was augmented 20 times: by random rotations in the $\left[-\frac{\pi}{7}; \frac{\pi}{7}\right]$ range along each axis, a random shearing in the $[-0.1; 0.1]$ range along each axis and the application of Gaussian blur with a kernel using a random standard deviation picked in the $[0.5; 2]$ range for half of the augmentations. The network outputs are the two left ventricular masks.

Post-processing. The network's outputs were thresholded at 0.5 to obtain binary 3D masks. In order to obtain spatially coherent masks, they were filtered to keep only the largest connected component and to forbid overlap between masks. Remaining holes in the masks were filled using the most frequent label in the hole neighbourhood. The masks were finally up-sampled to the original CT image resolution using a nearest neighbour interpolation.

2.2 Thickness Computation

Smooth and robust LV wall thickness estimation is not trivial, especially in three dimensions. It indeed requires to solve a partial differential equation using the endocardium and epicardium masks [12]. Such approach, which assign a thickness value to each voxel of the LV wall mask, is particularly adapted for simulations on regular grids; it has been previously used to such ends [1].

2.3 Electrophysiological Model

We used the thickness information to parameterise an Eikonal model previously described in [1]. Briefly, wall thinning is related to a macroscopic slowing of the activation front, due to a microscopic zig-zag course of activation in the infarcted tissue. A random pacing point was chosen in the healthy tissue, defined by a wall thickness superior to 5 mm, to initiate the propagation. The simulation was stopped at 500 ms.

2.4 Electrogram Simulation with the Eikonal Model

We propose an efficient way to simulate electrograms with the Eikonal model. It couples the activation map with a transmembrane action potential model and a propagation methodology based on the dipole formulation [3]. In this framework, every voxel is a considered a dipole with local current density:

$$\mathbf{j}_{eq} = -\sigma \nabla v, \tag{1}$$

where σ is the local conductivity and ∇v is the spatial gradient of the potential v. Using the chain rule, we can rewrite (1) as:

$$\mathbf{j}_{eq} = -\sigma \frac{\partial v}{\partial T} \times \frac{\partial T}{\partial X} = -\sigma \frac{\partial v}{\partial T} \nabla T, \tag{2}$$

where ∇T is the gradient of the activation map (output of the Eikonal model).

To compute $\frac{\partial v}{\partial T}$, we used a forward Euler scheme to solve the Mitchell-Schaeffer cardiomyocite action potential model [8] and stored its time derivative. We matched the activation time obtained with this model to the activation time obtained with the Eikonal model. To simulate bipolar electrograms, we placed two virtual electrodes (distance between them: 0.9 mm) inside the heart cavity and subtracted one signal to the other.

In order to handle the activation map gradient values at the mask boundaries, we ignored the gradient component(s) in the direction(s) flowing out of the mask. We adapted the volume used in the dipole moment computation accordingly [3].

2.5 Evaluation of the Automated Segmentation Impact

To evaluate the automated segmentation impact, we focused on 8 cardiac CT images (and their corresponding manual segmentation) of patients suffering of re-entrant VT. Theses images have been used neither for the automated segmentation training nor its validation. For 6 cases (patients 1–6) the original CT images were available; for the remaining 2 cases, resampled images aligned to the heart short axis were used.

Binary masks were resampled to the original CT image resolution when available, in order to compute a Dice score on the LV wall.

To compare thickness and activation maps, a mid-wall mesh was generated from the manual wall segmentation. Maps corresponding to manual and automated segmentation were then projected on this common frame of reference. A point-wise comparison was made possible this way, and median differences were computed.

To compare the electrograms obtained with both segmentation methods, they were compared with a Pearson correlation coefficient r.

3 Results

3.1 Segmentation

The training phase of the "high resolution" network reached a plateau at a Dice score of 0.96 for the LV wall (Fig. 3). Data augmentation was key to reaching this high score, especially adding the blur filter.

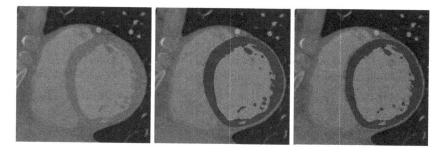

Fig. 3. Visual comparison of manual (middle) and automated (right) segmentations of the left ventricular wall shown on one slice of a CT that was not used during the DL network's fitting.

The median Dice score of the 8 images after up-sampling was lower: 0.90. There is no notable difference in the segmentation quality when using original CT images versus short-axis resampled images.

In zones of extreme thinning, the wall continuity was not always observed.

3.2   Thickness Computation

Across all cases, the median thickness of the manually segmented walls was
5.8 ± 3.2 mm versus 5.8 ± 3.0 mm for the automatic segmentation. Very similar thickness maps were obtained using the automated segmentation (median
differences across all cases: 0.7 mm, see Fig. 4).

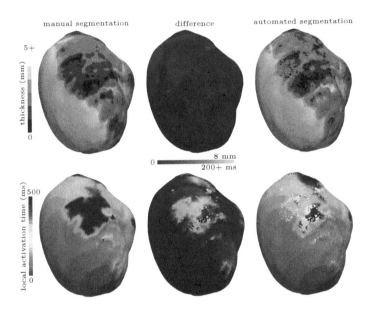

Fig. 4. Impact of the automated segmentation on the thickness and activation maps
for one example patient. The values are projected on the mid-wall meshes that were
used for quantitative comparisons.

3.3   Eikonal Model

As expected, given similar thickness maps and geometries, the "virtual pacing"
results were very close. The median activation time of the manual segmentation
models was 143 ms, with a median difference with automated segmentation as
little as 2 ms across all cases.

3.4   Electrogram Simulations

Bipolar signals generated with the automated and manual segmentations were
virtually identical, with a median Pearson's r correlation coefficient of 0.99. All
rs were above 0.94 except one (patient 8, 0.75), and all had p-values below 10^{-10}
(Fig. 5 and Table 1).

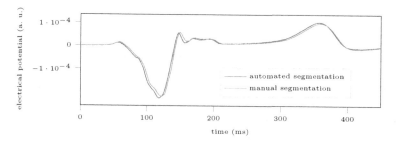

Fig. 5. Example comparison between simulated intra-cardiac bipolar electrograms using manual and automated segmentations (see Subsect. 2.4 for the methodology)

Table 1. [**Left**] **Segmentation quality.** DCS: dice score; HD: Hausdorff Distance; ADH: Average Hausdorff Distance; TD: Thickness Median Difference. [**Right**] **Model robustness:** difference between outputs from manual and automated segmentation. LATD: Local Activation Time Median Difference; EGr: Pearson's r coefficient between electrograms ($p < 10^{-10}$ for all patients)

Patient	DSC	HD (mm)	AHD (mm)	TD (mm)	LATD (ms)	EGr
1	0.88	19.15	0.10	0.7	5	0.99
4	0.90	17.47	0.09	0.8	2	0.99
5	0.87	13.25	0.09	0.8	3	0.99
6	0.91	5.00	0.05	0.5	1	1.00
7	0.91	16.44	0.06	0.6	1	1.00
8	0.88	7.57	0.09	1.4	29	0.75
9	0.89	22.86	0.11	0.7	2	0.94
10	0.91	12.71	0.08	0.7	1	0.99
Median	**0.90**	**14.85**	**0.09**	**0.7**	**2**	**0.99**

4 Discussion

4.1 Segmentation

The automated segmentation algorithm was shown to produce segmentations very close to those obtained by trained radiologists, even with different orientations of the input images. A perfect match between the algorithm and expert segmentation is not desirable anyway as there is also uncertainty in the manual segmentation. The available automated segmentation can probably be improved by training the neural networks on GPUs with more RAM, allowing a better input resolution and less loss of information in the down-sampling phase.

4.2 Thickness

The differences in the corresponding thickness maps are even smaller. Scar heterogeneity is preserved and comparable between manual and automated segmentation. In some particular cases of wall configuration the thickness computation could be problematic but it only happens in very localised cases which were easy to identify.

Ideally it would have been preferable to compare scar localization on MRI images, but they were not available for these patients.

4.3 Modelling

As expected, similarity of thickness maps leads to very similar simulation output between the automated and manual segmentation. The wall discontinuities are not problematic at all since they concern zones of extreme wall thinning that are considered non conductive anyway. These explain most of the discrepancy between the resulting activation maps.

5   Conclusion

We presented in this manuscript the automatic segmentation of the myocardial wall in CT images and the quantification of its impact on personalised models. We showed that most of the infarct related arrhythmia information extraction from CT images are not affected much by using an automated segmentation methodology, proving its robustness. This is another major step towards the future use of EP model personalisation in clinical practice.

Furthermore, we presented a novel methodology to generate electrograms from activation maps using the Eikonal model and the dipole formulation. Here we used the same action potential characteristics across the whole domain, but our formulation makes it possible to easily vary them, using the LV wall thickness for instance. Filtering the signals obtained in the same way they are filtered in the EP lab could further improve their realism.

Acknowledgements. The research leading to these results has received French funding from the National Research Agency grant IHU LIRYC (ANR-10-IAHU-04).

References

1. Cedilnik, N., et al.: Fast personalized electrophysiological models from CT images for ventricular tachycardia ablation planning. EP-Europace **20**, iii94–iii101 (2018)
2. Ghannam, M., et al.: Correlation between computer tomography-derived scar topography and critical ablation sites in postinfarction ventricular tachycardia. J. Cardiovasc. Electrophysiol. **29**(3), 438–445 (2018)
3. Giffard-Roisin, S., et al.: Estimation of Purkinje activation from ECG: an intermittent left bundle branch block study. In: Mansi, T., et al. (eds.) STACOM 2016. LNCS, vol. 10124, pp. 135–142. Springer, Cham (2017). https://doi.org/10.1007/978-3-319-52718-5_15

4. Grutta, L.L., Toia, P., Maffei, E., Cademartiri, F., Lagalla, R., Midiri, M.: Infarct characterization using CT. Cardiovasc. Diagn. Therapy **7**(2), 171–188 (2017)

5. Isensee, F., Kickingereder, P., Wick, W., Bendszus, M., Maier-Hein, K.H.: Brain tumor segmentation and radiomics survival prediction: contribution to the BRATS 2017 challenge (2018). arXiv:1802.10508 [cs]

6. Jia, S., et al.: Automatically segmenting the left atrium from cardiac images using successive 3D U-nets and a contour loss. In: Pop, M., et al. (eds.) STACOM 2018. LNCS, vol. 11395, pp. 221–229. Springer, Cham (2019). https://doi.org/10.1007/978-3-030-12029-0_24

7. Mahida, S., et al.: Cardiac imaging in patients with ventricular tachycardia. Circulation **136**(25), 2491–2507 (2017)

8. Mitchell, C.C., Schaeffer, D.G.: A two-current model for the dynamics of cardiac membrane. Bull. Math. Biol. **65**(5), 767–793 (2003)

9. Prakosa, A., et al.: Personalized virtual-heart technology for guiding the ablation of infarct-related ventricular tachycardia. Nat. Biomed. Eng. **2**, 732–740 (2018)

10. Yamashita, S., et al.: Myocardial wall thinning predicts transmural substrate in patients with scar-related ventricular tachycardia. Heart Rhythm **14**(2), 155–163 (2017)

11. Yamashita, S., et al.: Image integration to guide catheter ablation in scar-related ventricular tachycardia. J. Cardiovasc. Electrophysiol. **27**(6), 699–708 (2016)

12. Yezzi, A., Prince, J.: An eulerian PDE approach for computing tissue thickness. IEEE Trans. Med. Imaging **22**(10), 1332–1339 (2003)

Validation of Equilibrated Warping—Image Registration with Mechanical Regularization—On 3D Ultrasound Images

Lik Chuan Lee[1] and Martin Genet[2,3](✉)

[1] Department of Mechanical Engineering, Michigan State University,
East Lansing, MI, USA
lclee@egr.msu.edu
[2] Laboratoire de Mécanique des Solides, École Polytechnique/CNRS/Université
Paris-Saclay, Palaiseau, France
martin.genet@polytechnique.edu
[3] M3DISIM team, INRIA/Université Paris-Saclay, Palaiseau, France

Abstract. Image registration plays a very important role in quantifying cardiac motion from medical images, which has significant implications in the diagnosis of cardiac diseases and the development of personalized cardiac computational models. Many approaches have been proposed to solve the image registration problem; however, due to the intrinsic ill-posedness of the image registration problem, all these registration techniques, regardless of their variabilities, require some sort of regularization. An efficient regularization approach was recently proposed based on the equilibrium gap principle, named equilibrated warping. Compared to previous work, it has been formulated at the continuous level within the finite strain hyperelasticity framework and solved using the finite element method. Regularizing the image registration problem using this principle is advantageous as it produces a realistic solution that is close to that of an hyperelastic body in equilibrium with arbitrary boundary tractions, but no body load.The equilibrated warping method has already been extensively validated on both tagged and untagged magnetic resonance images. In this paper, we provide full validation of the method on 3D ultrasound images, based on the 2011 MICCAI Motion Tracking Challenge data.

Keywords: Image registration · Finite element method · Equilibrium gap regularization

1 Introduction

Image registration plays a very important role in quantifying cardiac motion from medical images, which has significant implications in the diagnosis of cardiac diseases [21] and the development of personalized cardiac computational

© Springer Nature Switzerland AG 2019
Y. Coudière et al. (Eds.): FIMH 2019, LNCS 11504, pp. 334–341, 2019.
https://doi.org/10.1007/978-3-030-21949-9_36

models [5,9] to understand the pathophysiological mechanisms of heart diseases [2,20] and the effects of treatments [14,16]. While image registration is still largely performed as a separate step in the personalization of models [6,9], it can be integrated and coupled with data assimilation techniques [13], also called integrated correlation [10], to estimate model parameters in a robust and efficient manner [2].

Many approaches have been proposed to solve the image registration problem, and they can be broadly categorized into local and global approaches. In the latter category, the different image registration techniques vary depending on *(i)* the displacement field support (image-based *vs.* mesh-based), *(ii)* the interpolation scheme (*e.g.*, linear *vs.* splines *vs.* polynomials), *(iii)* the chosen image similarity metric (*e.g.*, L2 *vs.* mutual information) and/or *(iv)* the minimization algorithm [1,17]. Due to the intrinsic ill-posedness of the image registration problem, however, all these registration techniques, regardless of their variabilities, require some sort of regularization [3,4,7,12,17–19]. Various regularizers have been proposed, from pure mathematical (laplacian smoothing) to geometrical to complex mechanical regularizer enforcing that the solution satisfies basic physical principles.

To address this issue, an efficient regularization approach was recently proposed based on the equilibrium gap principle [4], named equilibrated warping [7]. Compared to previous work [4,10], it has been formulated at the continuous level within the (nonlinear) finite strain hyperelasticity framework, discretized consistently, and solved using the finite element method. Regularizing the image registration problem using this principle is advantageous as it produces a realistic solution that is close to that of an hyperelastic body in equilibrium with arbitrary boundary tractions (but no body load). It could also be potentially formulated and integrated with data assimilation techniques to estimate boundary tractions (e.g., cavity pressure) that has important clinical implications [7]. Here, we propose to validate the equilibrated warping method based on the 2011 MICCAI Motion Tracking Challenge data [17] on both tagged (3DTAG) and untagged (SSFP) magnetic resonance and 3D ultrasound (3DUS) images.

2   Methods

2.1   Image Registration Problem

$\tilde{I}_0$ & $\tilde{I}$ are two images representing the same body $\mathcal{B}$ at two instants t_0 & t:

$$\tilde{I}_0 : \begin{cases} \Box_0 \to \mathbb{R} \\ \underline{X} \mapsto \tilde{I}_0\left(\underline{X}\right) \end{cases} , \qquad \tilde{I} : \begin{cases} \Box \to \mathbb{R} \\ \underline{x} \mapsto \tilde{I}\left(\underline{x}\right) \end{cases} , \tag{1}$$

where $\Box_0$ & $\Box$ are the image domains at t_0 & t, which are usually identical. The domains occupied by the body $\mathcal{B}$ at t_0 & t are denoted Ω_0 & Ω, respectively. The problem is to find the smooth mapping $\underline{\Phi}$ between materials points in the reference and deformed domains:

$$\underline{\Phi} : \begin{cases} \Omega_0 \to \Omega \\ \underline{X} \mapsto \underline{x} = \underline{\Phi}\left(\underline{X}\right) \end{cases} , \tag{2}$$

where $\underline{X}$ & $\underline{x}$ denote the position of a given material point in the reference and deformed configurations. Equivalently, one can search for the smooth displacement field $\underline{U}$:

$$\underline{U} : \begin{cases} \Omega_0 \rightarrow \mathbb{R}^3 \\ \underline{X} \mapsto \underline{U}(\underline{X}) = \underline{\Phi}(\underline{X}) - \underline{X} \end{cases}. \tag{3}$$

Due to its intrinsic ill-posedness, the problem is formulated as a regularized minimization problem:

$$\text{find } \underline{\Phi}^{\text{sol}} = \text{argmin}_{\{\underline{\Phi}\}} \left\{ (1 - \beta) \Psi^{\text{cor}}(\underline{\Phi}) + \beta \Psi^{\text{reg}}(\underline{\Phi}) \right\}, \tag{4}$$

where Ψ^{cor} is the image similarity metric, or "correlation energy", Ψ^{reg} is the regularization energy, and β defines the regularization strength. The correlation energy is assumed to be convex, at least in the neighborhood of the solution, though it is in general not quadratic. Similarly, the regularization energy is assumed to be convex in the neighborhood of the solution.

2.2 Image Intensity-Based Global Image Registration

In image intensity-based global approaches, the following correlation energy is generally used:

$$\Psi^{\text{cor}}(\underline{\Phi}) = \frac{1}{2} \int_{\Omega_0} \left(\tilde{I}(\underline{\Phi}(\underline{X})) - \tilde{I}_0(\underline{X}) \right)^2 d\Omega_0. \tag{5}$$

Other metrics have been proposed; however, we retain this one notably because it can be differentiated straightforwardly.

It can happen that the body to track is partly out of the images, at the reference frame and/or at the registered frames. This is especially true for cardiac echocardiography, where the ventricle barely fits within the imaging cone. In this case, in order to prevent the body parts outside the images to play a role in the registration, we propose the following augmentation of the correlation energy:

$$\Psi^{\text{cor}}(\underline{\Phi}) = \frac{1}{2} \int_{\Omega_0} \mathbb{1}(\underline{X}) \mathbb{1}(\underline{\Phi}(\underline{X})) \left(\tilde{I}(\underline{\Phi}(\underline{X})) - \tilde{I}_0(\underline{X}) \right)^2 d\Omega_0, \tag{6}$$

where $\mathbb{1}$ denotes the indicative function of the imaging domain, which corresponds to the entire image box in MRI, and only to the imaging cone in 3DUS.

2.3 Equilibrium Gap Regularization

In order to describe the equilibrium gap regularization that was first introduced in [4] and first formulated at the continuum level within the framework of nonlinear finite strain hyperelasticity in [7], let us first recall that mechanical equilibrium, i.e., conservation of momentum, in absence of body load and inertia, can be expressed as:

$$\begin{cases} \text{Div}\left(\underline{\underline{F}} \cdot \underline{\underline{S}}\right) = 0 \\ {}^t\underline{\underline{S}} = \underline{\underline{S}} \end{cases} \quad \forall \underline{X} \in \Omega_0, \tag{7}$$

where $\underline{S}$ is the second Piola-Kirchhoff stress tensor, and $\underline{F} = \frac{\partial \Phi}{\partial \underline{X}}$ is the transformation gradient [8]. These relations correspond to the conservation of linear and angular momentum, respectively. We also recall that the second principle of thermodynamics requires that:

$$\underline{S} = \frac{\partial \rho_0 \psi}{\partial \underline{E}}, \tag{8}$$

where $\rho_0 \psi$ is the free energy density, and $\underline{E} = \frac{1}{2} \left(\underline{C} - \underline{1} \right)$ the Green-Lagrange strain tensor, with $\underline{C} = {}^t\underline{F} \cdot \underline{F}$ the right Cauchy-Green dilatation tensor [8]. The Green-Lagrange strain tensor is symmetric, such that when computed through relation (8), the second Piola-Kirchhoff stress tensor is necessarily symmetric and the conservation of angular momentum is automatically verified. However, the conservation of linear momentum still needs to be enforced. In principle one could use $\Psi^{\mathrm{reg}} = \frac{1}{2} \left\| \mathrm{Div} \left(\underline{F} \cdot \underline{S} \right) \right\|^2_{L^2(\Omega_0)}$. However, Problem (4) is discretized using standard Lagrange finite elements, so that $\underline{F}$ and $\underline{S}$ belong to $L^2(\Omega_0)$, but not to $H(\mathrm{div}; \Omega_0)$. Thus, the following equivalent norm is used instead:

$$\Psi^{\mathrm{reg}} = \sum_K \frac{1}{2} \left\| \mathrm{Div} \left(\underline{F} \cdot \underline{S} \right) \right\|^2_{L^2(K)} + \sum_F \frac{1}{2h} \left[\!\left[\underline{F} \cdot \underline{S} \cdot \underline{N} \right]\!\right]^2_{L^2(F)}, \tag{9}$$

where K denotes the set of elements, F the set the interior faces with normal $\underline{N}$, and h a characteristic length of the finite element mesh. See [7] for more details.

2.4   Implementation and Workflow

The method has been implemented based on FEniCS[1][11] and VTK[2] [15] libraries, and the code is freely available[3]. In practice, to register the images, one only needs a segmentation of the object of interest, which can then be automatically meshed. From there the registration is entirely automatic.

3   Results and Discussion

In this section we present the validation of the equilibrated warping method for 3DUS cardiac images registration (results for SSFP & 3DTAG images were already presented in [7]), using the publicly available dataset of the Cardiac Motion Analysis Challenge that was held at the 2011 MICCAI workshop[4] and is described in details in [17]. Briefly, the dataset consists of untagged (SSFP) and tagged (3DTAG) magnetic resonance images as well as 3DUS images acquired from a dynamic phantom (PHANTOM) and fifteen healthy volunteers (V1-V2, V4-V16) with corresponding segmentations and markers that were manually

---

[1] https://www.fenicsproject.org.
[2] https://www.vtk.org.
[3] https://gitlab.inria.fr/mgenet/dolfin_dic.
[4] https://www.cardiacatlas.org/challenges/motion-tracking-challenge.

tracked by experts (*i.e.*, ground truth, GT). Results, in the form of tracked markers, from the challenge competitors (INRIA, IUCL, MEVIS, UPF) are also provided in the dataset.

Equilibrated warping was applied to all sets of 3DUS images. For all cases, a regularization strength of 0.1 was used, which was shown to be a good compromise in [7]. For the sake of illustration, Fig. 1 shows the result of the registration for V1. One can see the relatively good tracking of the ventricle despite the low image quality.

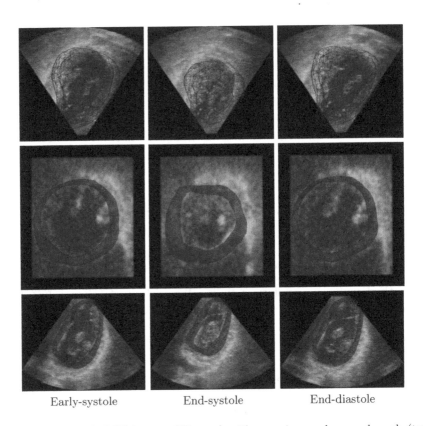

Early-systole   End-systole   End-diastole

Fig. 1. Sequence of 3DUS images (V1 case) with superimposed warped mesh (top), and sliced mesh in roughly short axis (middle) and long axis (bottom).

After successful computation of the displacement fields over the mesh at all time frames, the displacements were interpolated onto the markers, which were then warped as well. In order to quantitatively assess the quality of the registration, the following normalized error on the markers trajectory was used:

$$
\mathrm{err} = \frac{1}{n_{\mathrm{markers}}} \sum_{m=1}^{n_{\mathrm{markers}}} \frac{\sum_{f=1}^{n_{\mathrm{frames}}} \left\| \underline{X}^m (f) - \underline{X}^{m,\mathrm{GT}} (f) \right\|}{\sum_{f=1}^{n_{\mathrm{frames}}-1} \left\| \underline{X}^{m,\mathrm{GT}} (f+1) - \underline{X}^{m,\mathrm{GT}} (f) \right\|}, \tag{10}
$$

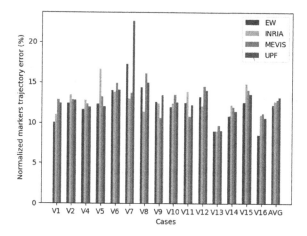

Fig. 2. Normalized markers error, for 3DUS images and for all cases, as well as normalized markers error mean over all cases (last column). EW stands for Equilibrated Warping [7], while the other columns represent challenge competitors [17].

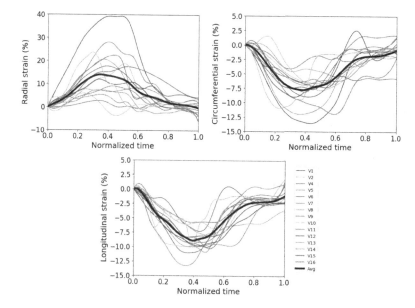

Fig. 3. Radial, circumferential and longitudinal strain components tracked from 3DUS images, for all cases as well as the average value.

where $n_{markers}$ is the number of "valid" markers (*i.e.*, markers that lie within the mesh in the reference configuration and can thus be tracked), n_{frames} is the number of frames, $\underline{X}^{m,GT}(f)$ is the ground truth position of marker m at frame f, and $\underline{X}^m(f)$ is the tracked position of marker m at frame f. See [7] for more details. A comparison of the normalized error between equilibrated warping and other challenge competitors is shown on Fig. 2. One can draw similar conclusions as for SSFP & 3DTAG images: equilibrated warping performs as well, if not better, than established registration methods.

For the sake of completeness, tracked strain components are also shown, on Fig. 3. Radial thickening, and circumferential and longitudinal shortenings are well captured, though underestimated, as already noted in [17].

4   Conclusion

Equilibrated warping provides a good registration in three different types of images (*i.e.* tagged and untagged magnetic resonance images as illustrated in [7], and 3DUS images as presented here for the first time) with a normalized marker error that is comparable, if not better, than other established methods from the challenge competitors. Besides its ability to correctly register material points in medical images as we have shown here, the main advantage of having a mechanically-sound regularization in equilibrated warping ensures that the displacement and strain fields are proper physical fields that can be used to derive physiologically relevant biomarkers [5, 21].

References

1. Bornert, M., et al.: Digital image correlation. In: Grédiac, M., et al.: Full-Field Measurements and Identification in Solid Mechanics. Wiley, Hoboken (2012). https://doi.org/10.1002/9781118578469.ch6
2. Chabiniok, R., et al.: Estimation of tissue contractility from cardiac cine-MRI using a biomechanical heart model. Biomech. Model. Mechanobiol. **11**(5), 609–630 (2012). https://doi.org/10.1007/s10237-011-0337-8
3. Christensen, G.E., et al.: Deformable templates using large deformation kinematics. IEEE Trans. Image Process. **5**(10), 1435–1447 (1996). https://doi.org/10.1109/83.536892
4. Claire, D., et al.: A finite element formulation to identify damage fields: the equilibrium gap method. Int. J. Numer. Methods Eng. **61**(2), 189–208 (2004). https://doi.org/10.1002/nme.1057
5. Finsberg, H., et al.: Efficient estimation of personalized biventricular mechanical function employing gradient-based optimization. Int. J. Numer. Methods Biomed. Eng. **34**(7), e2982 (2018). https://doi.org/10.1002/cnm.2982
6. Genet, M., et al.: A novel method for quantifying smooth regional variations in myocardial contractility within an infarcted human left ventricle based on delay-enhanced magnetic resonance imaging. J. Biomech. Eng. **137**(8), 081009 (2015). https://doi.org/10.1115/1.4030667
7. Genet, M., et al.: Equilibrated warping: finite element image registration with finite strain equilibrium gap regularization. Med. Image Anal. **50**, 1–22 (2018). https://doi.org/10.1016/j.media.2018.07.007
8. Holzapfel, G.A.: Nonlinear Solid Mechanics: A Continuum Approach for Engineering. Wiley, Chichester (2000)
9. Krishnamurthy, A., et al.: Patient-specific models of cardiac biomechanics. J. Comput. Phys. **244**, 4–21 (2013). https://doi.org/10.1016/j.jcp.2012.09.015

10. Leclerc, H., Périé, J.-N., Roux, S., Hild, F.: Integrated digital image correlation for the identification of mechanical properties. In: Gagalowicz, A., Philips, W. (eds.) MIRAGE 2009. LNCS, vol. 5496, pp. 161–171. Springer, Heidelberg (2009). https://doi.org/10.1007/978-3-642-01811-4_15

11. Logg, A., et al.: Automated Solution of Differential Equations by the Finite Element Method: The FEniCS Book. Lecture Notes in Computational Science and Engineering, p. 723. Springer, Heidelberg (2012). https://doi.org/10.1007/978-3-642-23099-8

12. Mansi, T., et al.: iLogDemons: a demons-based registration algorithm for tracking incompressible elastic biological tissues. Int. J. Comput. Vis. **92**(1), 92–111 (2011). https://doi.org/10.1007/s11263-010-0405-z

13. Moireau, P., et al.: Joint state and parameter estimation for distributed mechanical systems. Comput. Methods Appl. Mech. Eng. **197**(6–8), 659–677 (2008). https://doi.org/10.1016/j.cma.2007.08.021

14. Rausch, M.K., et al.: A virtual sizing tool for mitral valve annuloplasty. Int. J. Numer. Methods Biomed. Eng. **33**(2), e02788 (2017). https://doi.org/10.1002/cnm.2788

15. Schroeder, W., et al.: The Visualization Toolkit: An Object-Oriented Approach to 3D Graphics, 4th edn, p. 512. Kitware Inc, Clifton Park (2006)

16. Sermesant, M., et al.: Patient-specific electromechanical models of the heart for the prediction of pacing acute effects in CRT: a preliminary clinical validation. Med. Image Anal. **16**(1), 201–215 (2012). https://doi.org/10.1016/j.media.2011.07.003

17. Tobon-Gomez, C., et al.: Benchmarking framework for myocardial tracking and deformation algorithms: an open access database. Med. Image Anal. **17**(6), 632–648 (2013). https://doi.org/10.1016/j.media.2013.03.008

18. Veress, A.I., et al.: Measurement of strain in the left ventricle during diastole with cine-MRI and deformable image registration. J. Biomech. Eng. **127**(7), 1195–1207 (2005). https://doi.org/10.1115/1.2073677

19. Wang, H., et al.: Cardiac motion and deformation recovery from MRI: a review. IEEE Trans. Med. Imaging **31**(2), 487–503 (2012). https://doi.org/10.1109/TMI.2011.2171706

20. Xi, C., et al.: Patient-specific computational analysis of ventricular mechanics in pulmonary arterial hypertension. J. Biomech. Eng. **138**(11), 111001 (2016). https://doi.org/10.1115/1.4034559

21. Zou, H., et al.: Quantification of biventricular strains in heart failure with preserved ejection fraction patient using hyperelastic warping method. Front. Physiol. (2018). https://doi.org/10.3389/fphys.2018.01295

Ventricle Surface Reconstruction from Cardiac MR Slices Using Deep Learning

Hao Xu[1(✉)], Ernesto Zacur[1], Jurgen E. Schneider[2], and Vicente Grau[1]

[1] Institute of Biomedical Engineering, Department of Engineering Science,
University of Oxford, Oxford, UK
hao.xu@eng.ox.ac.uk
[2] Leeds Institute of Cardiovascular and Metabolic Medicine,
University of Leeds, Leeds, UK

Abstract. Reconstructing 3D ventricular surfaces from 2D cardiac MR data is challenging due to the sparsity of the input data and the presence of interslice misalignment. It is usually formulated as a 3D mesh fitting problem often incorporating shape priors and smoothness regularization, which might affect accuracy when handling pathological cases. We propose to formulate the 3D reconstruction as a volumetric mapping problem followed by isosurfacing from dense volumetric data. Taking advantage of deep learning algorithms, which learn to predict each voxel label without explicitly defining the shapes, our method is capable of generating anatomically meaningful surfaces with great flexibility. The sparse 3D volumetric input can process contours with any orientations and thus can utilize information from multiple short- and long-axis views. In addition, our method can provide correction of motion artifacts. We have validated our method using a statistical shape model on reconstructing 3D shapes from both spatially consistent and misaligned input data.

Keywords: Mesh reconstruction · Cardiac MRI · Deep learning

1 Introduction

Generating anatomically accurate 3D surface meshes is a key step in a wide range of applications including cardiac function analysis, interventional guidance and diagnosis [1–3]. Personalization of cardiac surfaces in 3D is also the first step required for computational simulations of cardiac electromechanics using the finite element method [4–6]. Cardiac MR (CMR) imaging provides accurate shape information of the heart non-invasively [2]. A standard clinical CMR study includes a stack of short-axis (SAX) slices, covering at least from the left/right ventricular (LV/RV) apex to the atrioventricular plane (base), plus at least two long-axis (LAX) views: horizontal long-axis (HLA, also known as 4 chamber view or 4CH) and vertical long-axis (VLA, also known as 2 chamber view or 2CH) [7]. Traditional isosurfacing algorithms cannot be directly used due to the sparsity

© Springer Nature Switzerland AG 2019
Y. Coudière et al. (Eds.): FIMH 2019, LNCS 11504, pp. 342–351, 2019.
https://doi.org/10.1007/978-3-030-21949-9_37

of the input data and because of the presence of motion artifacts (misalignment between slices caused by multiple breath holding and possible body movement during acquisition) [8], which make the task of reconstructing 3D structure from CMR data particularly challenging.

Reconstruction of 3D surfaces from CMR data is normally formulated as a 3D mesh adaptation problem to sparse contours or points [9–12], and solutions often incorporate shape priors during that process. The form of such prior could be a regular shape [10,11,13] or a statistical template with plausible variations [14], and deviations from regular and smooth geometries are penalized in a fitting process. These methods usually include an explicit smoothing term during fitting to further regularize the shape in addition to the use of shape priors, and both types of bias may compromise the accuracy of the reconstructed surfaces when handling pathological hearts [15,16]. In order to generate anatomically meaningful surfaces while also preserving the wide variability of the shape, reconstruction methods should incorporate a wide collection of plausible shape priors to refer to when fitting the input data. Deep learning algorithms have shown their capacity for storing a large number of accurate mappings between pair-wise data [17], and no explicit smoothing term is required by these methods, allowing the appearance of sharp edges and corners during reconstruction if necessary. However, for Convolutional Neural Networks (CNNs), input data are required to be highly structured (usually on a regular grid in 2D or 3D), and therefore most state-of-the-art deep learning methods for generating LV and RV 3D meshes using CMR data only take the SAX stack as input with the assumption that all SAX images are parallel [18,19]. This implementation limits the possibility of utilizing the LAX slices, which has been shown to improve the reconstruction of the ventricular meshes [12], as well as the possibility of incorporating out-of-plane geometric transformations for better correction of motion artifacts due to respiration [20,21].

In this paper, we propose to consider the sparse 3D information from contours in a volumetric form, and therefore tackle the problem similarly to volumetric image inpainting [22]. We transform the problem of mesh fitting from sparse input into a 3D volumetric mapping problem followed by isosurfacing from dense volumetric data. The sparse volumetric input can take contours in any positions within the volume, and therefore our method incorporates both SAX slices and LAX slices, and allows explicit out-of-plane motion artifact correction. Our method also has the capacity of reconstructing 3D meshes from misaligned input contours by itself.

1.1 Main Contributions

We have developed a novel bi-ventricular mesh reconstruction method through 3D volumetric mapping with a deep learning algorithm combining both SAX and LAX slices. To our best knowledge, this is the first deep learning method processing multiple views simultaneously without any assumption of image plane orientations.

We propose a method that can be used to reconstruct 3D meshes from both spatially consistent and misaligned input data, and our method is capable of processing intersecting slices with discrepancies among them, while most of the state-of-the-art methods rely on parallel contours.

2   Materials and Methods

Our 3D reconstruction method consists of three steps: 1. generating sparse volumetric input data from contours; 2. generating dense 3D volumetric predictions of LV myocardium and LV/RV cavities from input data using a variation of the 3D U-Net [23]; 3. generating 3D meshes from the predictions with an isosurfacing algorithm. The method is developed and evaluated by using synthetic data generated from a statistical shape model [14,24].

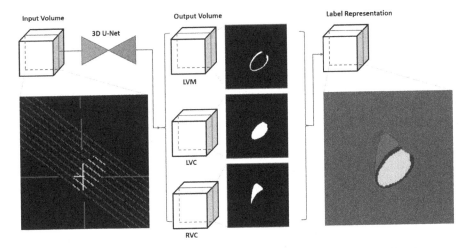

Fig. 1. Representation of the process carried out by the 3D U-Net. For illustration, cross-sectional views corresponding to specific slices (marked with dotted lines on the volumes) are shown; note that these do not correspond to original CMR slices. In the input volume, only voxels corresponding to the centre of CMR slices are assigned specific labels (2:red, 3:green, 4:blue and 1:cyan for RV cavity, LV cavity, LV myocardium and background, respectively), while all others are assigned the label "Unknown" (0). For each of the 3D output volumes, a value between 0 and 1 is assigned to each voxel representing the likelihood of belonging to the respective region. These 3D volumes are then combined to produce the label representation of the reconstructed volume, in which an average of 94.5% of the voxels are transformed from the original Unknown label. (Color figure online)

2.1   Data

We used the statistical shape model published by Bai et al. [14], in which the authors registered 1093 segmented hearts to a template space using rigid registration followed by the application of principal component analysis (PCA) to the surface meshes. The model is formed by labeled images with labels for background, LV myocardium and LV/RV cavities. We downloaded from the publicly available dataset (http://wp.doc.ic.ac.uk/wbai/data/) the mean shape model, the first 100 PCs and the corresponding eigenvalues or variances. We then used these to generate 120 different shapes simulating pathological and controlled cases from clinical data, limiting the variations to six standard deviations for any of the Principal Components.

To generate output references, we placed these shapes into a 3D volume with the size of $128 \times 128 \times 128$ and voxel size of $2 \times 2 \times 2 \, mm^3$, by aligning the centre of the smallest sphere enclosing the corresponding contours and the centre of the 3D volume, and labeled voxels enclosed by surfaces accordingly. To simulate input sparse volumes in conditions similar to real clinical acquisition, we used real image planes from clinical datasets (consisting of both SAX stack and two LAX slices). To generate spatially consistent input datasets, we first aligned one set of image planes to each of the shapes, and the voxels located within 0.5 mm away from the planes were then given their original labels acccording to the shape model. Voxels located more than 0.5 mm away from the plane, or those with inconsistent labels between slices were assigned the "Unknown" label. To mimic the misalignment caused by motion artifacts, we kept fixed image planes and applied 3D rigid transformations to the model (random rotations no larger than $10°$ and random translations of no more than 4 mm) before the calculation of the labels for each plane. An example of input and output volumetric data with cross-sectional views is shown in Fig. 1.

2.2   Volumetric Mapping

We adapted a variation of 3D U-Net for volumetric mapping. The schematic diagram of the network architecture is shown in Fig. 2.

The network consists of an encoder and a decoder, with skip connections between feature maps with the same resolution. The encoder starts from the input layer with the size of $128 \times 128 \times 128 \times 1$ and has three max-pooling stages where each pooling layer has a window size of $(2, 2, 2)$ and a stride of $(2, 2, 2)$, giving feature maps with four different resolutions. The decoder then upsamples the feature maps back to the original resolution also in three stages using upsampling layers with factors of $(2, 2, 2)$. The heavy duty of the computation within the network is carried out by convolutional blocks shown in Fig. 2, with each of them having two convolutional layers with kernel size of $(3, 3, 3)$ and two batch normalization (BN) layers. Other than the output layer, which has sigmoid activation functions, rectified linear units (ReLU) are used throughout the network. The output layer has 3 convolutional kernels with the size of $(1, 1, 1)$ giving the prediction in the form of $128 \times 128 \times 128 \times 3$ volume representing

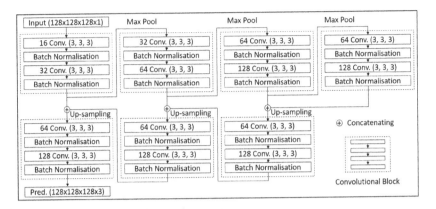

Fig. 2. A schematic diagram of our volumetric mapping network. For each convolutional layer, the number of filters is specified.

LV myocardium, LV cavity and RV cavity. The Dice coefficient loss was used during the training of the network to deal with imbalance between foreground and background voxels [25].

2.3 Isosurfacing

The prediction of the network was split into three $128 \times 128 \times 128$ volumes, and used to form the isosurfaces at 0.5 using marching cubes [26]. For each 3D volume, the largest object was then selected as the expected structure.

3 Experiments and Results

We performed two 4-fold cross-validation experiments with spatially consistent data and misaligned data respectively. For both experiments, the same number of bi-ventricular shapes were generated as described in Sect. 2.1, and they were randomly grouped into 4 sets of training data (80 shapes), validating data (10 shapes) and testing data (30 shapes) also in the same way, with the 4 testing datasets covering all generated shapes.

During the training phase of each cross-validation experiment, the neural network was randomly initiated and the network parameters were updated through back-propagation using the spatially consistent training data. The optimization was early stopped by evaluating the loss function of the validating data to prevent over-fitting. Once the training of the neural network for the spatially consistent cases was terminated, the learned parameters were used to initiate the network for reconstructing misaligned cases and fine-tuned using the misaligned training data. An overview of the training progress is shown in Fig. 3.

The testing data was only used for the evaluation of the method, and the predictions from the network for the testing data were then used to generate 3D meshes as described in Sect. 2.3. In both experiments, Dice coefficient and

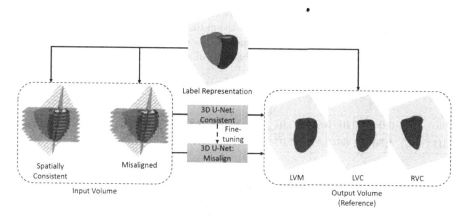

Fig. 3. Training progress overview. The blue, green and red objects respectively correspond to LV myocardium, LV cavity and RV cavity. For each case, both spatially consistent and misaligned input volumes are generated from the reference mesh according to Sect. 2.1. The output volume has voxels with values of 0 (light gray) and 1 (dark gray). (Color figure online)

Hausdorff Distance (HD) were the metrics used for quantitative analysis of the method. For the experiment with misaligned contours we registered the reconstructed shape to the reference shape with a rigid transformation before calculating the metrics. We also calculated the Euler characteristic for the generated meshes to evaluate the topology, and for all cases the value was expected to be equal to 2.

3.1 Statistical Shape Model

The Dice coefficient and HD results for experiments on spatially consistent and misaligned contours are shown in Table 1. For spatially consistent input data, the reconstructed LV/RV cavities shape matched the reference shape accurately with mean Dice coefficient values of 0.98. The LV myocardium has a more complicated shape with larger surface area, and therefore the Dice score is more sensitive to small changes, while the mean value is still 0.94. The voxel size is $2 \times 2 \times 2$ mm^3, and the mean values of HD are around 2 voxels. As the input data has only around 5.5% of voxels with known labels, the results suggest that our 3D U-Net stored a collection of plausible mappings between sparse input data and corresponding ventricular shapes during training and is able to utilize that information during the reconstruction of the testing cases.

The experiments on misaligned input data also achieved good accuracy, with above 0.93 mean Dice coefficient for LV/RV cavities and 0.83 for LV myocardium. As expected, comparing to the experiment on spatially consistent input data, the accuracy of these experiment is lower. This is unavoidable given the additional challenges involved in reconstructing the 3D shapes from misaligned input data, including discrepancies between slices, which results in the same set of input data

Table 1. Dice coefficient and Hausdorff Distance between reconstructed ventricular shape and reference ventricular shape for both spatially consistent and misaligned slices experiments. The Dice coefficient has values between 0 and 1 and Hausdorff Distance is in mm. The metrics are presented in the form of mean ± standard deviation.

	Consistent			Misaligned		
	LVM	LVC	RVC	LVM	LVC	RVC
Dice	0.94 ± 0.01	0.98 ± 0.01	0.98 ± 0.01	0.83 ± 0.04	0.95 ± 0.01	0.93 ± 0.02
HD (mm)	4.56 ± 1.66	3.17 ± 1.07	3.87 ± 2.28	6.30 ± 5.21	4.94 ± 3.61	5.91 ± 2.55

mapping to a much wider range of plausible output shapes. It is important to highlight the limitations of the validation method itself. The misalignment of all slices, combined, will result in an overall rigid transformation of the underlying three-dimensional shape. This rigidly transformed shape is most likely to be reconstructed by our network. If we compare this reconstruction with the original shape, the accuracy will be significantly affected, even if the shapes are exactly equal. We attempted at compensating for this unwanted effect by performing a rigid registration, but this still leaves residual errors that are partly responsible for the decrease of accuracy shown in Table 1.

For more than 85% of the 3D meshes directly generated from network predictions, the Euler characteristic is 2, suggesting a robust network output with no isolated false predictions, holes within the object or handles attached to it. However, the post-processing described in Sect. 2.3 is still needed and increases the rate by 8%.

3.2 Real Contours

The method is directly applicable to real clinical contours. One example of reconstructed meshes from clinical contours is shown in Fig. 3. Our method reconstructs plausible 3D meshes for LV myocardium and LV/RV cavities.

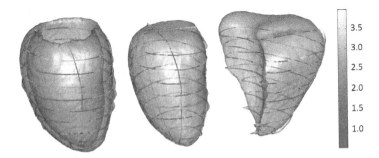

Fig. 4. Real contours aligned with the reconstructed meshes. The colours indicate the distance from contours to meshes. For distance smaller than 0.5 mm the colour is gray, and for distance larger than 4 mm the colour is purple. Colours corresponding to distances between 0.5 mm and 4 mm are shown in the color bar. (Color figure online)

The structure has no obvious distortions caused by misalignment, and the reconstructed meshes align well with the input contours. Quantitative validation in this case is complicated by the lack of gold standard 3D reconstructions (Fig. 4).

4   Conclusions

In this paper, we propose a bi-ventricular 3D mesh reconstruction method from CMR slices by transforming the mesh fitting problem into a volumetric mapping problem followed by isosurfacing. Our method takes advantage of a deep learning algorithm and is able to reconstruct anatomically realistic surfaces with a wide range of shapes from both spatially consistent and misaligned contours. The method has no constrains on the slice orientations and utilizes information from multiple SAX and LAX views simultaneously. It tolerates discrepancies between intersecting slices, and produces accurate 3D meshes from misaligned cases. We trained and evaluated our method using cases generated from a statistical shape model. We tested both the neural network predictions and reconstructed meshes using Dice coefficient and HD, and also evaluated topology with the Euler characteristic. A more detailed evaluation could be performed with more localized metrics, such as mesh quality metrics. We also applied our method to real contours and achieved good quality reconstructed meshes, suggesting the potential of the method to generate 3D meshes from CMR images directly. We showed that our network can compensate for motion artifacts such as the ones including out-of-plane misalignments. For future work, thanks to the flexibility of input slice orientations, our method can be developed to explicitly correct for image-plane rotation and translation in an iterative process to further minimize the disagreement between contours and reconstructed meshes.

Acknowledgments. We thank BHF Project Grant No. PG/16/75/32383 "Improving risk stratification in HCM through a computational anatomical analysis of ventricular remodelling" for support.

References

1. Vukicevic, M., Mosadegh, B., Min, J., Little, S.: Cardiac 3D printing and its future directions. JACC Cardiovasc. Imaging **10**, 171–184 (2017)
2. Suinesiaputra, A., et al.: Statistical shape modeling of the left ventricle: myocardial infarct classification challenge. IEEE JBHI **22**(2), 503–515 (2018)
3. Lehmann, H., et al.: Integrating viability information into a cardiac model for interventional guidance. In: Ayache, N., Delingette, H., Sermesant, M. (eds.) FIMH 2009. LNCS, vol. 5528, pp. 312–320. Springer, Heidelberg (2009). https://doi.org/10.1007/978-3-642-01932-6_34
4. Zacur, E., et al.: MRI-based heart and Torso personalization for computer modeling and simulation of cardiac electrophysiology. In: Cardoso, M.J., et al. (eds.) BIVPCS/POCUS -2017. LNCS, vol. 10549, pp. 61–70. Springer, Cham (2017). https://doi.org/10.1007/978-3-319-67552-7_8

5. Arevalo, H., et al.: Arrhythmia risk stratification of patients after myocardial infarction using personalized heart models. Nat. Commun. **7**, 11437 (2016)

6. Deng, D., Zhang, J., Xia, L.: Three-dimensional mesh generation for human heart model. In: Li, K., Li, X., Ma, S., Irwin, G.W. (eds.) ICSEE/LSMS -2010. CCIS, vol. 98, pp. 157–162. Springer, Heidelberg (2010). https://doi.org/10.1007/978-3-642-15859-9_22

7. AHA Writing Group on Myocardial Segmentation and Registration for Cardiac Imaging: Manuel D. Cerqueira, et al. "Standardized myocardial segmentation and nomenclature for tomographic imaging of the heart: a statement for healthcare professionals from the Cardiac Imaging Committee of the Council on Clinical Cardiology of the American Heart Association." Circulation **105**(4), 539–542 (2002)

8. Villard, B., Zacur, E., Dall'Armellina, E., Grau, V.: Correction of slice misalignment in multi-breath-hold cardiac MRI scans. In: Mansi, T., McLeod, K., Pop, M., Rhode, K., Sermesant, M., Young, A. (eds.) STACOM 2016. LNCS, vol. 10124, pp. 30–38. Springer, Cham (2017). https://doi.org/10.1007/978-3-319-52718-5_4

9. Zou, M., Holloway, M., Carr, N., Ju, T.: Topology-constrained surface reconstruction from cross-sections. ACM Trans. Graph. **34**, 128 (2015)

10. Young, A., et al.: Left ventricular mass and volume: fast calculation with guidepoint modelling on MR images. Radiology **2**, 597–602 (2000)

11. Medrano-Gracia, P., et al.: Large scale left ventricular shape atlas using automated model fitting to contours. In: Ourselin, S., Rueckert, D., Smith, N. (eds.) FIMH 2013. LNCS, vol. 7945, pp. 433–441. Springer, Heidelberg (2013). https://doi.org/10.1007/978-3-642-38899-6_51

12. Villard, B., Grau, V., Zacur, E.: Surface mesh reconstruction from cardiac MRI contours. J. Imaging **4**(1), 16 (2018)

13. Lamata, P., et al.: An accurate, fast and robust method to generate patient-specific cubic Hermite meshes. Med. Image Anal. **15**, 801–813 (2011)

14. De Marvao, A., et al.: Population-based studies of myocardial hypertrophy: high resolution cardiovascular magnetic resonance atlases improve statistical power. J. Cardiovasc. Magn. Reson. **16**, 16 (2015)

15. Zhang, X., et al.: Atlas-based quantification of cardiac remodeling due to myocardial infarction. PloS One **9**(10), e110243 (2014)

16. Alba, X., et al.: An algorithm for the segmentation of highly abnormal hearts using a generic statistical shape model. IEEE TMI **35**(3), 845859 (2016)

17. Zhang, C., et al.: Understanding deep learning requires rethinking generalization. arXiv preprint arXiv:1611.03530 (2016)

18. Oktay, O., et al.: Anatomically constrained neural networks (ACNNs): application to cardiac image enhancement and segmentation. IEEE TMI **37**(2), 384–395 (2018)

19. Duan, J., et al.: Automatic 3D bi-ventricular segmentation of cardiac images by a shape-refined multi-task deep learning approach. IEEE TMI (2019)

20. McLeish, K., Hill, D.L.G., Atkinson, D., Blackall, J.M., Razavi, R.: A study of the motion and deformation of the heart due to respiration. IEEE TMI **21**(9), 1142–1150 (2002)

21. Shechter, G., Ozturk, C., Resar, J.R., McVeigh, E.R.: Respiratory motion of the heart from free breathing coronary angiograms. IEEE TMI **23**, 1046–1056 (2004)

22. Bertalmio, M., Sapiro, G., Caselles, V., Ballester, C.: Image inpainting. In: Proceedings of the 27th Annual Conference on Computer Graphics and Interactive Techniques, pp. 417–424. ACM Press/Addison-Wesley Publishing Co. (2000)

23. Çiçek, Ö., Abdulkadir, A., Lienkamp, S.S., Brox, T., Ronneberger, O.: 3D U-Net: Learning dense volumetric segmentation from sparse annotation. In: Ourselin, S., Joskowicz, L., Sabuncu, M.R., Unal, G., Wells, W. (eds.) MICCAI 2016. LNCS, vol. 9901, pp. 424–432. Springer, Cham (2016). https://doi.org/10.1007/978-3-319-46723-8_49

24. Bai, W., et al.: A bi-ventricular cardiac atlas built from 1000+ high resolution MR images of healthy subjects and an analysis of shape and motion. Med. Image Anal. **26**, 133–145 (2015)

25. Milletari, F., Navab, N., Ahmadi, S.A.: V-net: fully convolutional neural networks for volumetric medical image segmentation. In: 4th 3DV, pp. 565–571 (2016)

26. Lorensen, W.E., Cline, H.E.: Marching cubes: a high resolution 3D surface construction algorithm. In: ACM Siggraph Computer Graphics, vol. 21, no. 4, pp. 163–169. ACM (1987)

FR-Net: Joint Reconstruction and Segmentation in Compressed Sensing Cardiac MRI

Qiaoying Huang[1](✉), Dong Yang[1], Jingru Yi[1], Leon Axel[2],
and Dimitris Metaxas[1]

[1] Department of Computer Science, Rutgers University, New Brunswick, NJ, USA
qh55@cs.rutgers.edu
[2] Department of Radiology, New York University, New York, NY, USA

Abstract. We provide a novel solution to the inverse problem in medical imaging that takes as input the undersampled k-space data from Magnetic Resonance Imaging (MRI) scans and outputs both the reconstructed images and the segmented myocardium. Previously, the undersampled k-space data is first transformed into a reconstructed MRI image. From this image, the myocardium is contours are subsequently extracted using a segmentation method. However, this sequential approach is not optimal and requires manual intervention. In order to automate and improve the results of these approaches, we propose a new method to solve the reconstruction and segmentation problems simultaneously. Our method is based on a novel deep learning approach we term "Joint-FR-Net", which consists of a reconstruction module derived from the fast iterative shrinkage-thresholding algorithm (FISTA) and a segmentation module. We test our approach on an undersampled short-axis (SAX) cardiac dataset and show the effectiveness of the Joint FR-Net in both image reconstruction and myocardium joint segmentation.

1 Introduction

Magnetic Resonance Imaging (MRI) is widely used for cardiac disease diagnosis and treatment monitoring. To achieve data acquisition efficiency, raw k-space data collected from MRI scanners are usually undersampled. Recently, deep neural networks have shown remarkable success in the field of MRI reconstruction. Different from Compressed Sensing (CS) based methods [7,9] that optimize the solution in an online manner, deep learning-based methods learn the mapping from input space to output space in an offline training process. The trained model can then be applied to reconstruct the test data in k-space. Deep learning based approaches are much faster and accurate than CS methods and several recent studies [3,4,6,11,13] have demonstrated their superiority.

Although these approaches solve the reconstruction problem, they do not simultaneously segment the myocardium and can't be used for automated cardiac disease diagnosis. In other words, these models are not designed to reconstruct images for specific applications such as myocardium segmentation [16].

© Springer Nature Switzerland AG 2019
Y. Coudière et al. (Eds.): FIMH 2019, LNCS 11504, pp. 352–360, 2019.
https://doi.org/10.1007/978-3-030-21949-9_38

Caballero *et al.* [2] treated both reconstruction and segmentation simultane-ously using a patch-based dictionary sparsity model and a Gaussian mixture model. However, it was an unsupervised learning technique that provided less accurate results compared to previous methods and also requires a large number of iterative steps. Other methods bypass the reconstruction of the MR image and only focus on segmentation part [12]. However, in clinical applications and disease diagnosis, it is necessary to devise an efficient automated approach that can simultaneously reconstruct MRI data and provide accurate segmentation of the myocardium.

In this paper, we devise a new approach that takes as input an undersam-pled k-space data and outputs both reconstructed MRI images and myocardium contour segmentation. Our approach is based on a deep learning based method called "Joint-FR-Net" that consists of two different modules: (a) the "FR-Net" emulates the FISTA algorithm to quickly reconstruct from k-space data into high-quality image, (b) a U-Net based segmentation network that segments the myocardium from the reconstructed image. Using an end-to-end training proce-dure, our proposed method achieves during testing state of the art results for the joint single-step based MRI image reconstruction and myocardium contour segmentation tasks from sparsely sampled k-space data.

2   Methodology

As shown in Fig. 1, our joint image reconstruction and myocardium segmenta-tion deep neural network based approach has the following structure: an image reconstruction network and a segmentation network.

2.1   Problem Formulation

The general objective of a CS-MRI reconstruction problem is to find the best reconstructed image x such that it minimizes the difference between undersam-pled data in k-space y, and undersampled data converted from x through the Fourier transform. The problem is formalized as follows:

$$x^* = \arg\min_x \frac{1}{2} \|E(F(x)) - y\|_2^2 + \lambda g(x), \tag{1}$$

where F is Fourier transform, E is an undersampled matrix and $\lambda g(\cdot)$ is a weighted regularization term. In MRI image reconstruction, the regularization function g usually takes ℓ_1-norm [7] or total variation norm [8].

Let $f(x) = \frac{1}{2} \|E(F(x)) - y\|_2^2$. According to FISTA [1], problem (1) is solv-able if f is a smooth convex function with a Lipschitz continuous gradient $L(f)$ and g is a continuous convex function and possibly nonsmooth. For any $L > 0$, consider the following quadratic approximation of Eq. (1) at a given point x'.

$$Q_L(x, x') = f(x') + (x - x') \cdot \bigtriangledown f(x') + \frac{L}{2} \|x - x'\|^2 + \lambda g(x), \tag{2}$$

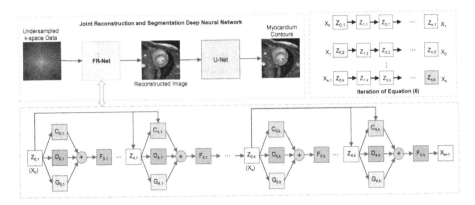

Fig. 1. Architecture of the proposed *Joint-FR-Net*. One submodule FR-Net transforms dynamic undersampled k-space data into a reconstructed image. The other submodule U-Net takes the reconstructed image as input and produces myocardium segmentation.

which admits a unique minimizer p_L. By ignoring constant terms related to x', we can rewrite p_L as:

$$p_L(x') = \arg \min \{Q_L(x, x')\} = \arg \min_x \left\{ \lambda g(x) + \frac{L}{2} \|x - (x' - \frac{1}{L} \triangledown f(x')\|^2 \right\}, \quad (3)$$

According to Eq. (3) and reconstruction problem equation (1), we can compute x_k iteratively using the following minimization approach:

$$x_k = \arg \min_x \left\{ \lambda g(x) + \frac{L_k}{2} \left\| x - (v_k - \frac{1}{L_k} \triangledown f(v_k)) \right\|^2 \right\}, \quad (4)$$

where v_k is a linear combination of x_{k-1} and x_{k-2} from the previous two iterations and L_k is a constant chosen by specific rule in the backtracking FISTA. This can generate a sequence of $\{x_1, x_2, \cdots, x_k\}$. In theory, if k goes to infinity, then x_k can approximate the optimal solution of equation (1). In order to speed up the converge, x_k is replaced by v_k, a linear combination of x_{k-1} and x_{k-2}. For further information, please refer to [1].

Note that in our MRI reconstruction problem, the gradient of $f(v_k)$ is defined as: $\triangledown f(v_k) = F^T(E(F(v_k)) - y)$, where F^T represents inverse Fourier transform. For specific forms of $g(\cdot)$ such as ℓ_1 norm, we can derive a closed form of x_k and then obtain the solution of x_k through gradient descent. More generally, we use gradient descent to generate a sequence of $\{z_{k,n}\}$ that gradually converges to x_k. Then we can generalize the problem to any form of $g(\cdot)$. Basically, from (4), $z_{k,n}$ is computed as follows:

$$z_{k,n} = z_{k,n-1} - \mu\{\lambda \nabla g(z_{k,n-1}) + L_k(z_{k,n-1} - v_k) + F^T(E(F(v_k)) - y)\}, \quad (5)$$

where μ denotes the gradient step and $\nabla g(\cdot)$ denotes the gradient of $g(\cdot)$. Figure 1 shows the relationship between z and x. Suppose the input is x_i. Using Eq. (5)

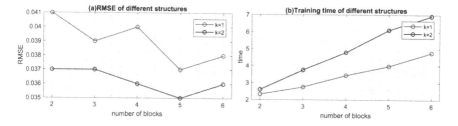

Fig. 2. RMSE and training time of different structures. The red line represents the result of one step. The blue line represents the result of two steps. (Color figure online)

iteratively n times, we can estimate x_{i+1}. Hence, there are n steps between x_i and x_{i+1}. $z_{i,j}$ denotes the jth step between x_i and x_{i+1}.

However, there are several drawbacks in this FISTA-based computational framework. First, it is very hard to derive a closed form of x_k when $g(\cdot)$ is one of such joint norms. Second, it requires a huge amount of computation to find a good solution using the iterative process. Furthermore, it is time-consuming to find a good step size μ, since it varies given different undersampling rates. Additionally, L_k is chosen according to rigorous conditions in backtracking FISTA, so it is sometimes hard to satisfy all conditions.

2.2 Deep Neural Network for MRI Reconstruction

In order to overcome the aforementioned difficulties, we first propose a deep neural network model called *FR-Net* for MRI reconstruction. Specifically, we create a corresponding neural network for x_k that can be solved via gradient descent. One advantage is that the neural network can be adapted to different forms of $g(\cdot)$ and hence closed form derivation is no longer required. Suppose at step k, the optimal value of Eq. (4) is x_k. At each iteration n in the gradient descent, a value $z_{k-1,n}$ is generated to approximate x_k. As n becomes larger, $z_{k-1,n}$ converges to x_k. Figure 1 shows how to iteratively compute z and x. In order to solve Eq. (5), we need to come out the close form of $\nabla g(z_{i,j})$. However, the regularization term g is always complicated. For example, it can be a joint norm whose gradient is hard to derive. Therefore, we replace $g(\cdot)$ with a convolution neural network G_k inspired by the work [15]. The intuition is that the convolutional layer has powerful ability of feature representation that it can generalize to any regularization norms. Specifically, G_k takes as input k-space data of a shape 256×256 and outputs feature vectors of the same shape. It consists of three convolution layers and each uses 3×3 kernels with 32 channels, followed by a Batch Normalization layer and a PReLU activation layer. The advantage of G_k over $g(\cdot)$ is that convolution neural networks can be generalized to any form of regularization term and hence derivation of closed form of $g(\cdot)$ is no longer needed. The second modification is that we substitute L_k with a fully connected layer called difference layer D_k since L_k is not easy to compute under some rigorous constraints in backtracking FISTA. Furthermore, $F^T(E(Fv_k) - y)$ is a

Table 1. Average reconstruction performance of 3-fold cross validation.

Method	RMSE↓	PSNR↑	RMSE↓	PSNR↑
	50%		25%	
ℓ_1-wavelet [14]	$0.0680_{\pm 0.0152}$	$23.5844_{\pm 2.0719}$	$0.0975_{\pm 0.0279}$	$20.6242_{\pm 2.7690}$
Total variation [7]	$0.0713_{\pm 0.0156}$	$23.1699_{\pm 2.0394}$	$0.0908_{\pm 0.0182}$	$21.0009_{\pm 1.6306}$
Low-rank [10]	$0.0508_{\pm 0.0135}$	$26.2022_{\pm 2.4367}$	$0.1014_{\pm 0.0284}$	$20.2634_{\pm 2.7302}$
RU-Net	$0.0331_{\pm 0.0081}$	$29.1018_{\pm 1.9535}$	$0.0546_{\pm 0.0091}$	$24.0784_{\pm 1.4528}$
FR-Net-2D	$0.0225_{\pm 0.0068}$	$32.8431_{\pm 2.3178}$	$0.0430_{\pm 0.0008}$	$26.5066_{\pm 1.6518}$
FR-Net-2D+time	$\mathbf{0.0220_{\pm 0.0061}}$	$\mathbf{33.0129_{\pm 2.2166}}$	$\mathbf{0.0411_{\pm 0.0079}}$	$\mathbf{26.9550_{\pm 1.7827}}$
Joint-FR-2D	$0.0236_{\pm 0.0070}$	$32.3520_{\pm 2.3128}$	–	–
Joint-FR-2D+time	$0.0241_{\pm 0.0067}$	$32.3455_{\pm 2.1326}$	–	–

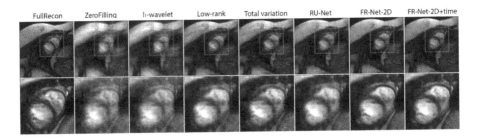

Fig. 3. Reconstruction results with k-space data at undersampled rate 25%.

constant value and we name it as C_k. The outputs of G_k, D_k and C_k are combined through element-wise addition and connected to another fully connected layer F_k. Finally, Eq. (5) is transformed to:

$$x_k = z_{n-1} - F_k\{G_{k-1}(z_{n-1}) + D_k(z_{n-1} - v_k) + C_k(v_k)\} \qquad (6)$$

2.3   Joint Learning of Reconstruction and Segmentation

Conventional MRI segmentation approaches consist of two separate steps. Basically, the first step is to reconstruct an image from undersampled k-space data and the second step is to feed the reconstructed image into an established segmentation model using automatic methods. For simplicity, we call this conventional approach *Two-step Model*. We propose a joint model called *Joint-FR-Net*. As shown in Fig. 1, *Joint-FR-Net* is an end-to-end deep neural network, which takes k-space data as input and directly learns segmentation mask as output. One significant difference from *Two-step Model* is that the "reconstructed image" is now a set of differentiable parameters that connect FR-Net and U-Net. The motivation behind this change is to bridge the gap between two originally-isolated models to become capable of learning parameters through backpropagation of shared gradients in the whole pipeline. From the perspective of model effectiveness and training efficiency, this mechanism allows two models to mutually benefit from each other. FR-Net passes to the input of U-Net the on-the-fly features of the "reconstructed image" instead of fixed noisy images so that features are shared and learned by both models. Meanwhile, U-Net directly backpropagates gradients to FR-Net to make the whole model more segmentation-oriented.

Technically speaking, let R_θ be the reconstruction network parameterized by θ and S_ϕ be the segmentation network parameterized by ϕ. In our application, R_θ and S_ϕ respectively represent FR-Net and U-Net. Let y denote the input of k-space data. The reconstructed image is $R_\theta(y)$ and the segmentation result is $S_\phi(R_\theta(y))$. Note that the input of the segmentation network is the output of the reconstruction network and hence the parameters ϕ and θ are updated simultaneously.

We also define a combination loss ℓ_{com} which is based on both the reconstruction loss function l_R and the segmentation loss function l_S. ℓ_{com} is defined as: $\ell_{com} = l_R(R^\theta(y), \hat{x}) + \beta \cdot l_S(S^\phi(R^\theta(y)), \tilde{x})$, where $\hat{x}$ is the ground truth of the reconstruction image and $\tilde{x}$ is the ground truth of the reconstruction mask. β is the hyper-parameter that balances between reconstruction and segmentation performances. In this paper, reconstruction loss l_R is defined as l_1 loss and segmentation loss l_S is defined as *dice* loss. During the training procedure, the key issue is to find an optimal equilibrium between reconstruction and segmentation such that they are mutually beneficial.

3 Experiments

In this section, we evaluate and analyze the performance of our model on both reconstruction and joint segmentation tasks.

Experimental Settings. All our experiments are conducted on a dynamic cardiac MRI dataset, which is collected from 25 volunteers and three of them are diagnosed with cardiac dyssynchrony disease. It contains $4,000$ 2D slices of SAX images with manually annotated LV contours over various spatial locations and cardiac phases. We randomly sample $2/3$ of the 25 subjects as the training set and the rest $1/3$ as the test set. We apply Fourier transform to convert SAX image to k-space data and undersample it using the same strategy proposed in [5]. Particularly, we keep eight lowest spatial frequencies and adopt Cartesian undersampling on k-space data along phase direction. Then undersampled k-space data can be simulated. Our model is assessed on the data at undersampled rates of 25% and 50%. Since the dataset is a dynamic sequence of SAX images, we consider both the 2D case and 2D+time case in the following experiments. In the reconstruction task, we quantitatively evaluate our method by root-mean-square error (RMSE) and peak signal-to-noise ratio (PSNR). In the segmentation task, we choose the commonly-used Dice's score and average perpendicular distance (APD) as our measure.

3.1   Reconstruction Task

As shown in Fig. 1, we initialize x_0 by Fourier transform on the k-space data. The number of steps (k in the Sect. 2.2) and the number of blocks between each step (n in the Sect. 2.2) are unknown and there could be infinite combinations. Therefore, we run an empirical experiment to decide the structure of the FR-Net. Figure 2 shows the corresponding RMSE and training time of different structures. We train different structures with undersampled data at rate of 50% and only run 15 epochs. Specifically, the red line means there is only one step, x_0 and x_1 ($k = 1$). The blue line means there are two steps x_0, x_1 and x_2. We increase the number of blocks between each step and see how their RMSE changes. Balancing between time efficiency and reconstruction quality, we choose five blocks respectively between x_0 and x_1, x_1 and x_2 ($k = 2$). x_2 is the output of reconstructed image.

We compare our model with both CS-based methods and the state-of-the-art deep learning approach. Specifically, we consider following three classic CS-based methods with different regularization terms: ℓ_1-wavelet [14], the total variation approach [7], and the low-rank approach [10]. We also consider one deep learning approach called RU-Net [4] that is a variation of U-Net. Both FR-Net and RU-Net are implemented in PyTorch and trained with ADAM optimizer on Tesla K80. In order to further improve the performance, we also adopt the same k-space correction strategy [4] by replacing zeros of the original k-space data with the values of the reconstructed image that in k-space. The average results are reported in Table 1. As we can see, our FR-Net-2D model achieves the best performance. And it also outperforms RU-Net in two cases with the lowest RMSE 0.0225 and 0.0430 respectively. For 2D+time data, we use 5 successive frames as input. The results show that the FR-Net-2D+time model which performs on dynamic data, achieves the best performance in terms of RMSE and PSNR in both cases. This validates that our model performs very well at different undersampled rates with and without temporal information. Figure 3 shows the reconstruction results at different undersampled rates. As we can see, the CS-based methods miss lots of details and the reconstructed images are blurry and fuzzy. In contrast, deep learning methods are more precise and efficient. Specifically, our FR-Net 2D and FR-Net 2D+time models yield much clearer and high-quality images.

Table 2. Average endo- and epicardium segmentation performance of test set. The input k-space data is at undersampled rate 50%.

Method	Dice↑	ADP(mm)↓	Dice↑	ADP(mm)↓
	Endo.		Epi.	
Fully-sampled	$0.7661_{\pm 0.2658}$	$4.1774_{\pm 5.0946}$	$0.7754_{\pm 0.2677}$	$5.7527_{\pm 10.8612}$
Zero-filling	$0.6357_{\pm 0.3009}$	$7.7575_{\pm 10.9018}$	$0.6570_{\pm 0.3040}$	$9.8359_{\pm 15.0050}$
Two-step	$0.6635_{\pm 0.3081}$	$9.1255_{\pm 21.1796}$	$0.7494_{\pm 0.2734}$	$8.2817_{\pm 18.4909}$
Joint-FR-2D	$0.7260_{\pm 0.2787}$	$4.9771_{\pm 5.7992}$	$0.7503_{\pm 0.2922}$	$6.9593_{\pm 12.3803}$
Joint-FR-2D+time	$\mathbf{0.7310}_{\pm 0.2603}$	$\mathbf{4.2884}_{\pm 4.9119}$	$\mathbf{0.7573}_{\pm 0.2763}$	$\mathbf{1.2628}_{\pm 1.6988}$

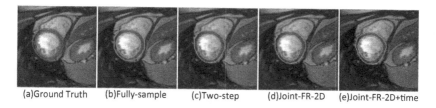

(a)Ground Truth   (b)Fully-sample   (c)Two-step   (d)Joint-FR-2D   (e)Joint-FR-2D+time

Fig. 4. Segmentation results of input data is at undersampled rate 50%.

3.2 Joint Segmentation Task

In this experiment, we train a U-Net with fully-sampled images and use it as the best achievable model. We also train another model with zero-filling images as the lower bound of segmentation performance. In order to show the mutual benefits of the Joint-FR-Net on solving the myocardium segmentation problem. We compare our method with the *Two-step* model which takes as input the images reconstructed from the FR-Net in Sect. 3.1. Our method respectively takes as input the 2D k-space data and the 2D+time dynamic data and we call them *Joint-FR-2D* and *Joint-FR-2D+time* in this experiment. The training and test data are based on the undersampled rate of 50%. Experiments are conducted on both endo- and epicardium segmentation. The average segmentation results are reported in Table 2. We observe that our two joint models outperform the Two-step model in terms of Dice and APD for both endo- and epicardium cases. The Two-step model only gains 0.6635 and 0.7494 Dice score in endo. and epi. case, respectively. We also find that the Joint-FR-2D+time model achieves better results than the Joint-FR-2D model. This highlights the generalization ability of our model to dynamic data. Moreover, we can see the Dice index of Joint-FR-2D+time model is very close to that of fully-sampled model. The reconstruction performance of the Joint-FR-2D and Joint-FR-2D+time models are also reported in Table 1. It achieves comparable results with the FR-Net-2D and FR-Net-2D+time models and outperforms other CS-Based methods. Thus we conclude that our joint model not only benefits the segmentation task but also achieves promising reconstruction results compared with the other CS-based models. We also plot the segmentation results in Fig. 4. Note that for better visualization, we use fully-sampled image as reconstructed image here. As the Joint-FR-Net models are trained with the raw k-space data, they compute similar contours to the ground truth.

4 Conclusion

We proposed a new deep learning method called "Joint FR-Net" that based on experimental results outperforms previous CS-based methods and the state-of-the-art deep learning methods on joint reconstruction and segmentation tasks.

References

1. Beck, A., Teboulle, M.: A fast iterative shrinkage-thresholding algorithm for linear inverse problems. SIAM J. Imaging Sci. **2**(1), 183–202 (2009)
2. Caballero, J., Bai, W., Price, A.N., Rueckert, D., Hajnal, J.V.: Application-driven MRI: joint reconstruction and segmentation from undersampled MRI data. In: Golland, P., Hata, N., Barillot, C., Hornegger, J., Howe, R. (eds.) MICCAI 2014. LNCS, vol. 8673, pp. 106–113. Springer, Cham (2014). https://doi.org/10.1007/978-3-319-10404-1_14
3. Huang, Q., Yang, D., Wu, P., Qu, H., Jingru, Y., Metaxas, D.: MRI reconstruction via cascaded channel-wise attention network. In: IEEE 16th International Symposium on Biomedical Imaging (ISBI 2019), pp. 1622–1626. IEEE (2019)
4. Hyun, C.M., Kim, H.P., Lee, S.M., Lee, S., Seo, J.K.: Deep learning for undersampled MRI reconstruction. arXiv preprint arXiv:1709.02576 (2017)
5. Jung, H., Ye, J.C., Kim, E.Y.: Improved k-t blast and k-t sense using focuss. Phys. Med. Biol. **52**(11), 3201 (2007)
6. Lee, D., Yoo, J., Ye, J.C.: Deep residual learning for compressed sensing MRI. In: ISBI, pp. 15–18 (2017)
7. Lustig, M., Donoho, D., Pauly, J.M.: Sparse MRI: the application of compressed sensing for rapid MR imaging. Magn. Reson. Med. **58**(6), 1182–1195 (2007)
8. Lustig, M., Donoho, D.L., Santos, J.M., Pauly, J.M.: Compressed sensing MRI. IEEE Signal Process. Mag. **25**(2), 72–82 (2008)
9. Ma, S., Yin, W., Zhang, Y., Chakraborty, A.: An efficient algorithm for compressed MR imaging using total variation and wavelets (2008)
10. Ravishankar, S., Bresler, Y.: MR image reconstruction from highly undersampled k-space data by dictionary learning. IEEE Trans. Med. Imaging **30**(5), 1028–1041 (2011)
11. Schlemper, J., Caballero, J., Hajnal, J.V., Price, A., Rueckert, D.: A deep cascade of convolutional neural networks for MR image reconstruction. In: Niethammer, M., et al. (eds.) IPMI 2017. LNCS, vol. 10265, pp. 647–658. Springer, Cham (2017). https://doi.org/10.1007/978-3-319-59050-9_51
12. Schlemper, J., et al.: Cardiac MR segmentation from undersampled k-space using deep latent representation learning. In: Frangi, A.F., Schnabel, J.A., Davatzikos, C., Alberola-López, C., Fichtinger, G. (eds.) MICCAI 2018. LNCS, vol. 11070, pp. 259–267. Springer, Cham (2018). https://doi.org/10.1007/978-3-030-00928-1_30
13. Sun, J., Li, H., Xu, Z., et al.: Deep ADMM-Net for compressive sensing MRI. In: Advances in Neural Information Processing Systems (NIPS), pp. 10–18 (2016)
14. Uecker, M., et al.: Berkeley advanced reconstruction toolbox. In: Proceedings of the International Society for Magnetic Resonance in Medicine, vol. 23, p. 2486 (2015)
15. Wu, D., Kim, K.S., Dong, B., Li, Q.: End-to-end abnormality detection in medical imaging. arXiv preprint arXiv:1711.02074 (2017)
16. Yang, D., Huang, Q., Axel, L., Metaxas, D.: Multi-component deformable models coupled with 2D-3D U-Net for automated probabilistic segmentation of cardiac walls and blood. In: IEEE 15th International Symposium on Biomedical Imaging (ISBI 2018), pp. 479–483. IEEE (2018)

SMOD - Data Augmentation Based on Statistical Models of Deformation to Enhance Segmentation in 2D Cine Cardiac MRI

Jorge Corral Acero[1]($\boxtimes$), Ernesto Zacur[1], Hao Xu[1], Rina Ariga[2],
Alfonso Bueno-Orovio[3], Pablo Lamata[4], and Vicente Grau[1]

[1] Department of Engineering Science, University of Oxford, Oxford, UK
jor.corral@eng.ox.ac.uk
[2] John Radcliffe Hospital, Oxford, UK
[3] Department of Computer Science, University of Oxford, Oxford, UK
[4] Biomedical Engineering Department, King's College London, London, UK

Abstract. Deep learning has revolutionized medical image analysis in recent years. Nevertheless, technical, ethical and financial constraints along with confidentiality issues still limit data availability, and therefore the performance of these approaches. To overcome such limitations, data augmentation has proven crucial. Here we propose SMOD, a novel augmentation methodology based on Statistical Models of Deformations, to segment 2D cine scans in cardiac MRI. In brief, the shape variability of the training set space is modelled so new images with the appearance of the original ones but unseen shapes within the space of plausible realistic shapes are generated. SMOD is compared to standard augmentation providing quantitative improvement, especially when the training data available is very limited or the structures to segment are complex and highly variable. We finally propose a state-of-art, deep learning 2D cardiac MRI segmenter for normal and hypertrophic cardiomyopathy hearts with an epicardium and endocardium mean Dice score of 0.968 in short and long axis.

Keywords: Data augmentation · Segmentation · Deep learning ·
Models of deformation · Cardiac MRI

1 Introduction

The expansion of big data has entered the cardiovascular medicine field in recent years [1, 2]. Machine learning approaches and deep learning techniques have been proven successful in medical imaging, and in particular in the cardiovascular field [3, 4]. This has had a strong impact on a wide range of applications, from classification, object detection and registration to phenotype clustering or risk prediction [3].

Among the uses of deep learning techniques in medical imaging, segmentation has been one of the most successful ones [3]. Since the U-net architecture was introduced by Ronneberger et al. in 2015 [5], a number of variations of that structure using

© Springer Nature Switzerland AG 2019
Y. Coudière et al. (Eds.): FIMH 2019, LNCS 11504, pp. 361–369, 2019.
https://doi.org/10.1007/978-3-030-21949-9_39

convolutional neural networks (CNN) have been explored, reporting excellent performance. A good example of this is the segmentation of both left and right ventricles (LV and RV) in cardiac magnetic resonance imaging (MRI) [6–8].

Despite these promising results, limitations and challenges remain [2], in particular concerning the availability, quantity and quality of training data [1]. Technical, ethical or financial constraints may limit data acquisition and confidentiality issues prevent access to data. Furthermore, manual data annotation is a costly and tedious process. Thus, the size of the training datasets in medical imaging is still a limiting factor. This scenario becomes even worse when dealing with specific diseases or applications. Since the performance of deep learning methods depends heavily on the available data, data augmentation and techniques such as transfer learning have become crucial for a better generalization and to avoid overfitting [9–12].

Image cropping, translation, rotation, scaling, flipping and even shearing are routinely applied as standard practice to enlarge datasets [11–14]. Other more complex approaches such as filtering, gamma correction, addition of Gaussian noise or even training a secondary network to perform the augmentation have been considered [9, 10, 12, 13]. Additionally, non-affine transformations, such as random elastic deformations or diffeomorphic transformations, have also been proposed [15, 16].

Statistical shape and deformation models have been used in a variety of medical image analysis tasks, including cardiac image segmentation but not, to the best of our knowledge, in combination with deep learning approaches [17]. We propose a new augmentation methodology based on models of deformation to segment 2D cine cardiac MRI scans and thus overcome the mentioned dataset size limitations. In brief, the shape variability of the training set space is modelled so new images with the gray level appearance of the original ones but unseen shapes within the space of plausible shapes are generated. As a result, the deformations are non-affine, follow a prior and are anatomically meaningful, in contrast with augmentation approaches based on random deformations [15]. We assess the performance of the proposed augmentation methodology versus standard augmentation approaches depending on data availability and on variability of the structure to be segmented. Finally, we build a state-of-art 2D cine MRI cardiac segmenter for both normal and Hypertrophic Cardiomyopathy (HCM) hearts, a condition that presents the largest amount of cardiac anatomical variability and represents the most common cause of sudden cardiac death in the young.

2   Materials and Methods

2.1   Dataset

We make use of two independent datasets containing cardiac MRIs for the assessment of our proposed augmentation methodology:

STACOM Dataset. This refers to the publicly available MICCAI 2018 LV Full Quantification Challenge STACOM dataset (https://lvquan18.github.io/). It contains 145 cine-MR mid-cavity short axis (SA) cardiac cycle sequences, 20 frames each. They were collected from 3 different hospitals, including diverse pathologies and a wide range of subject ages. The LV endocardium and epicardium have been manually contoured and revised by two cardiac radiologists in all frames. The original pixel size ranges from 0.68 mm/pixel to 2.08 mm/pixel. Pre-processing to align the images was performed, as described in the challenge.

HCM Dataset. This dataset, published at [18], contains 55 HCM phenotype positive, 18 HCM phenotype negative but genotype positive and 37 normal patients scanned at Oxford John Radcliffe Hospital. Only mid-ventricular SA (to facilitate inter-dataset comparisons), horizontal long axis (HLA) and vertical long axis (VLA) end-diastole instances are considered. The LV endocardium, epicardium and left atrium (LA) along with RV endocardium were contoured in the end-diastole frames and double checked by 2 experts. The pixel size ranges from 1.33 to 2.08 mm/pixel.

2.2   Statistical Models of Deformation

Our registration method consists of an initial rigid registration, minimizing the sum of square distances (SSD) of image intensities, followed by the application of Diffeo-morphic Log Demons [19] for non-rigid registration ($\sigma_{fluid} = 2$, $\sigma_{diff} = 1.8$ and $\sigma_i/\sigma_x = 0.82$). We use this method to build a model of deformation, in the following way: (a) we generate an atlas representing the specific dataset population; (b) each image is registered to the atlas to calculate one velocity field per image; (c) velocity fields are combined using Principal Component Analysis (PCA) and the deformation model is then used to generate new samples. Methodological details are provided below.

Generation of a Population-Specific Atlas. (1) The atlas is initialized to a randomly selected image among the training set, which we denote as A_0. (2) Each image from the dataset, I, is rigidly registered to A_0 to obtain the registered set I_{T0}. (3) The average of the images in the I_{T0} set is calculated, obtaining the first iteration of the atlas, A_1. (4) I_{T0} is non-rigidly registered to A_1 and averaged to obtain A_2. Step 4 is iterated until convergence is achieved. (5) The resulting atlas, A, is manually segmented to generate A_s.

Set of Velocity Fields to Describe the Shape Variability. The set of segmentations of the images, S, is non-rigidly registered to A_s, obtaining the set of velocity fields, $\{v_i\}$, to diffeomorphically bring each image to the atlas space. The set $\{v_i\}$ encodes the variability of shape of the set of images, I, with respect to the reference A, and can be exploited to generate new deformations, which result in new images.

Dimensionality Reduction and Image Generation. Each velocity of the set $\{v_i\}$ can be organized as the column of a matrix V, to apply PCA and thus reduce the dimensionality. A mode cut capturing 90% variability is considered and the random velocity fields are sampled as follows:

$$v_g = \bar{V} + U \cdot x \cdot d \qquad (1)$$

Where $\bar{V}$ is the mean of the velocity fields; U, the eigenvectors of the principal modes; d, the eigenvalues of the principal modes; and x, an array with random numbers following the distribution $N(0, \sigma)$. We set the value of σ to 1 using visual inspection to achieve a realistic but varied set of shapes.

Each of the images, i, is brought to the atlas space applying v_i and transformed back to the image space applying the inverse of the random velocity field, v_g (different from displacement fields, inverse transformations are obtained just inverting the sign of the velocity field). Thus, a new image with the appearance of image i but a random shape within the space of variability of the original images is obtained (see Fig. 1).

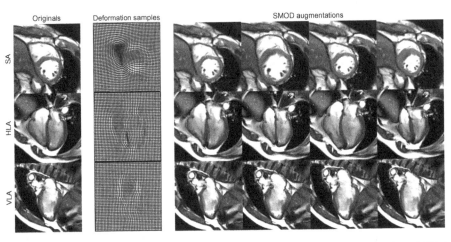

Fig. 1. Left column: Original mid-ventricular SA (top row), HLA (middle row) and VLA (bottom row) end-diastole MRI images. Middle column: Samples of the anatomically meaningful deformations obtained by applying SMOD on originals. Right Column: Augmentation samples from originals based on SMOD.

2.3   Augmentation Strategies

We consider three image augmentation strategies: (a) augmentation based on statistical models of deformations, SMOD, as described above; (b) standard augmentation, STD, based on random rotations (0–360°), flipping and translation (SA: –6 to 6 pixels in x and y, HLA and VLA: –10 to 10 pixels in x and y); and (c) a combination of both, SMOD+. For each cross-validation run, a model of deformations is learnt from scratch based on the particular training set to produce fair results. The augmentation process is "online", meaning that for each epoch the augmented training data fed to the network is different.

2.4   Neural Network

To segment the images, we used a Fully Convolutional Neural Network (FCNN) with skip connections following a U-Net architecture [5], as depicted in Fig. 2. The segmentation output layer has sigmoid activation functions and all the other FCN activations are rectified linear units (ReLU). Two convolutional layers (kernel size 3×3) and two batch normalization (BN) layers form a convolutional block, with the number of filters for each block specified in Fig. 2. Max-pooling and up-sampling layers have a stride of 2 in both dimensions. We used binary cross-entropy as loss function and early

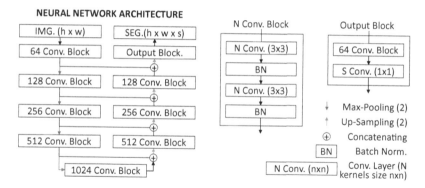

Fig. 2. Graphical overview of the proposed neural network structure.

stopping based on validation loss. The neural networks are implemented in Keras with Tensorflow as backend. The networks are trained with Adadelta optimizer on a NVIDIA GeForce GTX 1080 Ti 10 Gb. The Batch size is 8.

2.5   Assessment

We calculate Dice scores and Hausdorff distances between predicted and original segmentations to assess the performance of the augmentation strategies.

3   Experiments and Results

3.1   STACOM Dataset

We performed three batches of experiments considering different training set sizes to assess the impact of the augmentation strategy depending on data availability. In each batch, the three presented augmentation strategies were tested to segment the LV endocardium and epicardium in SA instances.

In the first batch, a five-fold cross-validation (29 subjects on each fold for testing) is performed. The remaining dataset is randomly split into 100 and 16 subjects, for training and validating data, respectively. In the second and third batches, the split into training/validation/testing data is 10/5/130 and 5/5/135 subjects, respectively. Cross fold validation was not considered in these 2nd and 3rd batches since the testing set, where the assessment is performed, already represents more than the 89% of the whole STACOM dataset in both scenarios. The images are cropped to 80 × 80 after augmentation and the whole cine sequence is considered in all experiments. That is, a training set of 100 subjects effectively means a training set of 20 × 100 segmented images.

The Dice score and Hausdorff distance results of the myocardium segmentations, calculated as the region between epicardium and endocardium, are depicted in Fig. 3. Similar results are obtained when evaluating endocardium and epicardium separately.

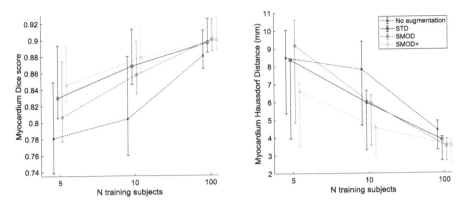

Fig. 3. STACOM dataset results. Mean Dice scores (left) and Hausdorff distances (right) in LV myocardium depending on the number of subjects to train with. Error-bars: 25th and 75th percentiles.

3.2   HCM Dataset

We assess the performance of the augmentation strategies depending on the intrinsic shape variability of the structure to be segmented on the HCM dataset, where not only the LV endocardium and epicardium are segmented but also the LA and the RV endocardium. Experiments to segment each of the three MRI planes were carried out with the three augmentation strategies. Five-fold cross validation is considered in all the experiments. The resulting split, in each of the folds, is: 22 testing, 78 training and 10 validation samples. After augmentation, the images are cropped to 80 × 80 in SA, and to 224 × 224 in HLA and VLA.

Figure 4 summarizes the segmentation Dice score results obtained for the different anatomical structures in the different MRI planes training under the three described augmentation strategies. Similar behavior and patterns are observed in Haussdorf distances (not shown). A representative example of the resulting segmentation using SMOD+ is presented in Fig. 5.

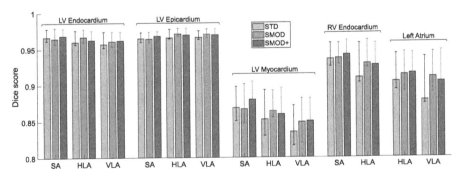

Fig. 4. HCM dataset results. Mean Dice scores and 25th and 75th percentiles grouped by segmented structure and MRI view.

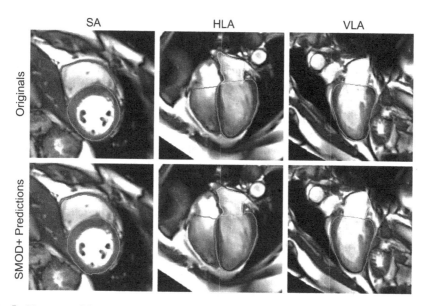

Fig. 5. Top row: LV endocardium (red), LV epicardium (green), LA (yellow) and RV endocardium (blue) ground truth segmentations of mid-ventricular SA (left), HLA (middle) and VLA (right) end-diastole MRI images from HCM dataset. Bottom row: SMOD+ predictions on these unseen instances. (Color figure online)

4 Discussion and Conclusions

Overall, our experiments confirm the importance of data augmentation, as reported in previous literature. Comparing to training with no augmentation, any of the three augmentation strategies significantly improves the results as it can be seen in Fig. 3.

The experiments completed on the STACOM dataset (Fig. 3) show that, while there is little difference among augmentation strategies when training with 100 subjects, SMOD+ outperforms the rest of strategies and becomes more relevant when the number of available subjects to train with decreases. We believe this occurs because the shape of the structures to segment (LV endocardium and epicardium) is reasonably simple and symmetrical, and thus when there are sufficient training samples the space of possible segmentations may be completely described with only rotations, translations and flipping. However, when the number of training subjects drops, standard augmentation is no longer enough to cover the space of variability and the models of deformations, learnt from the few samples, contribute filling the gaps and result in a better training. The observed pattern could be explained because the training network focuses more on local features rather than global ones to perform the segmentation task. The described behavior is confirmed both in Dice scores and Haussdorf distances.

In contrast to the STACOM data results when training with 100 subjects, the experiments on the HCM dataset show a more appreciable improvement in accuracy when using the proposed augmentation method (Fig. 4). SMOD+ reports better results than STD segmenting any structure in any plane, with difference being clearer in LV myocardium, LA and RV endocardium. We believe this confirms our appreciation as explained above: HCM cases have a larger variation in anatomy due to the disease, and the LA and RV have a less regular shape than the LV. We should also note that, even though the training set size (78 subjects) in the HCM experiments is comparable to STACOM one, only one frame of the cine sequences (end-diastole) is considered in HCM versus 20 instances for the STACOM dataset. Therefore, an alternative explanation for the difference in performance might be the difference in data availability.

A limitation of the work is that only one neural network architecture has been considered. Furthermore, different levels of deformation (σ value, Eq. 1) as well as training in larger and more heterogeneous datasets could be explored. Methods other than PCA, which assumes Gaussianity, could be considered to better describe the distribution of deformations. Moreover, comparison to augmentation based on completely random deformations (no anatomical meaning) could be tested.

We expect the developed augmentation methodology to have more of an impact in 3D since usually fewer samples are available. We also presume that it could be scaled up to other applications and anatomies, particularly those with less regular shapes. Additionally, models of deformation could be learnt from secondary larger datasets and applied to the actual training dataset, enlarging even further the space of plausible shapes and presumably leading to a better generalization. Even though we chose segmentation-to-segmentation registration since less noise is introduced, image-to-image registration could be used to obtain the velocity fields, opening the possibility for learning variability from unsegmented cohorts.

To conclude, (a) a new augmentation methodology based on statistical models of deformation has been developed; (b) it has been proven that it improves the standard augmentation training results when segmenting 2D cardiac MRI slices; (c) the improvement is remarkable especially when the training data available is very limited or the structures to segment are complex and variable; and (d) an accurate state-of-art segmenter for healthy and HCM cine MRI scans has been built.

Acknowledgements. This work was supported by the European Union's Horizon 2020 research and innovation program under the Marie Sklodowska-Curie (g.a. 764738) and by the British Heart Foundation (PG/16/75/32383). Authors are financially supported by a Wellcome Trust Senior Research Fellowship (to PL, 209450/Z/17/Z) and a BHF Intermediate Basic Science Research Fellowship (to ABO, FS/17/22/32644).

References

1. Rumsfeld, J.S., Joynt, K.E., Maddox, T.M.: Big data analytics to improve cardiovascular care: promise and challenges. Nat. Rev. Cardiol. **13**(6), 350–359 (2016)
2. Shameer, K., Johnson, K.W., Glicksberg, B.S., Dudley, J.T., Sengupta, P.P.: Machine learning in cardiovascular medicine: are we there yet? Heart **104**(14), 1156–1164 (2018)
3. Litjens, G., Kooi, T., Bejnordi, B.E., Setio, A.A.A., Ciompi, F., Ghafoorian, M., et al.: A survey on deep learning in medical image analysis. Med. Image Anal. **42**, 60–88 (2017)
4. Hosny, A., Parmar, C., Quackenbush, J., Schwartz, L.H., Aerts, H.J.W.L.: Artificial intelligence in radiology. Nat. Rev. Cancer **18**(8), 500–510 (2018)

5. Ronneberger, O., Fischer, P., Brox, T.: U-Net: convolutional networks for biomedical image segmentation. In: Navab, N., Hornegger, J., Wells, W.M., Frangi, A.F. (eds.) MICCAI 2015. LNCS, vol. 9351, pp. 234–241. Springer, Cham (2015). https://doi.org/10.1007/978-3-319-24574-4_28

6. Oktay, O., Ferrante, E., Kamnitsas, K., Heinrich, M., Wai, B., Caballero, J., et al.: Anatomically Constrained Neural Networks (ACNNs): application to cardiac image enhancement and segmentation. IEEE Trans. Med. Imaging 37(2), 384–395 (2018)

7. Poudel, R.P.K., Lamata, P., Montana, G.: Recurrent fully convolutional neural networks for multi-slice MRI cardiac segmentation. In: Zuluaga, M.A., Bhatia, K., Kainz, B., Moghari, M.H., Pace, D.F. (eds.) RAMBO/HVSMR -2016. LNCS, vol. 10129, pp. 83–94. Springer, Cham (2017). https://doi.org/10.1007/978-3-319-52280-7_8

8. Vigneault, D.M., Xie, W., Ho, C.Y., Bluemke, D.A., Noble, J.A.: Ω-Net (omega-net): fully automatic, multi-view cardiac mr detection, orientation, and segmentation with deep neural networks. Med. Image Anal. 48, 95–106 (2018)

9. Hussain, Z., Gimenez, F., Yi, D., Rubin, D.: Differential data augmentation techniques for medical imaging classification tasks. In: Proceedings American Medical Informatics Association Annual Symposium, vol. 2017, pp. 979–984 (2017)

10. Lemley, J., Bazrafkan, S., Corcoran, P.: Smart augmentation learning an optimal data augmentation strategy. IEEE Access 5, 5858–5869 (2017)

11. Asperti, A., Mastronardo, C.: The effectiveness of data augmentation for detection of gastrointestinal diseases from endoscopical images. In: Proceedings 11th International Joint Conference on Biomedical Engineering Systems and Technologies, pp. 199–205 (2018)

12. Goodfellow, I., Bengio, Y., Courville, A.: Deep Learning, vol. 13, no. 1. MIT Press, Cambridge (2017)

13. Perez, L., Wang, J.: The effectiveness of data augmentation in image classification using deep learning. CoRR, vol. abs/1712.0 (2017)

14. Dong, H., Yang, G., Liu, F., Mo, Y., Guo, Y.: Automatic brain tumor detection and segmentation using U-Net based fully convolutional networks. In: Valdés Hernández, M., González-Castro, V. (eds.) MIUA 2017. CCIS, vol. 723, pp. 506–517. Springer, Cham (2017). https://doi.org/10.1007/978-3-319-60964-5_44

15. Castro, E., Cardoso, J.S., Pereira, J.C.: Elastic deformations for data augmentation in breast cancer mass detection. In: IEEE EMBS International Conference on Biomedical & Health Informatics (BHI), pp. 230–234 (2018)

16. Arteaga, M.O., Sørensen, L., Cardoso, J., Modat, M., et al.: PADDIT: probabilistic augmentation of data using diffeomorphic image transformation, October 2018. arXiv:1810.01928

17. Alba, X., et al.: An algorithm for the segmentation of highly abnormal hearts using a generic stat. shape model. IEEE Trans. Med. Imaging 35(3), 845–859 (2016)

18. Raman, B., Ariga, R., Spartera, M., Sivalokanathan, S., Chan, K., Dass, S., et al.: Progression of myocardial fibrosis in hypertrophic cardiomyopathy: mechanisms and clinical implications. Eur. Heart J. Cardiovasc. Imaging 20(2), 157–167 (2019)

19. Lorenzi, M., Ayache, N., Frisoni, G.B., Pennec, X., Alzheimer's Disease Neuroimaging Initiative (ADNI): LCC-Demons: a robust and accurate symmetric diffeomorphic registration algorithm. Neuroimage 81, 470–483 (2013)

Comparing Subjects with Reference Populations - A Visualization Toolkit for the Analysis of Aortic Anatomy and Pressure Distribution

Sahar Karimkeshteh[1,2]([✉]), Lilli Kaufhold[1,2], Sarah Nordmeyer[3],
Lina Jarmatz[2], Andreas Harloff[4], and Anja Hennemuth[1,2]

[1] Fraunhofer MEVIS, Am Fallturm 1, 28359 Bremen, Germany
s.karimkeshteh@gmail.com
[2] Charité – Universitätsmedizin Berlin, 13353 Berlin, Germany
[3] Deutsches Herzzentrum Berlin,
Augustenburger Platz 1, 13353 Berlin, Germany
[4] Universitätsklinikum Freiburg,
Hugstetter Straße 55, 79106 Freiburg, Germany

Abstract. The analysis of anatomical and hemodynamic vessel parameters plays an important role in diagnosis and therapy planning for aortic diseases. Normal values and decision thresholds are usually based on global or local parameters provided by population studies. In order to enable a more holistic comparison of a single subject and a matching reference population we have developed a spatiotemporal normalization concept for the analysis of 4D PC MRI data of the thoracic aorta. This enables the comparison of geometric properties and pressure differences along the vessel course as well as in a sector model, which represents a cross-sectional value distribution. We tested the applicability of the presented approach by comparing subjects with aortic diseases to matching subgroups of a normal reference population. The presented framework enabled a visual and quantitative assessment of the local geometric and pressure distribution changes of different pathological alterations of the aorta. It will be extended to integrate further hemodynamic properties and larger reference cohorts to support clinical decision making based on hemodynamic information in near future.

Keywords: 4D PC MRI · Aortic disease · Pressure distribution ·
Population study

1 Introduction

Anatomic alterations as well as strong local changes in pressure are amongst others parameters used for decision making in congenital or acquired diseases of the aorta [1]. The information on the normal shape and function of the aorta in different genders and age groups is mainly based on average values for diameters, length and distance relations between anatomical landmarks such as the arch, the aortic valve, and branching vessels [2, 3]. As suggested by the European Society of Cardiology, recent studies also

© Springer Nature Switzerland AG 2019
Y. Coudière et al. (Eds.): FIMH 2019, LNCS 11504, pp. 370–378, 2019.
https://doi.org/10.1007/978-3-030-21949-9_40

explored the interplay between anatomical and hemodynamic properties based on 4D PC MRI [4, 5]. Although these studies provide insight into the parameter ranges of quantitative parameters, a direct comparison with existing patient data and the corresponding subpopulation is not easily possible. Landmark-related mapping approaches such as the dissection maps suggested by Mistelbauer et al. [6] and Behrendt et al. [7] and the spatiotemporal pressure maps by Lamata et al. [8] provide frameworks for exploring corresponding vascular regions of different subjects. The goal of this work is to provide a related approach, which enables the quantitative assessment of local anatomical properties and the corresponding pressure distribution.

Previous studies have confirmed that pressure difference maps can be derived from 4D flow MRI data [8–11]. Using the Pressure Poisson equation (PPE), the relative pressure difference to a reference location can be calculated at every spatial position. In [13] Meier et al. have presented an efficient algorithm to determine the pressure from velocity within a time-varying flow domain using the finite-element method for solving the PPE. The clinical applicability of this method was confirmed by comparison with catheter-based measurements for the assessment of pressure differences in the aorta [14]. We propose an exploration concept for the interplay of geometric parameters and the local pressure distribution. A landmark-based spatio-temporal normalization for the course of the aorta and the heart cycle as well as a sector concept for the normalized analysis of the cross-sectional pressure distribution will be combined. This concept will then be applied to compare different vessel pathologies with population-based normal values of relative pressure differences.

2 Method

For the generation of a reference dataset, 4D PC MRI of 117 healthy volunteers from the population study by Harloff et al. [5] have been further processed to provide pressure difference maps. Two linked concepts considering anatomical landmarks are then applied to calculate reference values in the normal population.

2.1 Data and Preprocessing

As described in [5], all 4D PC MRI reference datasets were acquired with a 3 T CMR system (TIM Trio, Siemens Healthineers, Erlangen, Germany). Parameters of 4D flow CMR were: TE/TR = 2.54/5 ms, flip angle = 7°, temporal resolution = 20 ms, matrix size = 340 × 255 × 75, bandwidth = 450 Hz/pixel, spatial resolution = 2.5 × 2.1 × 2.5 mm³, velocity sensitivity along all three directions = 150 cm/s, and parallel imaging (PEAK-GRAPPA) along the phase encoding direction (y) with an acceleration factor of $R = 5$ (20 reference lines). To calculate the PCMRA for interactive 3D segmentation, noise masking and phase unwrapping were applied. The PCMRA is calculated as follows:

$$PCMRA_{xyz} = \sqrt{\frac{1}{N}\sum_{t=1}^{N}\left(M_{xyz}^2(t)\cdot\left(vx_{xyz}^2(t)+vy_{xyz}^2(t)+vz_{xyz}^2(t)\right)\right)}$$

372   S. Karimkeshteh et al.

The interactive 3D segmentation is based on the watershed transform and an additional correction with a 3D brush tool [12]. The pre-processed velocity sequences and the 3D segmentation masks were used as input for the calculation of pressure difference maps with the method suggested by Meier et al. [13]. The Pressure Poisson equation (PPE)

$$-\Delta p = \nabla \cdot (\rho \frac{\partial}{\partial t} v + \rho v \cdot \nabla v - \eta \nabla^2 v - \rho g),$$

where ρ and η are the density and dynamic viscosity of blood and g is the gravity force density, is solved by the finite-element method (FEM) directly in the voxel grid with hexahedral elements using the conjugate gradient solver. The publication by Riesenkampff et al. [14] showed that after offset correction the pressure values calculated with this method were comparable to those measured with catheter sensors.

2.2 Analysis of Anatomical Properties and Pressure Differences Along the Vessel Course

The centerline is derived automatically from the given aorta mask through skeletonization. After resampling, the average and minimum vessel diameter are calculated at the intersections of the cross-sectional planes orthogonal to the centerline and the aorta mask. The average diameter is defined as $d_A = 2\sqrt{\frac{A}{\pi}}$, corresponding to the diameter of a circle with the same area A as the vessel mask cross-section. The minimum diameter d_{min} corresponds to the length of the shortest line that separates the cross-section area into equally sized $(A/2)$ parts.

In addition to the vessel diameter, the local curvature is calculated to assess the vessel bending. Curvature values at centerline points can be considered as the radius

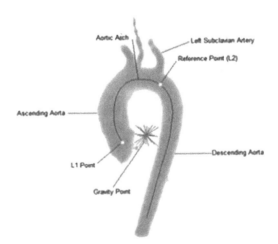

Fig. 1. Major landmarks for the normalized display of local aorta properties: L1 at the aortic valve and L2 at the end of the left subclavian artery branch are placed manually. They determine the distance of the reference planes and the vessel sections in the analysis. The gravity center of the centerline determines the orientation of the segment analysis of the cross-sections.

inversion of the osculating circle that is tangent to the centerline at this point ($k = 1/r$). For a parametric curve in a 3D $c(t) = (x(t), y(t), z(t))$, the curvature can be calculated as $k = \dfrac{\|c' \times c''\|}{\|c'\|^3}$ where $\times$ is the cross-product.

The pressure difference value at each centerline point is provided as a third parameter. As suggested by Meier et al. [13] and Mirzaee et al. [15], we choose the time point when the positive pressure difference between the values in the ascending aorta and the descending aorta is maximal for the initial visualization, assuming that this enables the detection of relevant pressure drops. The manually defined landmarks, which form the basis for the automatization of this calculation, are displayed in Fig. 1. The reference point $L2$ defines the offset for the pressure differences, which are calibrated such that at each voxel location $\vec{x}$, the relative pressure difference $p(\vec{x}) = p_{real}(\vec{x}) - p_{real}(L2)$ represents the difference of the real (unknown) pressure value from the one at $L2$. Corresponding parameters can then be shown as displayed in Fig. 2

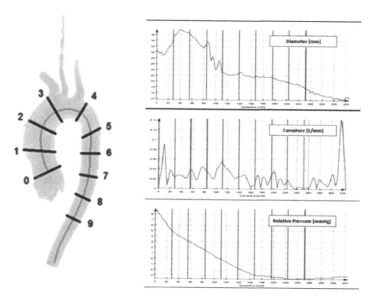

Fig. 2. Parameter curves and cross-section plane positions. The curves on the right represent the parameter values at the corresponding centerline positions. The cross-section planes 0 and 4 are placed at the locations of the landmarks L1 and L2. Their positions are displayed in the curve diagrams (note the zero value of the pressure difference curve at the position of plane 4).

2.3   Pressure Distribution in Cross-Sections

Our analysis concept uses 10 cross-sections but could easily be parameterized to work with alternative configurations. As displayed in Fig. 2, the locations and distances are defined by the positions of the manually chosen centerline points L1 and L2, which correspond to the positions of plane 0 and 4. Between plane 0 and 4, three planes are placed with equal distance along the centerline. The same distance is then used to place plane 5–9 in the descending aorta.

Based on the centerline and the given landmarks, we get the up vector, guide vector, plane normal vector and center point for the in-plane analysis as shown in Fig. 3. The scalar product is applied to determine the sector for each voxel. The vector between a voxel and the center point is considered for the calculation of the scalar product with all sector-separating rays. The highest value indicates the number of the corresponding sector, where this voxel is located. Statistical parameters (maximum, mean, minimum and standard deviation) of pressure are calculated per sector considering the values of all corresponding voxels. As suggested in previous publications comparing different flow and vessel wall parameters [16, 17], we segment cross-sections into 12 sectors.

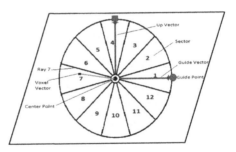

Fig. 3. The assignment is determined by the maximum of the scalar products of the voxel vector and all separating rays. The guide point is the projection of the connection between vessel center and centerline gravity point onto the cross-section plane. The up vector is the result of the cross product of the plane normal vector and the guide vector.

2.4   Exploration Concept for Population Analysis and Comparison with Individual Cases

In order to create a reference database, the extracted curves displayed in Fig. 2 as well as the sector-wise pressure values of the ten cross-section planes (0–9) are saved in Excel files for all subjects of the population. The presented framework enables the selection of gender and age intervals, and the exploration of the resulting population analysis of geometric and pressure parameters.

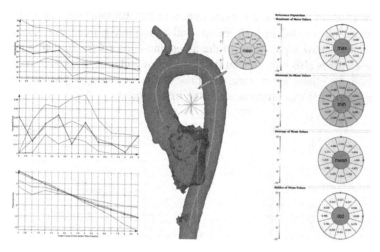

Fig. 4. The diagrams on the left show an example for the comparison of subject-specific parameters (red curve) with the parameters of the chosen reference group (min, max, mean). The sector diagrams on the right display the cross-section comparison. The sector diagram corresponding to plane 5 is shown in 3D to clarify its position. The 2D view enables a segment-wise comparison with the value intervals that occur in the reference population. (Color figure online)

As shown in Fig. 4 the average values of the centerline-related parameters are displayed with reference to the anatomical landmarks L1 and L2 and thereby in relation to the vessel segments defined by the cross-section planes. For the comparison of a specific case with population data, the population subgroup can be chosen similarly. The subject-specific curve is displayed together with the corresponding subgroup's average, minimum and maximum values. In addition to this overview, the pressure distribution per cross-section can be compared to the corresponding reference values and value ranges. The segment plots, which represent the analysis scheme, are described in Sect. 2.3.

Table 1. Characteristics of the population dataset. Numbers in parenthesis represent average and standard deviation.

Sex	Number	Age
Female	60	20–80 (50 ± 17)
Male	57	23–79 (47 ± 16)

3 Results

3.1 Application for Population Analysis

In order to illustrate a typical application scenario of a population analysis we chose to compare the centerline analysis for female and male subjects. Figure 5 displays the resulting normalized curves with the indicators of the analysis plane positions.

The average curvature values (black curves in left diagram) indicate an increasing trend between planes 0–2 in both groups and a decreasing trend after plane 4. The diameter curves show high values in the ascending aorta (plane 2) then a decreasing trend. These values also indicate that the diameter of aorta for the female volunteers is smaller than the diameter of aorta for males. The decrease in pressure along the descending aorta is more pronounced in the male group.

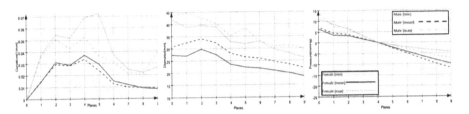

Fig. 5. Comparison of average parameters for male and female volunteers in the population described in Table 1

3.2 Comparison of Patient Datasets with the Corresponding Normal Population

We chose three datasets, a healthy volunteer, a patient with a stenosed bicuspid aortic valve, and a patient with aortic coarctation to test the comparison. Datasets were acquired with different scanners (Siemens Prisma and Philips Achieva) and settings. The aorta segmentation and vector fields resulting from expert pre-processing were imported and interactively analyzed with the presented framework. Results for the volunteer case are shown in Fig. 4. As expected, all values are within the parameter range of the age-matched reference population.

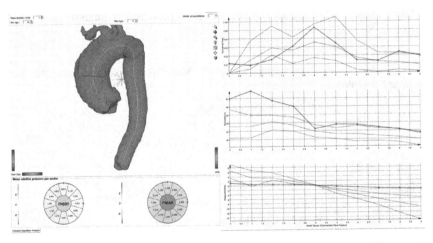

Fig. 6. Comparison of a female patient with aortic valve disease with the age- and sex-matched reference group. The higher curvature in the aortic root, a high diameter in the ascending aorta and a low pressure difference between ascending and descending aorta deviate clearly from the normal population. There is also a higher relative pressure in the extended wall sectors.

As shown in Fig. 6, for the patient with the stenosed bicuspid aortic valve, the differences of the parameters along the vessel course as well as the pressure distribution in the vessel cross-section are visible as well as quantitatively assessable. For the patient with aortic coarctation geometry and pressure course also differ from the normal values at several locations. Here even the shape of the curve differs strongly as shown in Fig. 7. The sector view shows a stronger variation of the pressure values in the cross-section after the strongly curved vessel segment with the coarctation.

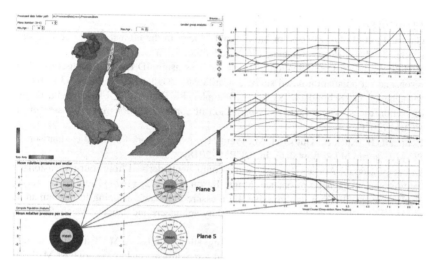

Fig. 7. Comparison of a female patient with aortic coarctation with the age- and sex-matched reference group. The patterns of curvature, diameter and pressure course differ from the reference of the normal population. There is a strong pressure drop between plane 3 and 5, and the sector analysis for plane 5 shows more pronounced differences between the inner and outer part of the cross-section.

4 Discussion and Conclusions

We have presented a solution for the visual and quantitative comparison of geometric and hemodynamic information in subgroups of a population as well as between single subjects and a reference group. In our application tests, it was possible to visually and quantitatively assess pathological deviations of anatomy and hemodynamics in patients with aortic diseases.

Major limitations of the approach arise from the fix normalization scheme, which on the one hand requires coverage of the anatomical region defined through the cross-section placement concept. On the other hand, the coarse normalization scheme does not enable a comparison of cross-sections related to patient-specific local pathologies, if these are not covered by one of the standard planes. Future work will focus on an extension of this concept towards an integration of additional parameters as well as an adaptable plane sampling scheme.

References

1. ESC Committee for Practice Guidelines: 2014 ESC Guidelines on the diagnosis and treatment of aortic diseases. Eur. Heart J. **35**(41), 2873–2926 (2014). https://doi.org/10.1093/eurheartj/ehu281. Epub 29 Aug 2014

2. Rylski, B., Desjardins, B., Moser, W., Bavaria, J.E., Milewski, R.K.: Gender-related changes in aortic geometry throughout life. Eur. J. Cardiothorac. Surg. **45**(5), 805–811 (2014). https://doi.org/10.1093/ejcts/ezt597

3. Redheuil, A., et al.: Age-related changes in aortic arch geometry: relationship with proximal aortic function and left ventricular mass and remodeling. J. Am. Coll. Cardiol. **58**(12), 1262–1270 (2011)

4. Garcia, J., et al.: Distribution of blood flow velocity in the normal aorta: effect of age and gender. J. Magn. Reson. Imaging **47**(2), 487–498 (2018)

5. Harloff, A., et al.: Determination of aortic stiffness using 4D flow cardiovascular magnetic resonance - a population-based study. J. Cardiovasc. Magn. Reson. **20**(1), 43 (2018)

6. Mistelbauer, G., Schmidt, J., Sailer, A., Bäumler, K., Walters, S., Fleischmann, D.: Aortic dissection maps: comprehensive visualization of aortic dissections for risk assessment. In: VCBM (2016)

7. Behrendt, B., Ebel, S., Gutberlet, M., Preim, B.: A framework for visual comparison of 4D PC-MRI aortic blood flow data. In: VCBM (2018)

8. Lamata, P., et al.: Aortic relative pressure components derived from four-dimensional flow cardiovascular magnetic resonance. Magn. Reson. Med. **72**, 1162–1169 (2014)

9. Bock, J., et al.: In vivo non-invasive 4D pressure difference mapping in the human aorta: phantom comparison and application in healthy volunteers and patients. Magn. Reson. Med. **66**(4), 1079–1088 (2011)

10. Ebbers, T., Wigstrom, L., Bolger, A.F., Engvall, J., Karlsson, M.: Estimation of relative cardiovascular pressures using time-resolved three-dimensional phase contrast MRI. Magn. Reson. Med. **45**(5), 872–879 (2001)

11. Tyszka, J.M., Laidlaw, D.H., Asa, J.W., Silverman, J.M.: Three-dimensional, time-resolved (4D) relative pressure mapping using magnetic resonance imaging. J. Magn. Reson. Imaging **12**(2), 321–329 (2000)

12. Hennemuth, A., et al.: Fast interactive exploration of 4D MRI flow data. In: Wong, K.H., et al. (eds.) SPIE Medical Imaging, vol. 7964, 79640E, pp. 1–11. SPIE (2011)

13. Meier, S., Hennemuth, A., Drexl, J., Bock, J., Jung, B., Preusser, T.: A fast and noise-robust method for computation of intravascular pressure difference maps from 4D PC-MRI data. In: Camara, O., Mansi, T., Pop, M., Rhode, K., Sermesant, M., Young, A. (eds.) STACOM 2012. LNCS, vol. 7746, pp. 215–224. Springer, Heidelberg (2013). https://doi.org/10.1007/978-3-642-36961-2_25

14. Riesenkampff, E., et al.: Pressure fields by flow-sensitive, 4D, velocity-encoded CMR in patients with aortic coarctation. JACC Cardiovasc. Imaging **7**(9), 920–926 (2014)

15. Mirzaee, H., et al.: MRI-based computational hemodynamics in patients with aortic coarctation using the lattice Boltzmann methods: clinical validation study. J. Magn. Reson. Imaging **45**, 139–146 (2017). https://doi.org/10.1002/jmri.25366

16. Stalder, A., Russe, M., Frydrychowicz, A., Bock, J., Hennig, J., Markl, M.: Quantitative 2D and 3D phase contrast MRI: optimized analysis of blood flow and vessel wall parameters. Magn. Reson. Med. **60**, 1218–1231 (2008)

17. Liu, J., Shar, J.A., Sucosky, P.: Wall shear stress directional abnormalities in BAV aortas: toward a new hemodynamic predictor of aortopathy? Front Physiol. **14**(9), 993 (2018)

Model-Based Indices of Early-Stage Cardiovascular Failure and Its Therapeutic Management in Fontan Patients

Bram Ruijsink[1](ID), Konrad Zugaj[1], Kuberan Pushparajah[1](ID),
and Radomír Chabiniok[1,2,3(✉)](ID)

[1] School of Biomedical Engineering and Imaging Sciences (BMEIS),
St. Thomas' Hospital, King's College London, London, UK
radomir.chabiniok@kcl.ac.uk
[2] Inria, Paris-Saclay University, Palaiseau, France
[3] LMS, Ecole Polytechnique, CNRS, Paris-Saclay University, Palaiseau, France

Abstract. Investigating the causes of failure of Fontan circulation in individual patients remains challenging despite detailed combined invasive cardiac catheterisation and magnetic resonance (XMR) exams at rest and during stress. In this work, we use a biomechanical model of the heart and Fontan circulation with the components of systemic and pulmonary beds to augment the diagnostic assessment of patients undergoing the XMR stress exam. We apply our model in 3 Fontan patients and one biventricular "control" case. In all subjects, we obtained important biophysical factors of cardiovascular physiology – contractility, contractile reserve and changes in systemic and pulmonary vascular resistance – which contribute to explaining the mechanism of failure in individual patients. Finally, we used the patient-specific model of one Fontan patient to investigate the impact of changes in pulmonary vascular resistance, aiming at in silico testing of pulmonary vasodilation treatments.

Keywords: Fontan circulation · Heart failure · Dobutamine stress ·
Pulmonary vascular resistance · Pulmonary vasodilation therapy

1 Introduction

Patients with Fontan circulation – a surgically established circulation for patients with a single functional ventricle in which the systemic and pulmonary circulations are coupled in series – experience a progressive decline in cardiovascular function, ultimately leading to heart failure [6]. Fontan failure (FF) is multifactorial and often involves both the heart and the circulation, e.g. a reduced myocardial contractility, contractile reserve, or an increased pulmonary vascular resistance (PVR). Identifying the exact cause of failure is therefore difficult. A potential treatment option for FF is pulmonary vasodilation (PVD) [4]. However, its efficacy remains inconclusive, especially in patients without elevated

© Springer Nature Switzerland AG 2019
Y. Coudière et al. (Eds.): FIMH 2019, LNCS 11504, pp. 379–387, 2019.
https://doi.org/10.1007/978-3-030-21949-9_41

PVR. Several specialised centres use combined cardiac MRI and catheterisation (XMR) at rest and during pharmacological stress to investigate the causes of FF, and to identify patients that might benefit from PVD [9]. These diagnostic exams generate rich datasets of combined ventricular and pulmonary pressures, blood flow and cardiac function. Despite this, understanding the role of each parameter in FF in individual patients remains challenging. We hypothesise that biomechanical modelling may improve the interpretation of XMR exams, and the augmented knowledge of patients' physiology could be considered in the PVD reactivity testing to possibly reduce its invasiveness.

In this work we build patient-specific models of the Fontan circulation (heart with systemic and pulmonary circulations in series) based on the stress XMR exam, which can provide values of myocardial contractility, SVR and PVR at rest and their changes during stress. Then, using such an "avatar" of rest and stress physiology from one patient, we investigate the impact of an increased and decreased PVR on the heart and circulation – the changes known to take place with aging and after PVD, respectively.

2   Methods

2.1   Data

Three patients with Fontan circulation (underlying diagnosis: hypoplastic left heart syndrome) and early-stage FF underwent an XMR dobutamine stress exam. During the exam, cine MRI, phase-contrast flow through aorta and branch pulmonary arteries (PAs), together with catheter pressure measurements in the aorta, systemic ventricle and PAs were obtained at rest and during pharmacological stress (infusion of dobutamine $10\,\mu g/kg/min$), see Fig. 1. None of the patients had significant collateral flow or fenestration shunts. In addition to the single-ventricular patients, one patient with a biventricular heart was included in the study. This patient suffered from biliary atresia and underwent a stress XMR during work-up for liver transplantation. This study was approved by our regional ethics committee, London UK (Ethics Number 09H0804062) and written informed consent was obtained for all subjects.

2.2   Model

The systemic ventricle is modelled using the multiscale biomechanical model described in [3], of which the geometry and kinematics were reduced to a sphere [2], while all other biophysical properties correspond to the full 3D model. Briefly, the passive part of the myocardium is modelled using the hyperelastic potential by Holzapfel and Ogden [5], and the active component by a system of ordinary differential equations representing chemically controlled actin-myosin interactions generating active stress, of which the asymptotic value σ_0 is directly related to myocardial contractility.

The ventricle is connected to the circulation system, represented by a 3-stage Windkessel (WK) model consisting of the very proximal part of the circulation

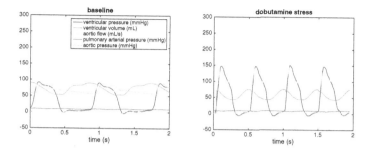

Fig. 1. Data of a selected Fontan patient acquired at rest and during stress.

(aorta and large arterial branches with resistance and capacitance R_{prox} and C_{prox}); main part of the systemic circulation (R_{syst}, C_{syst}); and the pulmonary circulation (R_{pulm}, C_{pulm}), see Fig. 2. The role of the proximal element is to capture the correct peak aortic pressure. Its effect on the overall ventricular afterload is limited, as R_{prox} and C_{prox} are ~1/10 of their systemic counterparts. Therefore, we associate systemic vascular resistance (SVR) directly with R_{syst}, while R_{pulm} represents PVR.

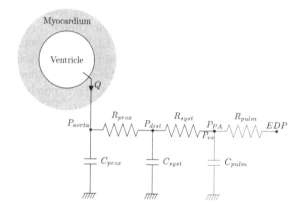

Fig. 2. Scheme of the model of systemic ventricle with the corresponding circulation: 2-stage Windkessel model for biventricular subject (terminating by central venous pressure P_{ve}); 3-stage Windkessel with the pulmonary component in blue (P_{PA} being pulmonary artery pressure, and ending by the end-diastolic ventricular pressure EDP). (Color figure online)

In the biventricular case, only a 2-stage WK model was used (without the pulmonary part). For the Fontan patients, in addition to the 3-stage WK, a circulation with SVR and PVR lumped into "total vascular resistance" (TVR) was calibrated. The role of these 2-stage WK models was to obtain a simpler representation of the afterload against which the systemic ventricle works.

The heart and circulation models where simulated using the MATLAB-based simulator CardiacLab developed at Inria in the team M≡DISIM (main contributor Philippe Moireau).

2.3   Model Calibration

The 3-stage WK model was calibrated sequentially – first the systemic, followed by the pulmonary part. The systemic and pulmonary resistances were initialised using the mean difference of respective pressures (mean aortic, pulmonary artery or terminal end-diastolic ventricular) and cardiac output (CO), as $R_i \sim \frac{\Delta P_i}{CO}$. The associated capacitances were pre-calibrated by assuming the time constants (given by multiplication of resistance and capacitance) being 1 s. For a detailed calibration, the measured aortic flow was imposed into the circulatory model firstly with the terminal pressure being the pulmonary artery pressure in the regime of a 2-stage (proximal+systemic) WK to calibrate the systemic part; then, the proximal resistance was adjusted to match the peak aortic pressure; lastly, the ventricular EDP was used as the terminal pressure in the 3-stage WK to calibrate the pulmonary part. For the biventricular "control" case (and to obtain TVR, the lumped SVR and PVR, in Fontan patients), the 2-stage WK was calibrated (i.e. steps 1 and 2 above). The calibration process was performed at rest and during stress.

The heart model was calibrated using the following sequence: First, the geometrical relations in the heart model (ventricular volume and wall thickness) were prescribed using the end-diastolic volume (EDV) and myocardial wall mass obtained from cine MRI. The volume of unloaded ventricle V_0 was assumed to be 50% of EDV [7]. The passive properties of the heart were adjusted by imposing the measured EDP and subsequently multiplying the hyperelastic potential by a "relative stiffness parameter" to obtain the measured EDV. Next, the active properties of the ventricle were adjusted by connecting the calibrated circulation WK model, imposing the timing of the activation (according to the measured ECG), and adjusting the myocardial contractility σ_0 to reach the end-systolic volume (ESV) and aortic flow as in the data. After this calibration at rest, the adjustments in the model during stress were obtained by connecting the WK calibrated using the stress data, adjusting the electrical activation according to the observed chronotropic effect, prescribing the preload pressure according to data, and lastly adjusting σ_0 to match the observed end-systolic volume (ESV) and aortic flow.

2.4   In Silico PVD Testing

Using the heart-circulation model of a selected Fontan patient, calibrated at rest and stress, we performed an in silico study to investigate the impact of modifying the PVR on pulmonary artery pressure (invasive indicator) and on stroke volume (SV, non-invasive indicator). The model was run with varying PVR (by factors 0.25, 0.5, 2 and 4), while keeping all other parameters as calibrated for

Table 1. Summary of the changes in contractility, systemic and pulmonary vascular resistances (SVR, PVR) and total pulmonary resistance (TVR) for Fontan patients (FP) and biventricular control case (CC) during dobutamine stress.

Patient	Contractility rest σ_0^{rest} (kPa)	Contractility stress ($\times \sigma_0^{rest}$)	SVR rest R_{syst}^{rest} ($\frac{MPa \cdot s}{m^3}$)	SVR stress ($\times$ rest)	PVR rest R_{pulm}^{rest} ($\frac{MPa \cdot s}{m^3}$)	PVR stress ($\times$ rest)	TVR rest ($\frac{MPa \cdot s}{m^3}$)	TVR stress
FP 1	65	2.4×	98	1.22×	11	0.91×	100	1.25×
FP 3	73	2.1×	108	1.62×	9	1.48 ×	125	1.48×
FP 5	73	2.0×	171	1.22×	19	0.76×	190	1.17×
CC 1	70	1.4×	130	0.90×	N/A	N/A	N/A	N/A

rest/stress). We evaluated the changes in SV and PA pressure in comparison to those in our original calibration.

3 Results

3.1 Patient-Specific Heart-Circulation Models at Rest and Stress

The quantitative values of parameters of the circulation and myocardial contractility are shown in Table 1. Figure 3 shows the simulated indicators compared with the data for a selected Fontan patient at rest and during stress. Figure 4 compares the systemic and pulmonary vascular resistances for all patients at rest and their adaptation during stress.

3.2 In Silico Test of PVD at Various Physiology Stages

Figure 5 shows the assumed effect of PVD in Fontan patient 5. The blue marker represents the PVR value obtained from the original calibrations. The mean pulmonary pressures were similar at rest and during stress. SV was higher at rest, compared to stress. With varying PVR, there was a significant change in SV at rest, see Fig. 5. This trend was much less pronounced during stress. Varying PVR did also result in a significant and expected change in average pulmonary pressures, both at rest and during stress.

4 Discussion

In this study, we explored the use of biomechanical models to investigate factors underlying heart failure in patients with Fontan circulation. We used XMR datasets obtained at rest and during stress to create patient-specific models to access the biophysical parameters that are not directly available from the data themselves. Good calibrations were obtained for all included cases at rest and during stress: the data-simulation error of maximum ventricular pressure averaged over all subjects and physiological states was 4.7 mmHg; error of mean pulmonary pressure 0.25 mmHg; and error of stroke volume 1.2 mL, see example

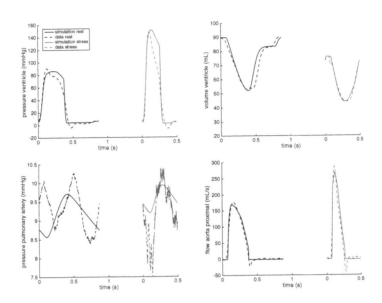

Fig. 3. Simulations at rest (black) and during stress (red) for Fontan patient 5. (Color figure online)

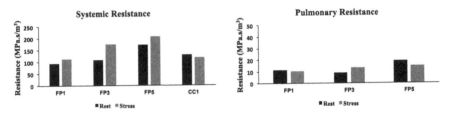

Fig. 4. Systemic and pulmonary vascular resistances at rest (black) and stress (red) in all Fontan patients (FP) and the control biventricular case (CC). (Color figure online)

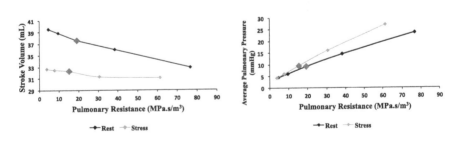

Fig. 5. Stroke volume (left) and mean pulmonary pressure (right) response to varying PVR in Fontan patient 5. The blue marker represents the original patient-specific calibrations. (Color figure online)

Fig. 3. This suggests that our proposed model, while simplifying both the heart and circulation, can well describe the different physiological conditions. With such close-to-real-time running models (computation time of one cardiac cycle being ~10 s on a standard laptop), we were able to obtain fast simulations using standard computers within a clinically relevant time-frame (i.e. compatible with ~3-h time-frame to clinically evaluate and report the outcomes of the exam).

At rest, myocardial contractility was slightly increased in Fontan patients compared to the biventricular case. However, all Fontan patients had a good contractile reserve ($> 2\times$ resting values) [1], see Table 1. These results suggest that FF in these particular Fontan patients was not due to systolic dysfunction.

Our coupled model also allowed to investigate the vascular response. At rest, we did not see significant differences in resistance between Fontan patients and the control case. However, during stress the vascular adaptations varied. In the control case, the resistance decreased – an expected response in normal physiology [10]. However, in our Fontan patients the total resistance of the system (TVR) increased. By looking at the changes in resistance of the two in-series-coupled beds (i.e. SVR and PVR), we could see that SVR was the main contributor to the effect, while there was only a relatively small change in PVR. This could reflect a vascular maladaptation of the systemic circulation to the chronic high systemic venous pressures [11] – the result of the circulations coupled in series.

While PVR is only a relatively small component of TVR, clinical and experimental studies have shown that it might have a large impact on cardiac output in the Fontan circulation [11]. This has formed the rationale behind PVD treatment. However, whether all patients benefit from PVD is unclear, especially when PVR is not elevated. We used our calibrated model in Fontan patient 5 to investigate the impact of hypothetical changes in PVR in silico, while keeping all other factors of the model as calibrated for rest and at stress (preload, heart rate, contractility and SVR). The model at rest suggests that varying PVR could result in a significant shift in stroke volume – i.e. a non-invasive index – despite the fact that the PVR was not significantly elevated in this patient [8].

The main limitation of our preliminary in silico PVD test is that preload was fixed while varying PVR, even though a change in PVR is likely to result in a change in flow towards the heart [8]. We aim to address this limitation by exploiting closed-loop heart-circulation models, once having data obtained during XMR exams that include in vivo PVD provocation both at rest and during stress. The approach nevertheless represents an initial step in studying some non-invasive indices, which might contribute to the assessment of PVD in individual patients.

Finally, we aim to further refine the circulatory model to include potential pathological systemic-to-venous or veno-venous shunts, which are often formed in Fontan patients, even though these were not present in our subjects.

5 Conclusion

In this paper, we showed that our models allow a detailed investigation of some key biophysical factors of the pathological stress response in Fontan patients. Moreover, we performed first steps in exploring the use of modelling to examine potential changes in PVR on the cardiovascular system. We aim to further investigate modelling in a larger group of patients, loop our models to include venous return and subsequently compare the results of in silico PVD experiments to data of prospectively examined patients. Ultimately, we hope to obtain the models that would contribute into the identification of patients who could benefit from PVD, in order to decrease the invasiveness of PVD-reactivity tests.

Acknowledgments. The authors acknowledge financial support from the Department of Health through the National Institute for Health Research (NIHR) comprehensive Biomedical Research Centre award to Guy's & St Thomas' NHS Foundation Trust in partnership with King's College London and the NIHR Cardiovascular MedTech Co-operative (previously existing as the Cardiovascular Healthcare Technology Co-operative 2012–2017), the support of Wellcome/EPSRC Centre for Medical Engineering [WT 203148/Z/16/Z], and from the Inria Associated team ToFMOD. The views expressed are those of the author(s) and not necessarily those of the NHS, the NIHR or the Department of Health.

In addition, the authors would like to acknowledge Philippe Moireau and Dominique Chapelle, Inria research team MΞDISIM, for providing the cardiac simulation software CardiacLab under a Royalty-fee software license.

References

1. Bussmann, W., Heeger, J., Kaltenbach, M.: Contractile and relaxation reserve of the left ventricle. I. Normal left ventricle. Z. Kardiol. **66**(12), 690–695 (1977)
2. Caruel, M., Chabiniok, R., Moireau, P., Lecarpentier, Y., Chapelle, D.: Dimensional reductions of a cardiac model for effective validation and calibration. Biomech. Model. Mechanobiol. **13**(4), 897–914 (2014)
3. Chapelle, D., Le Tallec, P., Moireau, P., Sorine, M.: An energy-preserving muscle tissue model: formulation and compatible discretizations. Int. J. Multiscale Comput. Eng. **10**(2), 189–211 (2012)
4. Ghanayem, N., Berger, S., Tweddell, J.: Medical management of the failing Fontan. Pediatr. Cardiol. **28**(6), 465–471 (2007)
5. Holzapfel, G., Ogden, R.: Constitutive modelling of passive myocardium: a structurally based framework for material characterization. Philos. Trans. R. Soc. Lond. A Math. Phys. Eng. Sci. **367**(1902), 3445–3475 (2009)
6. Khairy, P., Poirier, N., Mercier, L.A.: Univentricular heart. Circulation **115**(6), 800–812 (2007)
7. Klotz, S., et al.: Single-beat estimation of end-diastolic pressure-volume relationship: a novel method with potential for noninvasive application. Am. J. Physiol. Heart Circ. Physiol. **291**, H403–H412 (2006)
8. Kovacs, G., Berghold, A., Scheidl, S., Olschewski, H.: Pulmonary arterial pressure during rest and exercise in healthy subjects a systematic review. Eur. Respir. J. **34**, 888–894 (2009)

9. Pushparajah, K., et al.: Magnetic resonance imaging catheter stress haemodynamics post-Fontan in hypoplastic left heart syndrome. Eur. Heart J. Cardiovasc. Imaging 17(6), 644–651 (2015)
10. Ruffolo Jr., R.R., Messick, K.: Systemic hemodynamic effects of dopamine, (+)-dobutamine and the (+)- and (-)-enantiomers of dobutamine in anesthetized normotensive rats. Eur. J. Pharmacol. 109, 173–181 (1985)
11. Veldtman, G.R., et al.: Cardiovascular adaptation to the Fontan circulation. Congenit. Heart Disease 12(6), 699–710 (2017)

3D Coronary Vessel Tree Tracking
in X-Ray Projections

Emmanuelle Poulain[1,2($\boxtimes$)], Grégoire Malandain[2], and Régis Vaillant[1]

[1] GE Healthcare, 78530 Buc, France
epoulain93@gmail.com, regis.vaillant@ge.com
[2] Université Côte d'Azur, Inria, CNRS, I3S, Sophia Antipolis, France
gregoire.malandain@inria.fr

Abstract. CTA angiography brings potentially useful information for guidance in an interventional procedure. It comes with the challenge of registering this 3D modality to the projection of the coronary arteries which are deforming with the cardiac motion. A tree-spline i.e. a tree with a spline attached to each edge and shared control points between these points describes a 3D coronary tree and is able to represent its deformation along the time. We combine this description with a registration algorithm operating between the tree-spline and the angiographic projection of the coronary tree. It starts by the estimation of a rigid transformation for the iso cardiac phase time followed by a non-rigid deformation of the tree driven by the pairings formed between the projection of the edges of the tree-spline and the observed x-ray projection of the coronary arteries. The pairings are built taking into account the tree topology consistency. Anatomical constraints of length preservation is enforced when deforming the arteries. The proposed approach has been evaluated with clinical data issued from ten different clinical cases which enabling to form twenty three different experimental conditions. Encouraging results have been obtained.

Keywords: Deformable registration · Tracking · Coronary arteries · X-ray · Computed Tomography Angiography · CTA

1 Introduction

Coronary artery narrowing's are commonly treated in Percutaneous Coronary Intervention (PCI) procedures. These procedures are performed under the guidance of an x-ray imaging system and implies the use of iodinated contrast agent to opacify the lumen of the coronary arteries. Once the vessel, which is going to be the object of the procedure, has been imaged, the operator modifies the orientation of the image chain to place it such that vessel superimposition and projective foreshortening of this vessel is minimal. The angiography depicts the lumen of the artery with an excellent spatial and temporal resolution. It is limited in its capacity to depict the characteristics of the vessel wall as for example the

© Springer Nature Switzerland AG 2019
Y. Coudière et al. (Eds.): FIMH 2019, LNCS 11504, pp. 388–396, 2019.
https://doi.org/10.1007/978-3-030-21949-9_42

existence of calcifications. Computed Tomography Angiography (CTA) is more able to provide this additional information. Most of the times and for the benefit of limiting the dose delivered to the patient, it comes as a static image which depicts the vessel anatomy in diastolic phase. Physicians are eager to see both modalities in the same referential which is the one defined by the angiographic image in the configuration selected for the patient treatment. In this article, we propose and evaluate an algorithmic approach to bring these two imaging modalities in the same referential while keeping the dynamic of cardiac motion. [1,3,8] have proposed different strategies applicable solely to data in the same cardiac phase. The registration requires then the estimation of a rigid transformation followed by the conic projection with the x-ray source being the focal point. To extend to the cardiac cycle, [2] has proposed to adapt a generic model of the cardiac motion to end-diastolic CTA. This strategy raises the question of the validity of the generic model. [7] proposed to focus the attention of the algorithm to the vessel of interest and to track it along the cardiac sequences. It takes profit of a spline representation of the vessel of interest. The spline can be smoothly warped to follow the observed coronary artery deformation in the artery. The optimization of the control points is made with the double constraint of minimizing the distance to the projection of the vessel of interest and keeping the vessel length constant, which is anatomically meaningful. Encouraging results with an accuracy of 2 mm for the tracking of a landmark were observed. This approach is only applicable when the operator has avoided vessel superimposition over the vessel of interest. To further extend the concept, we explore here the benefit of doing the deformable registration over the whole coronary tree. This benefit is illustrated through tracking videos presented in https://3dvttracking.github.io/. In the following, we will describe the proposed method and explain the assessment strategy which includes metrics evaluating the registration and a specific metric related to the consistency of the position along the vessel.

2   Method

Before introducing the method, we first describe the data we have at hand. The 3D information is extracted from a CTA scan by a fully automated commercial product, *Auto-Coronary-Analyis* from GE Healthcare, providing a segmentation of the coronary vessel structure. The coronary vessels are separated between the right and the left coronary and the different branches are represented by their centerlines. So the anatomic structure is described by a tree $\mathcal{T}$. The aim of this work is to track $\mathcal{T}$ along the consecutive images of the x-ray record sequence. Even if the 3D model of the coronary vessels can be depicted by a tree, this may not be the case for the x-ray projection. Indeed, self superimpositions create crossings. The vessel segmentation may also cause over segmentation or miss some vessels. X-ray projections are segmented with an Hessian based vessel enhancement technique, and vessel like structures are extracted forming a set of curves which corresponds to the centerlines of the vessel. The segmented object is organized in a graph by applying standard processing methods to connect neighboring centerlines. Considering the consecutive images obtained in the sequence of N images by performing the acquisition after injection of the contrast agent, we obtain a set of graphs $\mathcal{G} = \{\mathcal{G}_1, \ldots, \mathcal{G}_N\}$.

We initiate the registration by identifying the initial rigid transformation, T° which maps $\mathcal{T}$ to the element $\mathcal{G}_1 \in \mathcal{G}$ corresponding to the same diastolic cardiac phase as the pre-operative CTA image [3].

The aim of the proposed tracking method is to track the tree $\mathcal{T}$ in all the consecutive phases of the cardiac cycle, which necessitates to deform it. A spline description [7] is a tool suited for a one vessel deformation which can be represented by the optimization of its control points. Nevertheless, the tree structure implies a vessel connectivity preservation. A tree-spline structure (as in [10]) makes this preservation possible. The registration itself is based on a two steps mechanism with first the determination of pairings between the projected edges of $\mathcal{T}$ and the centerlines represented through a graph structure. Second, the control points are adjusted by minimizing an energy depending on the distance between the paired points and constraints on edges.

2.1 Problem Modeling

The 3D temporal tracking requires a priori 3D model of the tree as introduced in [5,10]. The tree is represented as a set of centerlines which is a set of 3D curves. The different 3D curves are not completely independent: the extremity of a 3D curve is either a leaf of the tree and as such can move independently or is either the common extremity of other 3D curves. It is the case when the considered extremity is an anatomical bifurcation and in this case, it is the extremity of three different 3D curves. The representation of this tree shall support deformation of each 3D curves while keeping coherency between them when they share extremities. As seen in [7], the spline functions support a compact and smooth description of curves which can be continuously deformed by changing the position of the control points. We thus fit an approximating cubic (C^2 continuity) spline curve C^k for each edge A^k as in [6], using a centripetal method such that, $A^k \approx \{C^k(u) \mid u \in [0,1]\}$. More precisely the spline is defined as $C^k(u) = \sum_{i=1}^{n^k} N_{i,p}(u) P_i^k$, where $N_{i,p}$ is the ith B-spline of degree p, P_i^k the ith control point of C^k, u the spline abscissa (between 0 and 1), n^k the number of control points. To reach the goal of representing a tree, we assign a multiplicity of $p + 1$ to the first and last element of the associated knot vector U_k as below:

$$U_k = \{\underbrace{0, \ldots, 0}_{p+1}, u_{p+1}, \cdots, u_{m-p-2}, \underbrace{1, \ldots, 1}_{p+1}\}$$

with m the size of the vector. Given the description of the spline, the curve has to pass by the first (respectively last) control points (see [4] for details). Then we parametrize the set of splines such that the control points of 3D curves with a common extremity are also shared. Deformations are obtained by the optimization of the spline parameters. The set of control points to register the

3D vessel tree with the graph $\mathcal{G}_t$ is determined by solving this optimization problem:

$$\begin{cases} \hat{\mathcal{P}}_t = \mathrm{argmin}_{\mathcal{P}=\{P^1,\ldots,P^{n^a}\}\in\Omega} \sum_{k=1}^{n^a} E_d(C_{P^k}, \mathcal{G}_t) + \beta E_r(C_{P^k}) \\ \Omega = \{P \in \mathbb{R}^3 \mid P_{n_k}^k = P_0^j \text{ if } A^j \text{ is a daughter artery of } A^k\} \end{cases} \quad (1)$$

t denotes the temporal index of the frame, $E_d()$ and $E_r()$ are respectively the data attachment and the regularization energy terms, C_{P^k} the spline built with the control points P^k. In the following $\mathcal{P}_t$ denotes the set of control points for frame t while $\mathcal{P}_1^{init}$ denotes the set of control points for the 3D vessel tree after the pose estimation T° for frame 1. An initial position is used for the 3D vessel tree to build the data attachment term: it is the 3D vessels/splines $\mathcal{T}_{\mathcal{P}_1^{init}}$ issued from the pose estimation for the first frame $t = 1$ or $\mathcal{T}_{\mathcal{P}_{t-1}}$ for frame $t > 1$. For the sake of simplicity, t will be omitted in the following. Every 3D edges C^k are projected onto the angiographic frame and are denoted c^k. A 2D curve v^k corresponding to the projected 3D curve is extracted from the graph $\mathcal{G}$ as described in [3]. In the sequel, the registration between one couple of curves will be explained, therefore k will also be omitted.

Data Attachment Term. The data attachment term $E_d()$ is a sum of 3D residual distances issued from 3D to 2D pairings. The simplest method to build pairings is to use the closest neighbor scheme (as in the ICP). In [9], a variant of this approach is proposed: the idea is to represent the cardiac motion by covariance matrices on the different parameters describing the coronary tree. For this one, a generative 3D model is employed, i.e. a model including a probabilistic distribution of position for the arterial segment. The concept of distance is then extended from standard Euclidean distance to Mahalanobis distance. This geometrically oriented analysis does not include the constraint of ordered pairing as proposed in [3] where it is shown that a point pairing that respects the order along paired curves yields better results than the closest neighbor scheme. Such an ordered pairing was obtained by the means of the Fréchet distance, that allows *jumps* between paired points. In presence of vessel deformation, we observed that the coherency of the obtained pairings is sometimes lost. So we propose to constrain the pairing construction with a 2D elongation preservation. We first recall the Fréchet distance and its induced pairing [3]. Let $c = \{c_1, \ldots, c_{n^c}\}$ and $v = \{v_1, \ldots, v_{n^V}\}$ be the 2D curves to be paired. The points c_i are obtained as projection of points $C_{\mathcal{P}}(\bar{u}_i)$ from the 3D spline which represents the vessel. The points v_i are the discrete points forming the centerline of the vessel extracted from the angiographic images. The point pairings are entirely defined by a single injective function $F : \mathbb{N} \to \mathbb{N}$. The Fréchet distance is defined as:

$$\begin{cases} F(1) = \mathrm{argmin}_{i_v \in I_v} \|v_{i_v} - c_1\| \text{ with } I_v = \{1, \ldots, jump\} \\ F(i_c) = \mathrm{argmin}_{i_v \in I_v} \|v_{i_v} - c_{i_C}\| \text{ with } I_v = \{F(i_c - 1), \ldots, F(i_c - 1) + jump\} \end{cases}$$

with *jump* a parameter controlling the length of allowed jumps in pairings. Looking at the pairing produced by this metric (as in Fig. 1, left), we observed that the simple application of the criteria of minimizing consecutively the pairing length may lead to irregular pairings. Inspired by the Fréchet distance, we present a pairing function which aims to build a pairing function that advances at the same speed along the 2D curves to be paired. We define $d(p_1, p_2) = \sum_{i=p_1+1}^{p_2} \|c_i - c_{i-1}\|$ and F as:

$$\begin{cases} F(1) = \text{argmin}_{i_v \in I_v} \|v_{i_v} - c_1\|^2 + \lambda d(v_1, v_{i_v})^2 \text{ with } I_v = \{1, \ldots, jump\} \\ F(i_c) = \text{argmin}_{i_v \in I_v} \|v_{i_v} - c_{i_c}\|^2 + \lambda (d(v_{F(i_c-1)}, v_{i_v}) - d(c_{i_c-1}, c_{i_c}))^2 \\ \quad \text{with } I_v = \{F(i_c - 1), \ldots, F(i_c - 1) + jump\} \end{cases} \quad (2)$$

with λ proportional to the local distance between the neighborhood of i_c and i_v. This function favors point pairings between points which are approximately at the same distance from their respective neighborhoods.

Figure 1 shows the pairings obtained with this weighted Fréchet distance. It is especially useful when the two curves to be paired have experienced a non-rigid transformation with respect to each other. This is exactly the case as our objective is to follow the cardiac deformation. $F()$ provides 2D point pairings $(v_{F(i)}, c_i)$ between the 2D curves v and c. To compute 3D deformations, we have to define 3D point pairings. $c_i \in c$ is associated to its corresponding 3D point $C_\mathcal{P}(\bar{u}_i)$. The 3D point $V'_{F(i)}$ corresponding to $v_{F(i)}$ is the point from the backprojected line issued from $v_{F(i)}$ that is the closest to $C_\mathcal{P}(\bar{u}_i)$. The data attachment term is finally, $E_d(C_\mathcal{P}, \mathcal{G}_t) = \sum_{i=1}^{n_C} \|V'_{F(i)} - C_\mathcal{P}(\bar{u}_i)\|^2$

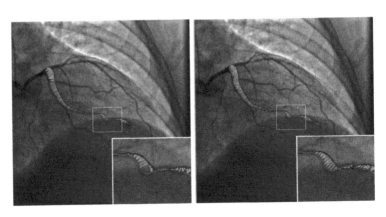

Fig. 1. Pairings (green) between a projected 3D vessel c (blue) and a 2D vessel v (red). Left, with the Fréchet distance; right, with the weighted Fréchet distance. (Color figure online)

Regularization Term. The regularization term aims at minimizing the 3D elongation of C: $E_r(C_\mathcal{P}) = \sum_{j=1}^{J} (\|C_\mathcal{P}(e_j) - C_\mathcal{P}(e_{j-1})\| - l_j)^2$ with, $e_j = \frac{j}{J}$, $l_j = \|C_{\mathcal{P}_1^{init}}(e_j) - C_{\mathcal{P}_1^{init}}(e_{j-1})\|$, J is the number of interval used to enforce the length constraint all along the vessel.

Energy Minimization. This global energy $E_d(C_\mathcal{P}, \mathcal{G}_t) + \beta E_r(C_\mathcal{P})$ is minimized via a gradient descent. Thanks to the spline description of the 3D curve, the analytic expression of the gradient is used for the gradient descent. The pairings are recomputed along the descent every 1000 iterations. The minimization is stopped when the gradient norm is below a threshold, whose value has been chosen in preliminary experiments.

3   Performance Evaluations

Qualitative evaluation of the performance can first be done by a visual control of the deformation of the projected deformed vessel of interest over the angiographic image along the cardiac cycle. We also propose two quantitative measures. The first derives from methodological expectations on the performance but does not cover directly the intended clinical application. The second one replicates more closely the expectations from a clinical standpoint.

Shape Preservation. For this analysis, we start from the idea that the vessel tree shall return to its initial state if the tracking is performed on a series of consecutive images which start and end by the same image. Let N the number of angiographic images in a sequence which covers a cardiac cycle, the tracking is done from the frame 1 to the frame N, resulting in N 3D trees corresponding to the same vessels temporally tracked, $\mathcal{T} = \{\mathcal{T}_1, \ldots, \mathcal{T}_N\}$. One can then generate the reverse sequence starting from image $N - 1$ down to image 1 and continue to apply the tracking algorithm. The result is an other set of 3D trees $\mathcal{T}' = \{\mathcal{T}'_{N-1}, \ldots, \mathcal{T}'_1\}$. To measure the similarity, we chose to compare the projections of $\mathcal{T}_1$ and $\mathcal{T}'_1$ on the vessel of interest W i.e. the vessel which is pathological and going to be fix in the intervention. Let I_w be the set of edge index which correspond to edges owing to the vessel of interest, we define the vessel as $W = \{C^j\}_{j \in I_w}$. We retain as a measure of shape preservation, the percentage of points which get close enough to their initial position sp and the percentage of points which get close enough to a point owing to W, sp^{cl}, these measures are defined as $sp = \frac{|E| * 100}{n^w}$ and $sp^{cl} = \frac{|E^{cl}| * 100}{n^w}$ with $E = \{i \in [1, n^w] \mid \|w_1(i) - w'_1(i)\| < l\}$, $E^{cl} = \{i \in [1, n^w] \mid \|w_1(i) - w'_1(k)\| < l\}$, with w_1 and w'_1 the projections of W_1 and W'_1, k the index of the closest point of $w_1(i)$ owing to w'_1, n^w the number of points in W and l a parameter which controls what close means here.

The point of computing this measure and not the distance between points is to detect when the tracking has followed a wrong vessel (as illustrated in the videos https://3dvttracking.github.io/), which in this case would give a low score of sp and sp^{cl}.

Landmark Tracking. The idea is to evaluate if a location defined along W is correctly tracked. A location in the vessel W is defined by its curvilinear abscissa. In the angiographic image, identifying a fixed point is more challenging.

For the purpose of the evaluation, we first manually point an easily identifiable landmark along the 2D vessel. Vessel bifurcations are natural candidates for such landmarks. To decide whether the same 3D point of the tracked vessel is paired to this landmark, we use the curvilinear abscissas u (along the vessel) of the paired 3D points to it. A perfect tracking (along with a perfect manual identification of the landmark) should yield the same curvilinear abscissa for all paired 3D points, thus the standard deviation of all curvilinear abscissas is an adequate measure to assess the tracking.

Formally, let $U = \{u_1, \ldots, u_N\}$ be the N abscissas of the paired 3D points, e.g. $W(u_t)$ is paired with the bifurcation/landmark in frame t, and $\bar{u}$ be the average value over U, the proposed measure is $lt = \sqrt{\frac{1}{N} \sum_{i=1}^{N} (u_i - \bar{u})^2}$.

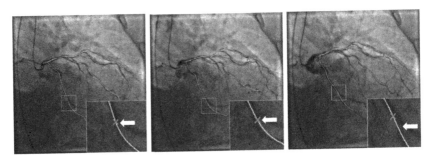

Fig. 2. Tracking results for one patient over one cardiac cycle. The yellow curve represents the projected 3D vessel, the blue cross represents the point tracked as the bifurcation, and the white arrow points to the bifurcation. Those images come from a 15 frames sequence. This figure shows the frames 1, 6, 15, from left to right. More tracking results are available on https://3dvttracking.github.io/ (Color figure online)

4   Results

To assess the performance of the proposed approach, we use anonymous data collected after informed patient consent for use in this type of investigation. These data come from ten different patients. Both the CTA and the angiographic images are available. We selected in the angiographic sequences a sub-sequence which covers a full cardiac cycle or a bit more depending on the patient case. The CT scans have been pre-processed to extract the coronary vessel trees as described above. Several x-ray projections with different angulation may have been selected for a given patient, yielding a total of 23 different tracking experiments. Each of them is analyzed separately from the other. Selection is based on the available angiographic views and the vessels are selected as the ones that could be the object of an interventional procedure. In the following, we propose to compare the trackings on the vessels of interest made thanks to the registration of the single vessel of interest as in [7], and the registration of the vessel of interest made thanks to the registration of the vessel tree as described in this work. Figure 2 is an example of the obtained results in one case. The method has been implemented through a $C++$ program on a standard modern processor, given an execution time of 18 seconds per image.

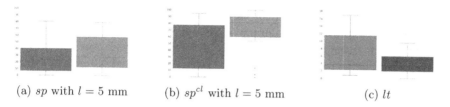

(a) sp with $l = 5$ mm   (b) sp^{cl} with $l = 5$ mm   (c) lt

Fig. 3. Each subfigure compares the one-vessel tracking (left, [7]) to the vessel-tree tracking (right). (a) and (b) present the shape preservation measures with respectively the point-to-point and the point-to-closest distances. (c) presents the landmark tracking measure.

Table 1. Average results for each evaluation, for the previous method presented in [7] and the actual method presented in this work.

Evaluation	Previous method	Actual method
sp	26%	31%
sp^{cl}	45%	70%
lt	6.7mm	4.3mm

Shape Preservation. Figure 3, (a) and (b), shows the results obtained thanks to the registration of only the vessel of interest (in blue), and the results obtained thanks to the registration of the entire tree (in orange), with $l = 5$ mm. In (a) the evaluation is computed point to point, in (b) point to closest point. For both, the method presented in this work clearly outperforms the method of [7]. [7] and our method obtained respectively in average for the first evaluation 26% and 33%, and 45% and 70% for the second evaluation (see Table 1).

Landmark Tracking. The results of this measure are shown in Fig. 3c. The values are the standard deviation of the curvilinear abscissa of points associated to the bifurcations for each projection of the vessel of interest. On the left, the measures (average 6.7 mm) are obtained with the approach presented in [7]. On the right, the measures (average 4.3 mm) are obtained thanks to the registration of the entire tree i.e. the measure described in this work (see Table 1).

5   Discussion and Conclusion

Fusing the anatomical content of a CTA scan with angiographic images of the same patient for guidance in interventional procedure is a common request from operators. One of the challenge is the management of the cardiac motion. In this paper, we have explored an approach involving the deformation of the 3D arterial tree represented by a tree-spline. The proposed approach involves several algorithmic steps: a rigid registration of the tree to an iso-cardiac phase projection followed by a deformation of the tree represented as a tree-spline. Encouraging results have been obtained and the use of the full tree in the different steps of the algorithms improves the accuracy of the approach. Indeed, the algorithm is more able to manage superimpositions of different branches over the vessel of interest which happens along the cardiac cycle. We observed 4.3 mm in average for the task of landmark tracking compared to 6.7 mm following the approach described in [7] which performs the deformation on the vessel of interest only. The considered dataset covers a larger set of clinical situation. Interestingly, the algorithm is able to perform a self-assessment of its performance using the shape preservation. Indeed this measure does not require any pre-defined ground truth. The obtained result shows that there is, in some cases, a "slipping" of the vessel of interest. This "slipping" is possible while keeping the 3D length of the vessel constant. The artery deforms itself in the direction of projection. Future works will be done to solve this. It may require improvement in the creation of the pairings.

References

1. Aksoy, T., Unal, G., Demirci, S., Navab, N., Degertekin, M.: Template-based CTA to x-ray angio rigid registration of coronary arteries in frequency domain with automatic x-ray segmentation. Med. Phys. 40(10) (2013)
2. Baka, N., et al.: Statistical coronary motion models for 2D+ t/3D registration of x-ray coronary angiography and CTA. Med. Image Anal. 17(6), 698–709 (2013)
3. Benseghir, T., Malandain, G., Vaillant, R.: A tree-topology preserving pairing for 3D/2D registration. Int. J. Comput. Assist. Radiol. Surg. 10(6), 913–923 (2015)
4. Groher, M., Zikic, D., Navab, N.: Deformable 2D–3D registration of vascular structures in a one view scenario. IEEE Trans. Med. Imaging 28(6), 847–860 (2009)
5. Heibel, T.H., Glocker, B., Groher, M., Paragios, N., Komodakis, N., Navab, N.: Discrete tracking of parametrized curves. In: CVPR, pp. 1754–1761. IEEE (2009)
6. Piegl, L., Tiller, W.: The NURBS Book. Springer Science and Business Media, Heidelberg (2012). https://doi.org/10.1007/978-3-642-97385-7
7. Poulain, E., Malandain, G., Vaillant, R.: 3D coronary vessel tracking in X-Ray projections. In: Pop, M., Wright, G.A. (eds.) FIMH 2017. LNCS, vol. 10263, pp. 204–215. Springer, Cham (2017). https://doi.org/10.1007/978-3-319-59448-4_20
8. Ruijters, D., ter Haar Romeny, B.M., Suetens, P.: Vesselness-based 2D–3D registration of the coronary arteries. Int. J. Comput. Assist. Radiol. Surg. 4(4), 391–397 (2009)
9. Serradell, E., Romero, A., Leta, R., Gatta, C., Moreno-Noguer, F.: Simultaneous correspondence and non-rigid 3D reconstruction of the coronary tree from single x-ray images. In: IEEE International Conference on Computer Vision (ICCV), pp. 850–857. IEEE (2011)
10. Shechter, G., Devernay, F., Coste-Maniere, E., McVeigh, E.R.: Temporal tracking of 3D coronary arteries in projection angiograms. In: Medical Imaging (2002)

Interactive-Automatic Segmentation and Modelling of the Mitral Valve

Patrick Carnahan[1,2(✉)], Olivia Ginty[2], John Moore[2], Andras Lasso[3], Matthew A. Jolley[4,5], Christian Herz[4], Mehdi Eskandari[6], Daniel Bainbridge[7], and Terry M. Peters[1,2,8]

[1] School of Biomedical Engineering, Western University,
London, ON, Canada
pcarnah@uwo.ca
[2] Imaging, Robarts Research Institute, Western University, London, ON, Canada
[3] Laboratory for Percutaneous Surgery, Queen's University, Kingston, ON, Canada
[4] Department of Anesthesiology and Critical Care Medicine,
Children's Hospital of Philadelphia, Philadelphia, PA, USA
[5] Division of Pediatric Cardiology, Children's Hospital of Philadelphia,
Philadelphia, PA, USA
[6] King's College Hospital, Denmark Hill, London, UK
[7] Department of Anesthesiology, London Health Sciences Centre Western University,
London, ON, Canada
[8] Departments of Medical Biophysics, Medical Imaging, Western University,
London, ON, Canada

Abstract. Mitral valve regurgitation is the most common valvular disease, affecting 10% of the population over 75 years old. Left untreated, patients with mitral valve regurgitation can suffer declining cardiac health until cardiac failure and death. Mitral valve repair is generally preferred over valve replacement. However, there is a direct correlation between the volume of cases performed and surgical outcomes, therefore there is a demand for the ability of surgeons to practice repairs on patient specific models in advance of surgery. This work demonstrates a semi-automated segmentation method to enable fast and accurate modelling of the mitral valve that captures patient-specific valve geometry. This modelling approach utilizes 3D active contours in a user-in-the-loop system which segments first the atrial blood pool, then the mitral leaflets. In a group of 15 mitral valve repair patients, valve segmentation and modelling attains an overall accuracy (mean absolute surface distance) of 1.40 ± 0.26 mm, and an accuracy of 1.01 ± 0.13 mm when only comparing the extracted leaflet surface proximal to the ultrasound probe. Thus this image-based segmentation tool has the potential to improve the workflow for extracting patient-specific mitral valve geometry for 3D modelling of the valve.

Keywords: Mitral valve · 3D echocardiography · Segmentation · Patient-specific modelling

© Springer Nature Switzerland AG 2019
Y. Coudière et al. (Eds.): FIMH 2019, LNCS 11504, pp. 397–404, 2019.
https://doi.org/10.1007/978-3-030-21949-9_43

1 Introduction

The mitral valve is an anatomically complex, dynamic structure integral for efficient blood flow and therefore healthy cardiac output. When it becomes dysfunctional, referred to as mitral regurgitation, patients can face declining cardiovascular health until cardiac failure and death. Furthermore, mitral regurgitation is the most common valvular disease affecting approximately 10% of those over 75 years old [2]. The preferred intervention for mitral regurgitation is repair, due to superior patient outcomes compared to replacement [1,9]. However, the repair must be tailored to the patient-specific anatomy and pathology, which requires expert training and experience. Consequently, there is a need for patient-specific models that can permit the training and procedure-planning of patient-specific repairs to minimize its learning curve and preventable errors [4,6]. Previous work has demonstrated the potential for patient-specific valve modelling for both surgical training as well as preoperatively predicting surgical outcomes [5]. The cause of mitral regurgitation varies across patients, as any failure of these structures can render the valve inefficient. Therefore, segmenting the mitral valve faces both the challenge of capturing anatomy that is complex, and in motion.

In order to prepare patient-specific models, the mitral valve must first be extracted from patient image data. There are several challenges specific to mitral leaflet segmentation in 3D TEE images. There is no intensity-based boundary between leaflets and adjacent heart tissue, and distinguishing between the anterior and posterior leaflets in the coaptation zone during systole is difficult due to the lack of an intensity-based boundary. Additionally, in diastole there can be signal dropout which appears as gaps in the leaflets. To facilitate clinical use and repeat-ability, several mitral leaflet segmentation methods have been proposed. These methods focus on varying goals between deriving quantitative valve measurements and extracting annular and leaflet geometry from 3D transesophageal echocardiograph (TEE) images. Burlina et al. proposed a semi-automatic segmentation method based on active contours and thin tissue detection for the purpose of computational modelling [3]. Scheinder et al. proposed a semi-automatic method for segmenting the mitral leaflets in 3D TEE over all phases of the cardiac cycle [13]. This method utilizes geometric priors and assumptions about the mechanical properties of the valve to model the leaflets through coaptation with a reported surface error of 0.84 mm. However, this method only represents the mitral leaflets as a single medial surface, rather than structures with thickness. Additionally, several fully automatic methods have been proposed that are based on population average atlases. Ionasec et al. describe a technique which uses a large database of manually labelled images and machine learning algorithms to locate and track valve landmarks [7]. While this method is fully automatic, the use of sparse landmarks potentially limits the patient-specific detail that can be extracted. Pouch et al. also describe a fully automatic method which utilizes a set of atlases to generate a deformable template which is then guided to the leaflet geometry using joint label fusion [11]. The surface error of this method is reported at 0.7 mm, however this is only achieved on healthy valves and performance is reduced when segmenting diseased valves. This method also may be limited by the quality of the input atlases, and could be biased towards the atlas geometry, limiting its potential for patient specific modelling. This is especially true considering the wide variety of pathologies and corresponding variety of morphologies possible with the MV.

Automatic 3D segmentation methods offer significant implications for the feasibility of patient-specific modelling in clinical use. Existing methods have demonstrated the ability to accurately segment the mitral valve structure, however remain highly time-intensive. Furthermore, some of these published methods show decreased performance when applied to highly diseased valves, demonstrating limitations in patient-specificity.

We aim to develop a segmentation method that can be applied to both normal and highly diseased valves, to extract patient-specific leaflet geometry. Our focus is on delineating the leaflet surfaces for the purpose of creating molds for our related patient-specific MV modelling project, where silicone is applied to the molds to create valves for use in surgical training and planning [5]. To accomplish this, we propose a semi-automatic segmentation method based on active contours that iterates using a user-in-the-loop strategy.

2   Methods

2.1   Image Acquisition and Data Sets

Fifteen patients with mitral valve regurgitation undergoing cardiac surgery were imaged preoperatively using Philips Epiq and iE33 systems as per clinical protocol. Of the fifteen patient datasets, six were acquired at King's College Hospital, London, UK, from patients with severe mitral regurgitation and nine were acquired at University Hospital, London, Canada. The 3D TEE images were exported into Cartesian format. The SlicerHeart module was used to import the Cartesian DICOM files into 3D Slicer [12]. Images at end-diastole were selected for image analysis. The exported Cartesian format images have an axial resolution of approximately 0.60 mm.

2.2   Semi-automated Image Analysis

Our software has been developed in the 3D Slicer[1] platform and utilizes the Insight Segmentation and Registration Toolkit (ITK)[2] software package. The iterative steps in our method do not use a fixed stopping point to account for the high variability in TEE data. Instead, the user runs the segmentation steps in increments until they are satisfied with the results. This leads to an ideal compromise between human judgement in ambiguous cases and guided automatic segmentation for ease of use and time efficiency. In addition, a user can view the result of the next step of the segmentation, compare it to the previous step, and

---

[1] www.slicer.org.
[2] www.itk.org.

make manual adjustments between active contour steps. We base our segmentation on the end diastole image where the leaflets are least likely to experience signal dropout, but where the anterior and posterior boundary is still clearly identifiable.

Before beginning the segmentation process, the user must define the valve annulus by placing a series of points. This is accomplished through the Slicer-Heart software, which facilitates the placing of the points and fits a smooth annulus curve [10]. This annulus definition is used throughout the automated process to provide context for the valve center, orientation and boundaries.

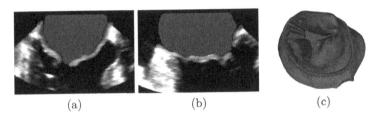

(a)   (b)   (c)

Fig. 1. Cross sectional views of a 3D TEE image and segmentation (a, b), with the blood pool segmentation shown in green and the leaflet segmentation shown in yellow. Rendering of extracted proximal surface mold (c) (Color figure online).

Blood Pool Segmentation. We first segment the atrial blood pool (BP). This blood pool segmentation provides context for the leaflet segmentation as well as the leaflet surface extraction. The image is first processed using a Gaussian filter with a variance of 1 mm, followed by a gradient magnitude filter. This creates a feature image highlighting the contrast edges, which determines the speed of the active contour growth. The center of the defined annulus is used to initialize a geodesic active-contour filter from ITK that grows to complete the blood pool segmentation, as pictured in Fig. 1. The active contour process is run with curvature, advection, and propagation scaling parameters of 1.2, 1.0, and 0.9 respectively. The curvature parameter controls boundary smoothing, while the advection parameter influences attraction to edges. The propagation scaling parameter applies an inwards or outwards force on the contour boundary creating a bias to either grow or shrink.

Shrinking Leaflet Segmentation. The boundary region of the blood pool segmentation within a distance of 11 voxels, or ~5 mm is taken as the initial estimate for the leaflet segmentation. This estimate is refined using another active contour approach which shrinks the segmentation down to the desired result. The active contour approach used here differs from the one used in the blood pool segmentation mainly in that it is biased to shrink. The parameters used for this phase are 0.9, 0.1, and −0.4 for curvature, advection and propagation scaling respectively. As the active contour process iterates, the segmentation pulls back to the leaflet boundaries as pictured in Fig. 1. Since our approach is interactive, the user is able to view adjacent image frames during the segmentation process to better inform their decision on the ideal stopping point. In addition, a volume rendered view can also be displayed alongside a 3D mesh of the segmentation, again providing more information to the user for completing the guided segmentation.

Proximal Surface Extraction. For manufacturing our physical MV models as described in previous work [5], we require a geometric model of the valve surface proximal to the TEE transducer. The proximal surface is extracted from the leaflet segmentation using the defined annulus for context. For points above the annulus plane, the surface is extracted by checking for self-intersection of the line between the annulus center and the point of the surface of the segmentation. This process finds all points on the inner surface, while excluding those on the outer surface. For points below the annulus, which includes the atrial wall, we compare the normal vector of the points on the leaflet segmentation surface and the corresponding closest point on the BP segmentation. Only points where the angle between the normal exceeds 100° are kept. When combined, we are left with the inner surface of the valve segmentation, which can then be used to create a mold for 3D printing pictured in Fig. 1.

3   Evaluation and Results

In order to evaluate the accuracy of the proposed segmentation method, we compared automated segmentations to expert manual segmentations for images from 15 subjects. The ground truth expert segmentations were created using manual segmentations performed by two clinical users. Our semi-automatic system was then used on the same images, and same reference frames. Our system was used with no manual user intervention between iterations in order to perform a baseline assessment of the algorithm independent of manual influence. Comparisons were made using the mean absolute surface distance (MASD) between the boundaries of the complete segmentations, as well as the MASD between the extracted proximal surfaces.

The results indicate a MASD for the proximal surface of 1.01 ± 0.13 mm, which is on the order of one to two image voxels along the depth of the image. The MASD for the complete valve boundary is higher at 1.40 ± 0.26 mm, or two to three image voxels. The maximum local error observed was 9.40 mm occurring at the boundary between the leaflet and the surrounding atrial tissue. The average completion time using the proposed semi-automated segmentation method was 8.93 ± 2.31 min, compared to the manual segmentation times which averaged 55.84 ± 12.87 min. There was an overall average speedup of 46.47 ± 10.64 min using the semi-automated method over performing manual tracings.

4 Discussion

The proposed semi-automatic guided segmentation method enables the extraction of mitral leaflet geometry from 3D TEE in a 3D printable form. This method allows a user to rapidly segment the mitral valve at end-diastole to extract its atrial surface for generating a patient-specific mold with minimal effort. The interactive, iterative nature of this segmentation system allows it to be used on a wide variety of pathological valves, as well as consistently work with a large range of image qualities from different systems.

Our results indicate similar overall performance to other semi-automatic methods and is on the same magnitude as previously reported inter observer variability of 0.60 ± 0.17 mm [8]. Our results are also consistent with previous studies which have observed that the greatest variability in manual and automatic segmentations occur at the boundaries of the model, rather than at the leaflet surfaces. As Fig. 2 shows, we see a boundary displacement of 1 mm or less for most of the leaflet surface, and see larger displacement only at the boundary of the leaflet as well as the chordae attachment points. This is a result of the somewhat arbitrary, non-contrast boundary between the leaflets, the atrial wall, and the surrounding tissue. In addition, our method shows improved results when only considering the atrial valve surface, the surface of the valve more proximal to the TEE probe. This region shows the best image contrast and is consequently the most consistently identified between manual and automatic segmentations. Furthermore, since we do not rely on any prior data or image atlases, our system does not demonstrate any biases or drop in performance for previously unseen valve geometries. This is a critical consideration when modelling highly diseased valves for preoperative planning, as a wide variety of valve geometries can be observed. For the purposes of patient-specific valve modelling, prior work has demonstrated the creation of physical valve models using silicone

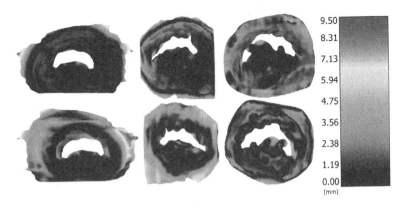

Fig. 2. Distance maps of the proximal segmentation surface (top) and the distal segmentation surface (bottom) for 3 cases. The largest regions of error are located where the leaflets meet the atrial wall, as well as near the valve commisures.

and 3D printed molds based on the proximal valve surface [5]. Our method is well suited to the task of extracting mitral valve geometry for the creation of patient-specific models such as these.

4.1   Limitations and Future Work

In cases of very poor image quality, the automated segmentation may miss regions of dropout, or fail to capture the exact valve geometry. Expert users performing these same segmentations often use other points in the cardiac cycle to inform their segmentation. Adapting this software to utilize additional time points may help to improve the robustness of the method in very difficult cases. In addition, color Doppler images are often captured in order to diagnose mitral valve regurgitation. This additional information could be used to increase the accuracy of segmentations as it contains information about the flow of blood between the leaflets. Further work validating the effectiveness of this system may be necessary to ensure it accurately captures patient-specific detail across healthy, mildly diseased and severely diseased valves. Our semi-automatic segmentation technique is currently being used as part of an ongoing prospective study which aims to validate the effectiveness of our dynamic silicone based valves for predicting surgical outcomes.

5   Conclusion

We present a technique that enables the extraction of mitral leaflet geometry from 3D TEE for creating 3D printed models used in creating accurate patient specific physical models. Segmentations from our software successfully replicate gold-standard MV segmentations within reasonable tolerance with respect to image resolution. The overall mean surface distance analysis demonstrates that our software can extract the proximal surface of the MV to within approximately one millimeter. This level of accuracy of the is suitable for patient-specific mitral valve modelling applications. This segmentation software reduces the time required for completing an accurate mitral valve segmentation and improves the workflow of the mitral valve modelling process.

Acknowledgements. This work was funded by the Canadian Foundation for Innovation (20994), the Ontario Research Fund (IDCD), and the Canadian Institutes for Health Research (FDN 201409).

References

1. Ailawadi, G., et al.: Is mitral valve repair superior to replacement in elderly patients? Ann. Thorac. Surg. **86**(1), 77–86 (2008). https://doi.org/10.1016/j.athoracsur.2008.03.020
2. Benjamin, E.J., et al.: Heart disease and stroke statistics–2018 update: a report from the American heart association. Circulation **137**(12), E67–E492 (2018). https://doi.org/10.1161/CIR.0000000000000558

3. Burlina, P., et al.: Patient-specific modeling and analysis of the mitral valve using 3D-TEE. In: Navab, N., Jannin, P. (eds.) IPCAI 2010. LNCS, vol. 6135, pp. 135–146. Springer, Heidelberg (2010). https://doi.org/10.1007/978-3-642-13711-2_13

4. Eleid, M.F., et al.: The learning curve for transcatheter mitral valve repair with MitraClip. J. Intervent. Cardiol. 29(5), 539–545 (2016). https://doi.org/10.1111/joic.12326

5. Ginty, O., Moore, J., Peters, T., Bainbridge, D.: Modeling patient-specific deformable mitral valves. J. Cardiothorac. Vasc. Anesth. 32(3), 1368–1373 (2018). https://doi.org/10.1053/j.jvca.2017.09.005

6. Holzhey, D.M., Seeburger, J., Misfeld, M., Borger, M.A., Mohr, F.W.: Learning minimally invasive mitral valve surgery. Circulation 128(5), 483–491 (2013). https://doi.org/10.1161/circulationaha.112.001402

7. Ionasec, R.I., et al.: Patient-specific modeling and quantification of the aortic and mitral valves from 4-D cardiac CT and TEE. IEEE Trans. Med. Imaging 29(9), 1636–1651 (2010). https://doi.org/10.1109/tmi.2010.2048756

8. Jassar, A.S., et al.: Quantitative mitral valve modeling using real-time three-dimensional echocardiography: technique and repeatability. Ann. Thorac. Surg. 91(1), 165–171 (2011). https://doi.org/10.1016/j.athoracsur.2010.10.034

9. McNeely, C.A., Vassileva, C.M.: Long-term outcomes of mitral valve repair versus replacement for degenerative disease: a systematic review. Curr. Cardiol. Rev. 11(2), 157–62 (2015). http://www.ncbi.nlm.nih.gov/pubmed/25158683

10. Nguyen, A.V., et al.: Dynamic three-dimensional geometry of the tricuspid valve annulus in hypoplastic left heart syndrome with a Fontan circulation. J. Am. Soc. Echocardiogr. (2019). https://doi.org/10.1016/j.echo.2019.01.002

11. Pouch, A., et al.: Fully automatic segmentation of the mitral leaflets in 3D transesophageal echocardiographic images using multi-atlas joint label fusion and deformable medial modeling. Med. Image Anal. 18(1), 118–129 (2014). https://doi.org/10.1016/j.media.2013.10.001

12. Scanlan, A.B., et al.: Comparison of 3D echocardiogram-derived 3D printed valve models to molded models for simulatedrepair of pediatric atrioventricular valves. Pediatr. Cardiol. 39(3), 538–547 (2017). https://doi.org/10.1007/s00246-017-1785-4

13. Schneider, R.J., Tenenholtz, N.A., Perrin, D.P., Marx, G.R., del Nido, P.J., Howe, R.D.: Patient-specific mitral leaflet segmentation from 4D ultrasound. In: Fichtinger, G., Martel, A., Peters, T. (eds.) MICCAI 2011. LNCS, vol. 6893, pp. 520–527. Springer, Heidelberg (2011). https://doi.org/10.1007/978-3-642-23626-6_64

Cardiac Displacement Tracking with Data Assimilation Combining a Biomechanical Model and an Automatic Contour Detection

Radomír Chabiniok[1,2,3], Gautier Bureau[1,2], Alexandra Groth[4],
Jaroslav Tintera[5], Jürgen Weese[4], Dominique Chapelle[1,2],
and Philippe Moireau[1,2(✉)]

[1] Inria, Université Paris-Saclay, Palaiseau, France
philippe.moireau@inria.fr
[2] LMS, Ecole Polytechnique, CNRS, Université Paris-Saclay, Palaiseau, France
[3] School of Biomedical Engineering and Imaging Sciences (BMEIS),
St Thomas' Hospital, King's College London, London, UK
[4] Philips, Research Laboratories, Hamburg, Germany
[5] Institute for Clinical and Experimental Medicine, Prague, Czech Republic

Abstract. Data assimilation in computational models represents an essential step in building patient-specific simulations. This work aims at circumventing one major bottleneck in the practical use of data assimilation strategies in cardiac applications, namely, the difficulty of formulating and effectively computing adequate data-fitting term for cardiac imaging such as cine MRI. We here provide a proof-of-concept study of data assimilation based on automatic contour detection. The tissue motion simulated by the data assimilation framework is then assessed with displacements extracted from tagged MRI in six subjects, and the results illustrate the performance of the proposed method, including for circumferential displacements, which are not well extracted from cine MRI alone.

Keywords: Biophysical heart modeling · Data assimilation · cine MRI

1 Introduction

It is now widely accepted that data assimilation strategies are the fundamental ingredient for personalizing biophysical cardiac models, as they have the capability to decrease the discrepancy between model and data, while estimating values of key biophysical parameters [2]. However, one major limitation in the practical use of such modeling and data assimilation strategies in clinical applications lies in the difficulty of formulating and effectively computing adequate data-fitting terms, especially where imaging data are concerned. In some specific imaging techniques such as tagged MRI, displacements fields can be extracted

© Springer Nature Switzerland AG 2019
Y. Coudière et al. (Eds.): FIMH 2019, LNCS 11504, pp. 405–414, 2019.
https://doi.org/10.1007/978-3-030-21949-9_44

from image sequences [5,16], in which case comparison with models is straight-forward, but the question arises of assessing the accuracy of these extracted displacements. However, for more standard image data such as cine MRI, CT, or ultrasound, data processing does not readily provide such displacements fields. Some previous proof-of-concept works have successfully used pre-segmented cine MR images [1] with data-fitting terms based on distances between the simulated model boundaries and the segmentation contours [1,9,13]. The segmentation itself however represents an additional step that complexifies the data assim-ilation pipeline – with possible additional errors. Our objective here is to use a previously-proposed method for automatic contour detection [4] in cardiac images to design an integrated data-fitting term in a complete data assimilation framework with a cardiac biomechanical model.

2 Methods

The section presents the clinical data, the biophysical model, the cine MRI data-fitting term, the data assimilation method, and the assessment strategy based on tagged MRI.

2.1 Clinical Data

Four healthy volunteers and two patients with a suspected cardiomyopathy were involved in the study as *test subjects*. For each subject, the following MR cardiac images were acquired by using a Philips Achieva 1.5T MR system (for healthy volunteers), and Siemens Aera 1.5T (for the patients):

- Time-resolved cine MRI (multiple 2D cine bSSFP) in breath-hold and ret-rospective ECG gating, with the following parameters: temporal resolution 25–30 ms, field of view (FOV) 350 × 350 mm, parallel imaging using acceler-ation factor of 2, acquired spatial resolution ∼2.3 × 2.3 × 8 mm (inter-slice spacing 10 mm).
- Tagged MRI in prospective ECG triggering: 3D tagged MRI [16] for Philips scanner (FOV 100 × 100 × 100 mm, spatial resolution 3.4 × 7.7 × 7.7 mm, temporal resolution ∼35 ms); 2D tagged (CSPAMM) for Siemens scanner for which 3D tagged sequence was not available (FOV 350 × 350 mm, spatial resolution 3 × 3 × 7 mm, temporal resolution ∼35 ms).
- Whole-heart 3D bSSFP sequence [6] acquired in free breathing by using breath-navigator, acquisition matrix 212 × 209 × 200, acquired voxel size 2 × 2 × 2 mm, repetition time 4.5 ms, echo time 2.2 ms, echo train length 26 and flip angle 90°.

In addition, cine MRI together with the 3D whole-heart sequence was acquired in ten *training subjects* (with Philips Achieva 1.5 T scanner). The train-ing group included five healthy volunteers and five patients with dilated car-diomyopathy. There was no intersection between the *test* and *training* groups.

Cine sequences covered the whole ventricles by a set of parallel short-axis (SA) slices (the so-called "SA cine stack" or "SA stack"). Spatial mis-registration of the slices within the SA cine stack (caused by acquiring the stack in 5–15 breath-hold periods) was corrected by the rigid registration function from the Image Registration Toolkit IRTK [18], while using the high-resolution 3D whole-heart image as a template.

2.2   Biomechanical Heart Model

Any data assimilation procedure relies on an underlying physical model. By this, we mean a patient-specific geometry with additional microstructure information such as the myocardial fiber directions, but also a mechanical formulation combining a constitutive law and boundary conditions in a continuum mechanics framework.

Geometry – A biventricular geometry is obtained by manual segmentation of the end-diastolic time frame of the SA cine MR stack of images covering the whole heart. The segmentation produces a surface triangle mesh $\Gamma_0^{h,k}$, $1 \leq k \leq 6$ where k denotes the subject number, and h means that this is a meshed surface. Then, a tetrahedral mesh $\Omega_0^{h,k}$ is generated from the boundary using the GHS3D software[1], namely such that $\partial \Omega_0^{h,k} = \Gamma_0^{h,k}$. The myocardial fiber directions τ are prescribed analytically at each integration point of the volume mesh according to a rule-based criterion with the elevation angle being $-60/60°$ for the LV (from epicardium to endocardium) and $-75/75°$ for RV. Note that defining the reference configuration from an initial geometry can be improved by solving a static mechanical equilibrium involving the passive constitutive law and an internal ventricular pressure, see [15] and references therein.

Biomechanical Model – We rely on the biomechanical model described in [3]. We consider a total Lagrangian formulation allowing to compute at every time t the deformation mapping φ with respect to the reference configuration $\Omega_0^{h,k}$, namely

$$\varphi : \begin{vmatrix} \Omega_0^{h,k} \to \Omega_{h,k}(t) \\ \mathbf{x} \mapsto \boldsymbol{x}(t) = \varphi(\mathbf{x},t) = \mathbf{x} + \boldsymbol{u}(\mathbf{x},t) \end{vmatrix}$$

where $\boldsymbol{u}$ stands for the displacement field. We recall the following kinematics tensors definitions: the deformation gradient $\mathsf{F} = \nabla \boldsymbol{u}$ with J its determinant, the right Cauchy-Green deformation tensor $\mathsf{C} = \mathsf{F}^\mathsf{T} \cdot \mathsf{F}$ and $\mathsf{e} = \frac{1}{2}(\mathsf{C} - 1)$ the Green-Lagrange strain tensor. The constitutive law defining the second Piola-Kirchhoff stress tensor Σ consists of an active part – based on Huxley's law – along the fiber directions $\sigma_a(\mathsf{F},t)\boldsymbol{\tau}(\mathbf{x}) \otimes \boldsymbol{\tau}(\mathbf{x})$, a hyperelastic part $\Sigma^p(\mathsf{C})$ and a viscoelastic part $\Sigma^v(\partial_t \mathsf{C}, \mathsf{C})$ [3]. The heart geometry is fixed on the apex and anterior wall (simplifying the contact between epicardium and thorax and diaphragm)

---

[1] Inria, Project-Team Gamma.

by using visco-elastic boundary conditions with stiffness and viscosity parameters α, β [13]. Similar albeit more compliant boundary conditions are prescribed at the basal boundary as a substitute for atria and large vessels. Finally, the systemic and pulmonary circulations are modeled by Windkessel models, each containing a combination of proximal and distal resistance-capacitance systems [17]. A valve model connects the left and right ventricular pressures $p_{v,i}$ to the systemic and pulmonary circulations [17], respectively. The principle of virtual work in the space of admissible test displacements $\mathcal{V}_{\mathrm{ad}}$ reads

$$\int_{\Omega_0^{h,k}} \rho \partial_{tt} \boldsymbol{u} \cdot \boldsymbol{w} \, \mathrm{d}\Omega + \int_{\Omega_0^{h,k}} \Sigma : \mathrm{De}[\boldsymbol{u}](\boldsymbol{w}) \, \mathrm{d}\Omega + \int_{\Gamma_{0,b}^{h,k}} (\alpha \boldsymbol{u} + \beta \partial_t \boldsymbol{u}) \cdot \boldsymbol{w} \, \mathrm{d}\Gamma$$

$$= -\sum_{i=l,r} \int_{\Gamma_{0,\mathrm{endo}_i}^{h,k}} p_{v,i} \mathsf{J} \mathsf{F}^{-1} \boldsymbol{n} \cdot \boldsymbol{w} \, \mathrm{d}\Gamma, \quad \boldsymbol{w} \in \mathcal{V}_{\mathrm{ad}},$$

where $\mathrm{De}[\boldsymbol{u}]$ denotes the differential of the strain tensor e around the displacement $\boldsymbol{u}$ and $\boldsymbol{n}$ the outward unit normal vector. The model is discretized in time using a Newmark scheme and in space using finite elements. It is implemented in the MPI-based MoReFEM library[2] and we follow [1] to pre-calibrate the model to patients' data.

2.3   Distance Computation in Cine MRI

In order to provide a data-fitting term directly based on cine MRI, we use the framework of [4,14] – trained using all ten *training* subjects – to detect endo- and epicardial boundaries in the SA cine images. Then, for any computed deformation $\varphi(\mathbf{x}, t)$, the boundary detection mechanism is used to find target points in the image for each triangle $T^{h,k}$ from the endo- $\Gamma_{\mathrm{endo}_i}^{h,k}(t)$, $i = l, r$, and epicardial $\Gamma_{\mathrm{epi}_i}^{h,k}(t)$ surfaces of the deformed model. From these target points, we can compute a signed distance between all points of $\Gamma_{\mathrm{ext}}^{h,k}(t) = \cup_{i=l,r} \Gamma_{\mathrm{endo}_i}^{h,k}(t) \cup \Gamma_{\mathrm{epi}_i}^{h,k}(t)$ and the target points $\boldsymbol{x}_{\mathrm{Im}}(T^{h,k}(\boldsymbol{x}))$.

$$\mathrm{dist}_{\mathrm{Im}} : \left| \begin{array}{l} \Gamma_{\mathrm{ext}}^{h,k}(t) \to \mathbb{R}^3 \\ \boldsymbol{x} \mapsto \boldsymbol{x}_{\mathrm{Im}}(T^{h,k}(\boldsymbol{x})) - \boldsymbol{x} \end{array} \right.$$

Note that this signed distance contains (1) the information of being in or out of the mesh boundaries, (2) the distance itself $\|\boldsymbol{x}_{\mathrm{Im}}(T^{h,k}(\boldsymbol{x})) - \boldsymbol{x}\|$ to the target point, and (3) the direction pointing to the target point $\nu_{\mathrm{Im}} = (\boldsymbol{x}_{\mathrm{Im}}(T^{h,k}(\boldsymbol{x})) - \boldsymbol{x})/\|\boldsymbol{x}_{\mathrm{Im}}(T^{h,k}(\boldsymbol{x})) - \boldsymbol{x}\|$. This signed distance together with the confidence of this measure $\delta_{\mathrm{Im}} : \Gamma_{\mathrm{ext}}^{h,k}(t) \to \mathbb{R}$ (locally for each surface element) will be integrated in our data assimilation framework as the data-fitting term.

---

[2] https://gitlab.inria.fr/MoReFEM.

2.4 Data Assimilation Strategy

We follow a sequential strategy initially proposed in [12] and further refined in [9] for the time integration procedure. This strategy has proven to be a very effective data assimilation procedure with real data for both recovering the model trajectory – hence the deformation mapping – and jointly identifying some uncertain biophysical parameters [1,13]. Our work is here restricted to the trajectory estimation, but its originality lies in that we rely on a state-of-the-art data-fitting distance directly computed in the image sequence, instead on the distance to segmentations used in the previous papers. The principle of our data assimilation procedure is as follows. Assuming that we are at a time t^n when data are available, and that we have computed a deformation mapping $\varphi^n(x) = x + u^n(x)$, we then introduce a correction step based on the data, which consists in seeking $\hat{u}^n$ such that for all admissible test displacement fields $w \in \mathcal{V}^{\mathrm{ad}}$,

$$\kappa \int_{\Omega_0^{h,k}} \nabla(\hat{u}^n - u^n) : \mathsf{A}[u^n] : \nabla w \, d\Omega$$

$$+\gamma \int_{\Gamma_{\mathrm{ext},0}^{h,k}} \delta_{\mathrm{Im}}(\varphi^n)[(\hat{u}^n - u^n) \cdot \nu_{\mathrm{Im}}(\varphi^n)][w \cdot \nu_{\mathrm{Im}}(\varphi^n)] \, d\Gamma$$

$$= \gamma \int_{\Gamma_{\mathrm{ext},0}^{h,k}} \delta_{\mathrm{Im}}(\varphi^n)\mathrm{dist}_{\mathrm{Im}}(\varphi^n) \cdot w \, d\Gamma$$

where γ is a scaling parameter for our overall data confidence, κ is a regularization parameter scaled to balance the model overall stiffness with respect to the data confidence, and $\mathsf{A}[u^n]$ is the tangent stiffness tensor around the current displacement u^n. We point out that this variational formulation is linear with respect to $\hat{u}^n$, with a coercive bilinear form in the left-hand side. Then the model time integration is restarted from $\hat{u}^n$ instead of u^n. Computing successively $u^{n-1} \to u^n \to \hat{u}^n$ is called a prediction-correction scheme. The prediction is made by the model dynamics time scheme whereas the correction is performed using the data. Note that the correction fundamentally corresponds to computing the first iteration of a quasi-Newton based minimizing procedure involving the predicted displacement field and the data-fitting term. Indeed, our sequential data assimilation procedure aims at a compromise between predicting the next displacement using the biophysical model, and minimizing the distance with respect to the new data.

2.5 Assessment Based on Displacements Extracted from Tagged MRI

Displacements from tagged MRI are extracted by using the Image Registration Toolkit library (IRTK) [18], implemented within the visualisation software Eidolon [10]. This provides full 3D displacement vectors u_{3Dtag} in the case of 3D tagged sequences, or their projection in the short-axis plane u_{SAtag} for 2D tagged MR images. By t_d we denote the time corresponding to the first frame of the tagged MR sequence (typically between 30 and 60 ms from the R wave of QRS complex, due to the prospectively triggered tagged MRI acquisition), and relate the extracted displacements to this time instance, namely $u_{\text{3Dtag}}(\boldsymbol{x}, t_d) = u_{\text{SAtag}}(\boldsymbol{x}, t_d) = 0$. The model simulated displacements $\boldsymbol{u}(\mathbf{x}, t)$ are computed in the reference configuration, namely at points $\mathbf{x} \in \Omega_0^{h,k}$. Therefore, to compare them with the 3D tag, we first compute for all the model time steps t_n, the model displacement field with respect to the first tagged frame in deformed configuration:

$$
\tilde{u}^n : \left|
\begin{array}{l}
\Omega^{h,k}(t) \to \mathbb{R}^3 \\[6pt]
\boldsymbol{x} \mapsto \boldsymbol{u}^n((\boldsymbol{\varphi}^n)^{-1}(\boldsymbol{x})) - \boldsymbol{u}^d((\boldsymbol{\varphi}^n)^{-1}(\boldsymbol{x}))
\end{array}
\right.
$$

A global discrepancy between displacements obtained from the model and data assimilation and those extracted from the 3D tagged MRI is then computed by

$$
d_{\text{3D}} = \left[\sum_{n \in T_{\text{3D}}} \int_{\Omega_0^{h,k}} ||\tilde{u}^n(\boldsymbol{\varphi}^n(\mathbf{x}), t_n) - \boldsymbol{u}_{\text{3Dtag}}(\boldsymbol{\varphi}^n(\mathbf{x}), t_n)||^2 \frac{\Delta t}{T|\Omega|} \mathrm{d}\Omega \right]^{\frac{1}{2}},
$$

where T_{3D} is the set of indices associated with each time when a tagged MRI is available, Δt is the time step between two tagged MR images while T is the total acquisition time, and $|\Omega|$ the total volume of ventricle. Likewise, for the subjects with only 2D tagged MRI available, we use the similarly defined measure d_{SA}. The time integration is performed only up to end-systole as the tag lines are progressively fading in the diastole.

For the quantitative comparison, the components of the errors d_{3D} or d_{2DSA} in the radial and circumferential directions (with respect to the prolate coordinate system local to each mesh node, defined in the model reference configuration) are used.

3   Results

Figure 1 shows the end-systolic cine MR image of a few representative subjects in the study, together with contours of two models: the initial model simulation without the data-based correction in blue, and the result of model simulation with the data-based correction in red. This qualitative comparison demonstrates that the data corrected simulation is very close to the manually segmented surfaces. Figure 2 demonstrates quantitatively the impact on errors in predicted displacements in radial and circumferential directions evaluated with respect to the displacements extracted from tagged MRI. The decrease of error in radial component confirms the qualitative analysis made from Fig. 1. The decrease of error in the circumferential component demonstrates that the corrected model was able to capture some component of the circumferential direction, which is not directly accessible from cine MRI.

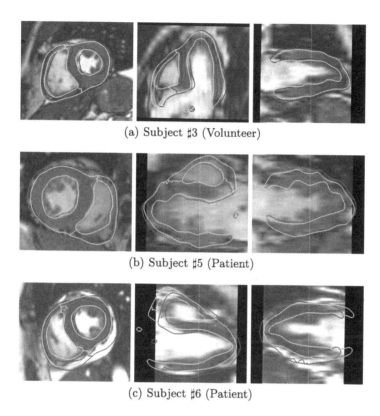

(a) Subject ♯3 (Volunteer)

(b) Subject ♯5 (Patient)

(c) Subject ♯6 (Patient)

Fig. 1. Qualitative assessment of selected subjects at end-systole: comparison of the contours of the model without taking into account the data (blue) and the model corrected by the data-fitting term (red). (Color figure online)

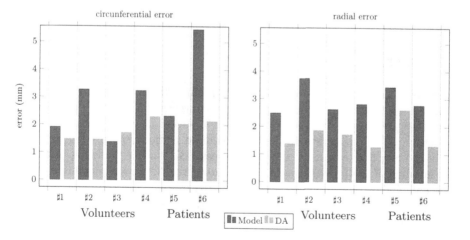

Fig. 2. Errors assessed with respect to the displacements extracted from tagged MRI. Data-free model simulation (blue) and model with data assimilation (red) (Color figure online)

4 Discussion

The datasets were acquired by two 1.5T MR systems from different vendors –
Philips and Siemens – in two different clinical environments. Even though the
scanning protocols were adapted to be consistent as much as possible, this nevertheless brought some challenges. First, the distance computation in cine MRI
was trained only on the datasets acquired by the Philips scanner. Nevertheless,
the performance was actually very good in all subjects. This is a sign that the
current implementation may be directly suitable even for multi-platform and/or
multi-center projects. Of course, a detailed assessment would be needed when a
new type of scanner is to be included.

While 3D tagged MRI was used during the acquisition by the Philips system, only "more standard" 2D tagged MRI were acquired with Siemens (as the
3D tagged sequence was not implemented on the platform). Even though we
had to adjust the quantitative verification accordingly, the comparisons of local
displacements with the pseudo-ground truth (Fig. 2) showed that the model is
improved in all cases (except for the circumferential component in Subject ♯3,
where the initial model error was already small).

Additionally to MRI, the same strategy of automatic contour detection may
be trained also for cine CT, and for 3D echo data. This would further increase the
accessibility of 3D patient-specific cardiac models, and would be suitable even for
patients contraindicated for MRI exam (e.g. non-MRI-conditional pacemakers).

The modeling setup used in this work was as in a previous project [1]. Even
though the model may be improved – typically by using a more refined passive
law [7,8], and a reference configuration closer to experimental data [11] – our
work demonstrates that the strategy of incorporating image data in a data assimilation strategy is successful. Including more sophisticated modeling ingredients
may bring further improvements, albeit we expect that this would mostly help
in the estimation of myocardial properties – a natural subsequent step in our
work.

5 Conclusion

A complex 3D heart model interacting with cine MRI was used to reconstruct
clinically important motion indicators, which would be normally inaccessible
from the given type of sequence. Secondly, a successful personalization of the
model by using image data as they are acquired without any pre-segmentation
paves the way for a wider use of patient-specific 3D cardiac modeling. Although
the number of included subjects was limited, the data were acquired in two
different clinical environments and using MRI systems from different vendors.
Overall, the performed proof-of-concept work demonstrates a good potential that
would justify a wider-scale pre-clinical study.

Acknowlegement. The authors acknowledge financial support from the Department of Health through the National Institute for Health Research (NIHR) comprehensive Biomedical Research Centre award to Guy's & St Thomas' NHS Foundation Trust in partnership with King's College London and the NIHR Cardiovascular MedTech Co-operative (previously existing as the Cardiovascular Healthcare Technology Co-operative 2012–2017), the support of Wellcome/EPSRC Centre for Medical Engineering [WT 203148/Z/16/Z]. The views expressed are those of the author(s) and not necessarily those of the NHS, the NIHR or the Department of Health.

References

1. Chabiniok, R., Moireau, P., Lesault, P.F., Rahmouni, A., Deux, J.F., Chapelle, D.: Estimation of tissue contractility from cardiac cine MRI using a biomechanical heart model. Biomech. Model Mechanobiol. **11**(5), 609–30 (2012)
2. Chapelle, D., Fragu, M., Mallet, V., Moireau, P.: Fundamental principles of data assimilation underlying the Verdandi library: applications to biophysical model personalization within euHeart. Med. Biol. Eng. Comput. **51**(11), 1221–1233 (2013)
3. Chapelle, D., Le Tallec, P., Moireau, P., Sorine, M.: An energy-preserving muscle tissue model: formulation and compatible discretizations. Int. J. Multiscale Comput. Eng. **10**(2), 189–211 (2012)
4. Ecabert, O., et al.: Automatic model-based segmentation of the heart in CT images. IEEE Trans. Med. Imaging **27**(9), 1189 (2008)
5. Genet, M., Stoeck, C.T., von Deuster, C., Lee, L.C., Kozerke, S.: Equilibrated warping: finite element image registration with finite strain equilibrium gap regularization. Med. Image Anal. **50**, 1–22 (2018)
6. Giorgi, B., Dymarkowski, S., Maes, F., Kouwenhoven, M., Bogaert, J.: Improved visualization of coronary arteries using a new three-dimensional submillimeter MR coronary angiography sequence with balanced gradients. Am. J. Roentgenol. **179**(4), 901–910 (2002)
7. Hadjicharalambous, M., et al.: Analysis of passive cardiac constitutive laws for parameter estimation using 3D tagged MRI. Biomechan. Model. Mechanobiol. **14**(4), 807–828 (2015)
8. Holzapfel, G., Ogden, R.: Constitutive modelling of passive myocardium: a structurally based framework for material characterization. Philos. Trans. Roy. Soc. London: Math. Phys. Eng. Sci. **367**(1902), 3445–3475 (2009)
9. Imperiale, A., Routier, A., Durrleman, S., Moireau, P.: Improving efficiency of data assimilation procedure for a biomechanical heart model by representing surfaces as currents. In: Ourselin, S., Rueckert, D., Smith, N. (eds.) FIMH 2013. LNCS, vol. 7945, pp. 342–351. Springer, Heidelberg (2013). https://doi.org/10.1007/978-3-642-38899-6_41
10. Kerfoot, E., et al.: Eidolon: visualization and computational framework for multi-modal biomedical data analysis. In: Zheng, G., Liao, H., Jannin, P., Cattin, P., Lee, S.-L. (eds.) MIAR 2016. LNCS, vol. 9805, pp. 425–437. Springer, Cham (2016). https://doi.org/10.1007/978-3-319-43775-0_39
11. Klotz, S., et al.: Single-beat estimation of end-diastolic pressure-volume relationship: a novel method with potential for noninvasive application. Am. J. Physiol. Heart Circ. Physiol. **291**, H403–H412 (2006)
12. Moireau, P., Chapelle, D., Le Tallec, P.: Filtering for distributed mechanical systems using position measurements: perspectives in medical imaging. Inverse Prob. **25**(3), 035010 (2009). (25pp)

13. Moireau, P., et al.: Sequential identification of boundary support parameters in a fluid-structure vascular model using patient image data. Biomech. Model. Mechanobiol. **12**(3), 475–496 (2013)

14. Peters, J., Ecabert, O., Meyer, C., Kneser, R., Weese, J.: Optimizing boundary detection via simulated search with applications to multi-modal heart segmentation. Med. Image Anal. **14**(1), 70–84 (2010)

15. Rausch, M.K., Genet, M., Humphrey, J.D.: An augmented iterative method for identifying a stress-free reference configuration in image-based biomechanical modeling. J. Biomech. **58**, 227–231 (2017)

16. Rutz, A., Ryf, S., Plein, S., Boesiger, P., Kozerke, S.: Accelerated whole-heart 3D CSPAMM for myocardial motion quantification. Magn. Reson. Med. **59**, 755–763 (2008)

17. Sainte-Marie, J., Chapelle, D., Cimrman, R., Sorine, M.: Modeling and estimation of the cardiac electromechanical activity. Comput. Struct. **84**(28), 1743–1759 (2006)

18. Shi, W., et al.: A comprehensive cardiac motion estimation framework using both untagged and 3D tagged MR images based on non-rigid registration. IEEE Trans. Med. Imaging **31**(6), 1263–1275 (2012)

An Adversarial Network Architecture Using 2D U-Net Models for Segmentation of Left Ventricle from Cine Cardiac MRI

Roshan Reddy Upendra[1(✉)], Shusil Dangi[1], and Cristian A. Linte[1,2]

[1] Center for Imaging Science, Rochester Institute of Technology, Rochester, NY, USA
{ru6928,sxd7257,calbme}@rit.edu
[2] Biomedical Engineering, Rochester Institute of Technology, Rochester, NY, USA

Abstract. Cardiac magnetic resonance imaging (CMRI) provides high resolution images ideal for assessing cardiac function and diagnosis of cardiovascular diseases. To assess cardiac function, estimation of ejection fraction, ventricular volume, mass and stroke volume are crucial, and the segmentation of left ventricle from CMRI is the first critical step. Fully convolutional neural network architectures have proved to be very efficient for medical image segmentation, with U-Net inspired architecture as the current state-of-the-art. Generative adversarial networks (GAN) inspired architectures have recently gained popularity in medical image segmentation with one of them being SegAN, a novel end-to-end adversarial neural network architecture. In this paper, we investigate SegAN with three different types of U-Net inspired architectures for left ventricle segmentation from cardiac MRI data. We performed our experiments on the 2017 ACDC segmentation challenge dataset. Our results show that the performance of U-Net architectures is better when trained in the SegAN framework than when trained stand-alone. The mean Dice scores achieved for three different U-Net architectures trained in the SegAN framework was on the order of 93.62%, 92.49% and 94.57%, showing a significant improvement over their Dice scores following stand-alone training - 92.58%, 91.46% and 93.81%, respectively.

Keywords: Image segmentation - Deep learning ·
Cine magnetic resonance image · Cardiac image analysis ·
Left ventricle segmentation

1 Introduction

Cardiac magnetic resonance imaging (CMRI) is considered as the benchmark for analysis of the cardiac function and quantification of the ventricular volume [6]. The analysis of the function of the ventricles, described by their ejection fraction (EF), stroke volume, mass and wall thickness are crucial in clinical cardiology for diagnosis, evaluating risks and planning therapy [2,14]. In particular, the function of left ventricle is a very good predictor of myocardial damage, cardiac

© Springer Nature Switzerland AG 2019
Y. Coudière et al. (Eds.): FIMH 2019, LNCS 11504, pp. 415–424, 2019.
https://doi.org/10.1007/978-3-030-21949-9_45

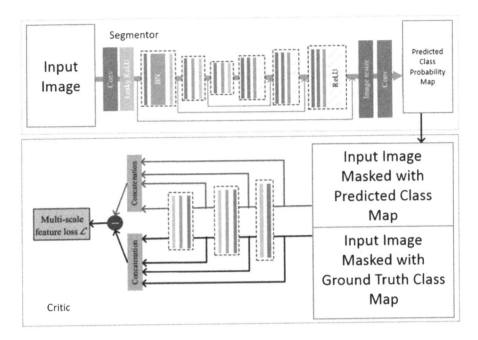

Fig. 1. SegAN architecture inspired from GAN [15]

failure, etc. Therefore, accurate and robust segmentation of left ventricle from the MRI data plays an important role in a large number of cardiac problems.

Manual segmentation can be a very laborious task prone to significant user variability (Fig. 1). Therefore, semi-automatic or fully automated segmentation methods would be very useful to cardiologists in the decision making process [8]. Some of the challenges in automated segmentation of the left ventricles are delineation between the myocardium of the left ventricle and other surrounding chambers, high contrast plus brightness heterogeneity in the ventricular cavity due to the presence of blood, the presence of papillary muscles, noise due to motion artifacts, and the variable structure of the heart [2].

A number of methods for automatic segmentation of the left ventricle from MRI images have been proposed. Traditional algorithms such as thresholding, edge detection, region growing, clustering, etc., were proposed initially [8]. These algorithms work decently for mid-ventricle slices, but often fail in the basal and apical slices. Also, they require considerable user-intervention. In graph based segmentation algorithms [7], graphs are created and a cost is assigned to each pixel or node. A minimum cost path is found using a graph searching algorithm to segment the left ventricle. These methods fail in complex cardiac structures, like papillary and trabecular muscles (PTMs). In [11], active shape models (ASM) are used to segment the left ventricles using the energy minimization of rigidity and elasticity internally, and edges externally. It is difficult to segment left ventricle from low contrast images using ASMs. In spite of all these research in left ventricle segmentation from MRI images, the accuracy of existing algorithms is not sufficient for clinical applications.

With the increase in popularity of deep learning due to the availability of resources for training, medical image segmentation has benefited immensely. Convolutional neural networks (CNNs) work outstandingly well for image classification where the output of the CNN to an image is a class label. The availability of large number of CMRI images enabled the use of deep learning for left ventricular segmentation. Several international challenges have been organized in the past few years to develop and evaluate segmentation algorithms for both ventricles [9,12,13]. The top ten results of the 2017 automated cardiac diagnosis challenge (ACDC) have all used CNNs for segmentation of the ventricles [2].

Ronneberger et al. [10] proposed U-Net, an elegant network architecture using fully convolutional network. U-Net is void of any fully connected layers and the convolutional layer labels each pixel in the image allowing segmentation with fewer training images [10].

Generative adversarial networks (GAN) [3] are becoming increasingly popular for medical image segmentation as training these generative models enable latent representations, which can serve as useful features. Eule et al. [4] show that segmentation of epithelial tissue using cycle-GAN outperforms the state-of-the-art U-Net. Xue et al. [15] proposed a GAN inspired end-to-end architecture, called segmentation adversarial network (SegAN), for semantic segmentation. They achieved better Dice score and precision compared to the state-of-the-art U-Net architecture in the segmentation of the MICCAI BRATS (2013 and 2015) brain tumor segmentation challenge [15].

In this paper, we combine three U-Net models with SegAN adversarial architecture to segment the left ventricle on the 2017 automated cardiac diagnosis challenge (ACDC) dataset. The objective of this paper is to test if SegAN, when combined with different U-Net architectures, produces better segmentation results than when stand-alone U-Net architectures are trained.

2 Methodology

Inspired by GAN, Xue et al. have come up with SegAN, an adversarial network that has two networks, segmentor and critic, analogous to generator and discriminator in GAN, respectively. The segmentor, a fully convolutional neural network, takes in raw images as input and outputs a probability label map. The critic network, which is the encoder part of the fully convolutional neural network needs two inputs - the masked image by the ground truth labels and the masked image by predicted labels obtained from the segmentor. The aim of the segmentor network is to minimize the L_1 loss function and that of critic network is to maximize the L_1 loss function [15].

2.1 Conventional GAN Models

In GANs, the loss function is defined as -

$$\min_{\theta_G} \max_{\theta_D} L(\theta_G, \theta_D) = E_{x \sim P_{data}}[log D(x)] + E_{z \sim P_z}[log(1 - D(G(z)))]. \quad (1)$$

In the above equation, θ_G and θ_D are the parameters of generator G and discriminator D, respectively. x and z are real image from unknown distribution P_{data} and random input for G from probability distribution P_z, respectively. The generator G outputs a high dimensional vector which is the input to the discriminator D. The discriminator D is trained to maximize the probability of assigning the correct label to the training data and the data generated from G. The generator G is simultaneously trained to minimize the objective function $log(1 - D(G(z)))$ to generate images that are difficult to differentiate for D [3]. The aim of the generator is to produce images that are as similar as possible to the real image and the aim of the discriminator is to successfully distinguish between the real image and the fake image produced by the generator.

2.2   Loss Function in SegAN

In SegAN, the aim is to solve the mapping between input images and their segmentation masks. The loss function L for SegAN is given by -

$$\min_{\theta_S} \max_{\theta_C} L(\theta_S, \theta_C) = \frac{1}{N} \sum_{n=1}^{N} l_{mae}(f_C(x_n \circ S(x_n)), f_C(x_n \circ y_n)). \quad (2)$$

In this equation, θ_S and θ_C are the parameters of segmentor S and critic C, respectively and N represents the number of training images. $(x_n \circ S(x_n))$ and $(x_n \circ y_n)$ are input images masked with segmentor predicted label map and ground truth, respectively. $f_c(x)$ are the features extracted from image x by critic and l_{mae} is the mean absolute error (MAE) given by -

$$l_{mae}(f_C(x), f_C(x')) = \frac{1}{L} \sum_{i=1}^{L} ||f_C^i(x) - f_C^i(x')||_1, \quad (3)$$

with L representing the number of layers in the critic network [15].

The segmentor and critic networks are trained alternatively, just like GAN. The difference between GAN and SegAN is that GAN has two separate losses for generator and discriminator, while, the SegAN has only one multi-scale L_1 loss function for both segmentor and critic.

2.3   Segmentor and Critic

We use three different segmentor networks to predict the segmented mask and compare their results. The first one is the original *U-Net* [10]. The second one is a U-Net architecture with skip connection used in [15] (*U-Net A*). The third segmentor used is a modified version of the U-Net architecture inspired from [5] (*U-Net B*). The input to all these three networks are raw images and the output is a predicted mask.

For the critic network, we used a similar structure to the downsampling part of the corresponding segmentor network to extract hierarchical features from multiple layers of the network. We then concatenated all these features extracted across multiple layers and computed the overall L_1 loss using the concatenated feature vector [15]. The input to the critic network are two images - input image masked with predicted class map and input image masked with the ground truth class map; and output is a feature vector.

2.4   Training U-Net and SegAN Models

Our experiments involve three different U-Net architectures - the original *U-Net* from [10], the encoder-decoder network used as segmentor in [15] (*U-Net A*), and a modified U-Net inspired from [5] (*U-Net B*), which is the current state-of-the-art for left ventricle segmentation in the ACDC 2017 dataset. The U-Net models are trained with cross entropy loss function. We experimented with Dice loss but cross entropy loss gave us better results. This is supported by results in [1], too.

The segmentor and the critic network are trained alternately using back-propagation and the loss function. First, the segmentor outputs a predicted class map. Then, the segmentor is fixed and the critic is trained in the next step using gradients calculated from the loss function. After that, the critic is fixed and the segmentor is trained using gradients from the loss function passed to the segmentor from the critic [15]. As explained in GANs, this process resembles a min-max game, where the segmentor aims to minimize the loss and the critic tries to maximize it. Provided additional data and more epochs, the segmentor will produce segmented masks i.e. labelled maps that are similar to the ground truth. For each U-Net model we use as segmentor, we use the encoder part of that particular U-Net model as critic.

We train the U-Net and SegAN models by resizing each slice to a 224×224 image and feeding it into the network with a learning rate of 0.0008, a batch size of 8, a decay of 0.5, a beta value of 0.5, one GPU and 50 epochs.

2.5   Dataset

The Automated Cardiac Diagnosis Challenge (ACDC) dataset was released during the MICCAI 2017 conference in conjunction with the STACOM workshop. The images were acquired using two different MRI scanners with different magnetic strength - 1.5 T and 3.0 T. The short axis slices cover the left ventricle from base to apex such that we get one image every 5 mm to 10 mm. A complete cardiac cycle is usually covered by 28 to 40 images. Their spatial resolution is 1.37 to 1.68 mm^2/pixel [2]. The training dataset is composed of 100 subjects and the test dataset is composed of 50 subjects.

The image dataset corresponding to each subject consists of two image volumes, one at end-diastole and one at end-systole, with each containing 10 slices, therefore leading to a total of $1,902$ images. Since we do not have the ground truth for the 50 test subjects, we divide the training dataset into 80 subjects for training and 20 subjects for validation. The Dice score and the IoU in this paper are the result of 5-fold cross validation of the training dataset.

Table 1. Segmentation evaluation, mean (std-dev) for end diastole (ED) and end systole (ES) left ventricle segmentation in the 2017 ACDC dataset. Statistical significance (T-test) of the results of SegAn architecture compared against U-Net models are represented by * for p < 0.1 and ** for p < 0.05. The best Dice values achieved are labeled in **bold**.

	Dice (ED) (%)	IoU (ED) (%)	Dice (ES) (%)	IoU (ES) (%)
U-Net	93.41 (4.23)	87.25 (3.12)	91.75 (2.26)	83.64 (4.01)
SegAN + U-Net	*94.71 (1.24)***	*89.55 (2.46)***	*92.54 (3.89)*	*84.91 (5.75)*
U-Net A	92.62 (2.75)	85.27 (1.81)	90.30 (7.11)	81.58 (5.60)
SegAN + U-Net A	*93.88 (2.86)**	*88.54 (1.12)*	*91.10 (4.15)**	*82.74 (5.71)*
U-Net B	94.91 (2.40)	91.55 (3.23)	92.72 (4.71)	87.44 (3.81)
SegAN + U-Net B	***95.87 (1.71)****	***92.94 (3.27)****	***93.14 (2.56)****	***88.94 (3.92)***

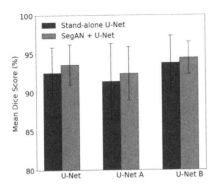 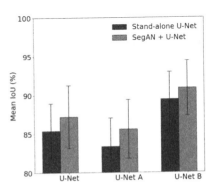

Fig. 2. Comparison of (a) mean Dice scores and (b) mean IoU values U-Net models and its corresponding SegAN architecture (Color figure online)

3   Experiments and Results

The focus of our experiment is to compare the results of a stand-alone 2D U-Net architecture with a SegAN architecture. For example, we obtain segmentation results using U-Net [10] with cross entropy loss as cost function. Then, we use this U-Net [10] as segmentor and the downsampling part of the U-Net as critic in the SegAN architecture with multi-scale L_1 loss as cost function. The results of these two networks are compared to determine if the SegAN architecture improves the segmentation results of the U-Net model. Experiments are performed with three variants of 2D U-Net architectures - an original **U-Net** [10], the encoder-decoder network used as segmentor in [15] (**U-Net A**) and a modified U-Net inspired from [5] (**U-Net B**), the current state-of-the-art for left ventricle segmentation in the ACDC 2017 dataset.

Table 1 summarizes the segmentation performance of the investigated frameworks with and without the SegAN integration. The results are obtained using 80% of the training subjects as training dataset and the 20% of the training subjects as validation dataset.

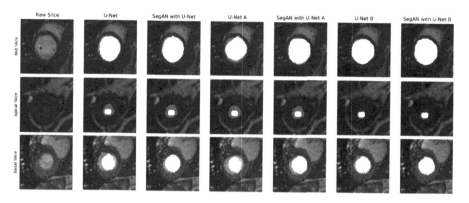

Fig. 3. Examples of segmentation of the left ventricle in mid, apical and basal slice (top to bottom). The white, red and blue regions represent true positives, false positives and false negatives, respectively. (Color figure online)

In Table 1, we can observe that the mean Dice scores and mean IoU values of the three SegAN architectures are higher than their corresponding U-Net models. To compare the performance the three stand-alone U-Net models with their SegAN frameworks, we conducted a statistical significance (T-test) test. The mean Dice score showed significant improvement ($p < 0.05$) from 93.41% (*U-Net*) to 94.71% (*SegAN + U-Net*), ($p < 0.1$) from 92.62% (*U-Net A*) to 93.88% (*SegAN + U-Net A*) and ($p < 0.05$) from 94.91% (*U-Net B*) to 95.87% (*SegAN + U-Net B*) in end diastole, and ($p < 0.1$) from 90.30% (*U-Net A*) to 91.10% (*SegAN + U-Net A*) and ($p < 0.1$) from 92.72% (*U-Net B*) to 93.14% (*SegAN + U-Net B*) in end systole. The mean IoU values showed significant improvement ($p < 0.05$) from 87.25% (*U-Net*) to 89.55% (*SegAN + U-Net*) and ($p < 0.1$) from 91.55% (*U-Net B*) to 92.94% (*SegAN + U-Net B*) in end diastole.

The highest mean Dice score and mean IoU in our experiments are obtained using the SegAN architecture with U-Net B as its segmentor network and the U-Net B's encoder as the critic network. The *SegAN + U-Net B* outperforms *U-Net* by 2.46% (Dice) in end diastole and 1.40% in end systole.

In Fig. 2, it can be observed that the segmentation performance of SegAN frameworks (shown in red) is better than the performance of the corresponding stand-alone U-Net architectures (shown in blue). When we use these U-Net models as segmentor, we see significant improvement in both Dice score and IoU, for ED and ES.

Figure 3 shows examples of mid, apical and basal slices of the heart and the corresponding segmented masks using the six architectures. The white regions represent the overlap between the ground truth mask and the tested mask. The red and blue regions represent the false positive (pixels predicted as left ventricle by the tested algorithm, but not annotated in the ground truth), and false negative (pixels not predicted as left ventricle by the tested algorithm, but annotated in the ground truth) regions, respectively.

4   Discussion

In this paper, the integration of three different U-Net models into the SegAN framework is evaluated on the 2017 ACDC segmentation challenge dataset. Our goal was to investigate if the SegAN framework improves the segmentation performance of U-Net models. Our experiments reveal that U-Net models, when trained in the SegAN framework, produces significantly better segmentation results than when trained stand-alone, consistently. The features extracted across multiple layers of the critic network and concatenated into the feature vector used to compute the multi-scale L_1 loss captures pixel-, low-, mid- and high-level features. This multi-resolution approach to feature extraction enables the SegAN model to learn the dissimilarities between the generated and the ground truth segmentation maps across the multiple layers of the critic network.

We use cross entropy loss as cost function for training U-Net models. We also experimented with training all three U-Net variants with a Dice loss cost function, however the results indicated a consistently lower performance than that achieved using cross entropy loss. We also experimented with multi-scale L_2 loss as cost function for training the SegAN models, however, the results were not very consistent. Further investigation with multi-scale L_2 loss as cost function will be conducted, to determine if it can outperform multi-scale L_1 loss as cost function.

To evaluate our method, we used a 5-fold cross-validation strategy, in which we employed five different combinations of 80 training and 20 testing datasets from the available 100 datasets. This is a common approach used to validate novel deep learning techniques, as it enables testing the robustness of the method across different training datasets, while also removing the bias associated with a single 80 training - 20 testing data split.

One of the disadvantage of SegAN is that it needs more computational time than the stand-alone U-Net models. It takes around 110 s to run each epoch of a regular U-Net model, but when the same U-Net model is used in the SegAN framework, it needs around 320 s.

5   Conclusion and Future Work

SegAN is a promising algorithm that was recently shown to outperform the current state-of-the-art methods in brain tumor segmentation [15]. Moreover, in this work we showed that the SegAN integration alongside three different U-Net variants led to significant improvement in Dice score for left ventricle segmentation.

In light of the improved performance of the SegAN addition in both brain tumor and left ventricle segmentation, further investigation into these methods and their further refinement is justified. As such we will employ the current state-of-the-art methods from the leaderboard of the 2017 ACDC segmentation challenge as segmentor and show their performance using adversarial regularization. One of the variants of SegAN that we would like to experiment in our future work would be to use cross entropy loss as cost function for training the segmentor and multi-scale L_1 loss to train the critic network, instead of using only the multi-scale L_1 loss for both segmentor and critic.

Lastly, we will also aim to answer the question whether this adversarial regularization method may improve the segmentation results of any fully convolutional network when employed as a segmentor along with a critic network.

Acknowledgement. Research reported in this publication was supported by the National Institute of General Medical Sciences of the National Institutes of Health under Award No. R35GM128877 and by the Office of Advanced Cyber-infrastructure of the National Science Foundation under Award No. 1808530.

References

1. Baumgartner, C.F., Koch, L.M., Pollefeys, M., Konukoglu, E.: An exploration of 2D and 3D deep learning techniques for cardiac MR image segmentation. In: Pop, M., et al. (eds.) STACOM 2017. LNCS, vol. 10663, pp. 111–119. Springer, Cham (2018). https://doi.org/10.1007/978-3-319-75541-0_12
2. Bernard, O., et al.: Deep learning techniques for automatic MRI cardiac multi-structures segmentation and diagnosis: is the problem solved? IEEE Trans. Med. Imaging **37**(11), 2514–2525 (2018)
3. Goodfellow, I., et al.: Generative adversarial nets. In: Advances in Neural Information Processing Systems, pp. 2672–2680 (2014)
4. Haering, M., Grosshans, J., Wolf, F., Eule, S.: Automated segmentation of epithelial tissue using cycle-consistent generative adversarial networks. bioRxiv p. 311373 (2018)
5. Isensee, F., Jaeger, P.F., Full, P.M., Wolf, I., Engelhardt, S., Maier-Hein, K.H.: Automatic cardiac disease assessment on cine-MRI via time-series segmentation and domain specific features. In: Pop, M., Sermesant, M., Jodoin, P.-M., Lalande, A., Zhuang, X., Yang, G., Young, A., Bernard, O. (eds.) STACOM 2017. LNCS, vol. 10663, pp. 120–129. Springer, Cham (2018). https://doi.org/10.1007/978-3-319-75541-0_13
6. La, A.G., et al.: Cardiac MRI: a new gold standard for ventricular volume quantification during high-intensity exercise. Circ. Cardiovasc. Imaging **6**(2), 329–338 (2013)
7. Lin, X., Cowan, B., Young, A.: Model-based graph cut method for segmentation of the left ventricle. In: 27th Annual International Conference of the Engineering in Medicine and Biology Society, IEEE-EMBS 2005, pp. 3059–3062. IEEE (2006)
8. Nasr-Esfahani, M., et al.: Left ventricle segmentation in cardiac MR images using fully convolutional network. arXiv preprint arXiv:1802.07778 (2018)
9. Petitjean, C., et al.: Right ventricle segmentation from cardiac MRI: a collation study. Med. Image Anal. **19**(1), 187–202 (2015)
10. Ronneberger, O., Fischer, P., Brox, T.: U-Net: convolutional networks for biomedical image segmentation. In: Navab, N., Hornegger, J., Wells, W.M., Frangi, A.F. (eds.) MICCAI 2015. LNCS, vol. 9351, pp. 234–241. Springer, Cham (2015). https://doi.org/10.1007/978-3-319-24574-4_28
11. Santiago, C., Nascimento, J.C., Marques, J.S.: A new ASM framework for left ventricle segmentation exploring slice variability in cardiac MRI volumes. Neural Comput. Appl. **28**(9), 2489–2500 (2017)

12. Suinesiaputra, A., et al.: A collaborative resource to build consensus for automated left ventricular segmentation of cardiac MR images. Med. Image Anal. **18**(1), 50–62 (2014)

13. Suinesiaputra, A., et al.: Left ventricular segmentation challenge from cardiac MRI: a collation study. In: Camara, O., Konukoglu, E., Pop, M., Rhode, K., Sermesant, M., Young, A. (eds.) STACOM 2011. LNCS, vol. 7085, pp. 88–97. Springer, Heidelberg (2012). https://doi.org/10.1007/978-3-642-28326-0_9

14. White, H.D., Norris, R.M., Brown, M.A., Brandt, P.W., Whitlock, R., Wild, C.J.: Left ventricular end-systolic volume as the major determinant of survival after recovery from myocardial infarction. Circulation **76**(1), 44–51 (1987)

15. Xue, Y., Xu, T., Zhang, H., Long, L.R., Huang, X.: Segan: adversarial network with multi-scale l 1 loss for medical image segmentation. Neuroinformatics **16**, 383–392 (2018)

Analysis of Three-Chamber View Tagged Cine MRI in Patients with Suspected Hypertrophic Cardiomyopathy

Mikael Kanski[1], Teodora Chitiboi[1,2], Lennart Tautz[3,4],
Anja Hennemuth[3,4], Dan Halpern[5], Mark V. Sherrid[6],
and Leon Axel[1(✉)]

[1] Department of Radiology, NYU Langone Health, NYU School of Medicine,
New York, NY, USA
Leon.Axel@nyumc.org
[2] Siemens Healthineers, Princeton, NJ, USA
[3] Charité University Hospital, Berlin, Germany
[4] Fraunhofer Institute for Medical Image Computing - MEVIS,
Bremen, Germany
[5] Division of Cardiology, NYU Langone Health, NYU School of Medicine,
New York, NY, USA
[6] Hypertrophic Cardiomyopathy Program, Division of Cardiology,
NYU Langone Health, NYU School of Medicine, New York, NY, USA

Abstract. Cardiovascular magnetic resonance imaging (CMR) is frequently used for patients with suspected hypertrophic cardiomyopathy (HCM). However, although magnetization-tagged MRI can provide quantitative assessment of regional function in HCM, it is not routinely used in clinical CMR of HCM. Therefore, we investigated the added value of acquiring a single three-chamber (3Ch) tagged cine image sequence to our clinical HCM protocol. Forty-eight patients with HCM, five patients with "septal knuckle" (SK), and 20 healthy volunteers underwent CMR at 1.5T, including a 3Ch tagged cine sequence. The images were analyzed for regional deformation (strain) and myocardial wall thickness, and results were compared to data from healthy volunteers. In HCM, we found a reduced tangential strain ($p < 0.003$) with decreased relaxation in hypertrophied segments, and mildly reduced strain ($p < 0.035$) with decreased relaxation in segments without hypertrophy. There was an inverse correlation between principal strain magnitude and local septal wall thickness ($p < 0.001$). Patients with SK had a higher tangential strain in the basal septum compared to patients with asymmetrical septal hypertrophy due to HCM, which may aid in differentiating focal HCM from SK. Thus, adding a 3Ch tagged MRI sequence to clinical CMR examinations of suspected HCM patients can provide functional information, aiding in differentiating SK from HCM, with minimal additional time cost.

Keywords: MRI tagging · Hypertrophic cardiomyopathy · Strain

© Springer Nature Switzerland AG 2019
Y. Coudière et al. (Eds.): FIMH 2019, LNCS 11504, pp. 425–432, 2019.
https://doi.org/10.1007/978-3-030-21949-9_46

1   Introduction

Hypertrophic cardiomyopathy (HCM) is a relatively common cardiac disease (prevalence $\sim 0.2\%$ of the general population [1]), and is the most common cause of sudden death in the young. HCM is linked to mutations in genes encoding elements of the myocardial contractile apparatus and is usually defined as left ventricular (LV) hypertrophy (LV thickness ≥ 15 mm, or 13–14 mm (borderline) with compelling evidence such as family history). While this definition is suitable in most cases, apical hypertrophies can be missed; an apex measuring 10 mm in thickness is already relatively thick. The disease has a wide range of phenotypes, all sharing the manifestation of an asymmetrical LV hypertrophy. The basal septum is affected in about 70% of the cases [2], which may results in an asymmetrical septal hypertrophy leading to LV outflow tract (LVOT) obstruction. LVOT obstruction is clinically manifested as limited exercise tolerance, and may increase the risk of sudden death [3]. Due to its severity, it is crucial to detect and initiate treatment of HCM at an early stage. The therapy is today limited to either invasive treatment such as septal myectomy or alcohol ablation, or medical treatment with negative inotropes, such as Disopyramide [4]. Mavacamten, a myosin ATPase inhibitor, is a promising alternative now undergoing clinical trials [5]. The three main mechanisms that promote obstruction include (a) the septal wall hypertrophy, (b) anterior displacement of the papillary muscles [6], and (c) elongated or malformed mitral valve (MV) leaflets. In this work we will focus on strain evolution in myocardial hypertrophy. We investigate the potential utility of a single three-chamber (3Ch) tagged cardiovascular magnetic resonance (CMR) slice in conjunction with the standard 3Ch cine. We: (1) propose a practical quantitative approach to the evaluation of regional wall strain and morphological characterization in obstructive and non-obstructive HCM, and (2) investigate the possibility of distinguishing between focal HCM and so-called septal knuckle (SK).

2   Materials and Methods

We retrospectively analyzed tagged and conventional 3Ch cine MRI image series acquired in 60 consecutive patients (15 women) referred for clinical CMR for suspected HCM, and 20 healthy volunteers (14 women). While there are not enough women to study sex-based differences, the expected magnitudes are likely not large enough to affect our results. The 3Ch view is defined as a long-axis slice bisecting the apex, and the mitral and aortic orifices (the name derives from that three "chambers" are seen in the image: left ventricle [LV], right ventricle [RV], and left atrium [LA]). The tagged images were acquired in addition to our standard clinical CMR protocol, during one breath hold. All subjects were retrospectively studied under a research protocol approved by the local Internal Review Board (IRB).

Conventional steady-state free precession (SSFP) and magnetization-tagged MRI [7] cine images were acquired (1.5T Aera or Avanto systems, Siemens, Erlangen, Germany) in 3Ch orientation during breath hold. Typical imaging parameters included: in-plane resolution: 1.25×1.25 mm, slice thickness: 6 mm, temporal resolution: 30–50 ms (depending on breath-hold capability); tag spacing: 6 mm. The 3Ch view plane was defined to bisect the centers of the MV, the aortic valve, and the LV apex. The SSFP images were used for measurement of wall-thickness and MV structures.

2.1   Strain Analysis

A flow diagram of the following steps is shown in Fig. 1. The deformation field between each two consecutive temporal frames, which provides a local motion estimation of the myocardium over the cardiac cycle, was computed using a multiresolution non-rigid registration algorithm based on local phase, Morphon [8].

The tagged images were semi-automatically segmented using an in-house developed script in MeVisLab (v2.8.2). A single manual myocardial contour was drawn on a mid-systolic image frame and was then automatically tracked throughout the cardiac cycle in forward and backward directions, using the previously extracted deformation field [9]. The segmentation propagation method alternates between a tracking step using Runge-Kutta integration of the displacement field and a Laplacian smoothing step of the displaced discrete contour points, for each consecutive temporal frame. We also calculated the approximate mid-endocardial centerline of the LV wall by morphological erosion and opening with a 3×3 pixel kernel of the segmented LV mask, for every time point in the cine MRI series. Using this, the displacement vectors were

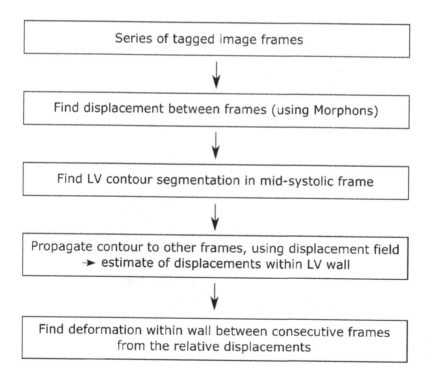

Fig. 1. Flow diagram of strain analysis from tagged images.

projected to the normal (radial displacement, positive towards the LV center) and tangent (tangential displacement, negative towards the LV apex) directions, with respect to the local myocardium centerline. To analyze the rate of relaxation of the myocardium during early diastole, we normalized the strain values to the maximum strain occurring during systole, in order to obtain a fractional relative strain measure.

To analyze the rate of relaxation of the myocardium during early diastole, we normalized the strain values to the maximum strain occurring during systole, in order to obtain a fractional relative strain measure.

2.2   Wall Thickness

The myocardium was divided into seven segments (basal septum defined as segment 1: midlevel septum 2; apical septum 3; apex 4; apical free wall 5; midlevel free wall 6; and basal LV free wall 7), which were tracked over the cardiac cycle, using the extracted motion field (Fig. 2). For each segment, the mean wall thickness at end-diastole (ED) was manually measured. For each individual segment, we subdivided the group of HCM patients into patients with local hypertrophy and patients with normal wall thickness of that particular segment. For example, in a patient with asymmetrical septal hypertrophy only, the septal parts are hypertrophied, but the basal free wall had normal thickness; this patient's LV free wall (segment 7) was defined as HCM normal segment. Hypertrophy was defined as 13 mm for basal segments, 12 mm for mid LV segments, and 10 mm for apical segments.

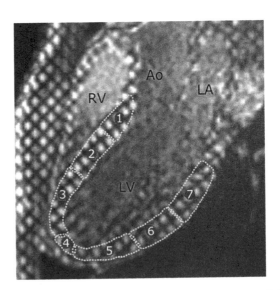

Fig. 2. The 3Ch view at end-diastole (with tags still visible in the blood), divided in 7 segments tracked throughout the cardiac cycle. Segment 1: basal septum; 2: midventricular septum; 3: apical septum; 4: apex; 5 apical free wall; 6: midventricular free wall; and 7: basal free wall. LV = left ventricle; RV = right ventricle; LA = left atrium; Ao = aorta.

2.3   Statistical Methods

We compared he measurements of the patient and normal volunteer subgroups, using the Mann-Whitney test for statistical assessment of differences and the Pearson correlation (scipy.org). Differences with $p < 0.05$ were considered as statistically significant.

3   Results

Of the 60 patients, 48 (80%) were diagnosed as HCM, and five (6%) as having septal knuckle (SK) but not HCM. Six patients showed normal findings on MRI and were therefore excluded, and one additional subject was excluded from analysis due to image artifacts related to mechanical MV, leaving 53 patients for analysis. Patients with HCM were classified as having obstructive or non-obstructive HCM. Subject characteristics are presented in Table 1.

Table 1. Subject characteristics. BMI: body-mass index; LVOT: left ventricular outflow tract; SAM: systolic anterior motion of the mitral valve; SK: septal knuckle.

	Normals	SK	HCM	
			Obstructive	Non-obstructive
Number of subjects [women]	20 [14]	5 [1]	30 [9]	18 [4]
Age (years)	36.9 ± 11.9	54 ± 9.9	55.4 ± 12.7	50.7 ± 11.5
Weight (kg)	71.3 ± 14.4	87.2 ± 15.8	86.8 ± 19.4	88.3 ± 20.4
BMI (kg/m^2)	24.4 ± 4.6	28.4 ± 7.6	29.2 ± 6.1	28.2 ± 5.5
LVOT obstruction	0	0	27	N/A
Non-LVOT obstruction	0	0	3	N/A
SAM	0	0	24	2
Mitral regurgitation	0	3	21	5
Mild	0	2	11	5
Moderate	0	1	10	0

Of the 48 confirmed HCM patients, 27 were found to have LVOT obstruction (nine women), and three had non-LVOT obstruction (2 midventricular, 1 apical). Most HCM patients had qualitatively reduced local cardiac deformation in the tagged image series, especially in the hypertrophied areas (Fig. 3).

3.1   Myocardial Strain

The HCM patients with local hypertrophy had significantly reduced maximal magnitude of the tangential strain and reduced maximal shear strain for all segments ($p < 0.003$ and $p < 0.035$, Fig. 3 top panels). Figure 3 bottom panels shows the average tangential strain over time for the for all subjects in the basal septal segment (left) and the basal free-wall segment (right). Figure 3 suggests that in patients with

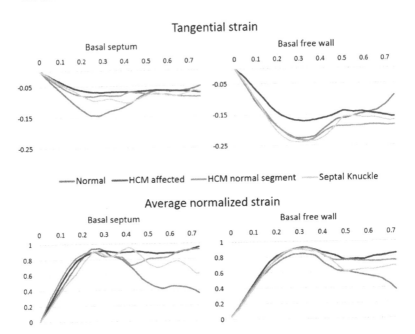

Fig. 3. Tangential strain (top panels) and average normalized strain (bottom panels) in the basal septum (left graphs) and the basal free wall (right graphs). Note that the tangential strain tends to remain increased (more negative) during diastole compared to healthy volunteers in the free wall, and that the average normalized strain is higher (reduced diastolic function) in the basal septum as well as in remote myocardium. Time is represented as fraction of the cardiac cycle.

HCM, even myocardium segments without any visible hypertrophy were nevertheless functionally affected, showing a reduced diastolic function (slower relaxation) compared with normal volunteers.

In the principal strain angle representation, the average angle between the principal strain direction and the normal to the myocardium wall was significantly larger in patients with HCM compared to normal volunteers (22.5 ± 5 vs. 16.7 ± 3.8, $p < 0.001$). However, the computed angle is somewhat influenced by the relative position of each 3Ch image. The magnitudes of maximal and minimal principal strains (P1 and P2, respectively) were significantly reduced for the hypertrophied HCM segments compared to normal-thickened segments in HCM in most cases (Table 2).

Table 2. P1 and P2 in hypertrophied and normal-thickened segments in patients with HCM. * $p < 0.05$; ** $p < 0.01$; *** $p < 0.001$

Segment	1	2	3	4	5	6	7
P1 ± SD (%)							
Hyper	12.4 ± 3.3**	14.9 ± 4.9*	17.0 ± 4.4***	18.6 ± 7.6*	13.4 ± 6.3**	19.6 ± 9.0	28.4 ± 9.4
Norm, thick	17.5 ± 4.4	18.9 ± 5.0	24.6 ± 6.9	24.3 ± 9.7	21.9 ± 5.9	23.7 ± 7.3	36.8 ± 10.1
P2 ± SD(%)							
Hyper	−5.8 ± 1.6**	−8.8 ± 3.7	−8.9 ± 3.7***	−8.8 ± 3.7*	−5.6 ± 2.8***	−9.0 ± 5.4	−16.6 ± 5.4
Norm, thick	−8.2 ± 1.7	−11.5 ± 5.1	−14 ± 4.6	−12.2 ± 5.0	−10.7 ± 3.8	−12.3 ± 4.4	−21.9 ± 6.7

3.2   Septal Knuckle

The data in Fig. 3 suggests that the systolic strain magnitude in patients with SK was slightly reduced compared to normal subjects, but less so when compared to the average hypertrophied HCM patients (Fig. 4). In particular, the strain magnitude was reduced in the thickened basal segments of the patients with SK to a degree comparable to HCM patients with similar wall thickness, for both principal strains. The patients with SK also had qualitatively faster diastolic relaxation in their normal thickness segments, e.g., the basal free wall (Fig. 3, upper-right) compared to the segments with normal wall thickness from the HCM patients. Figure 3 (bottom panels) shows the average relative strain, as it progresses throughout the cardiac cycle, for the basal septum and basal free wall. Around 40–50% into the cardiac cycle, the normal segments have a significantly smaller (p < 0.01) strain than both the hypertrophied and non-thickened corresponding segments of the HCM patients. The fractional strain of thickened basal septum of SK segments is similar to the HCM segments, but the normal-width segments of patients with SK behave more similarly to the normal subjects' segments, suggesting a potentially useful way to help differentiate SK from focal basal-septal HCM.

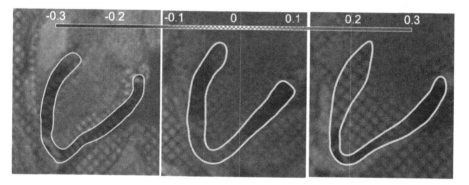

Fig. 4. Minimum principal strain (P2) at end-systole for a normal volunteer (left), septal knuckle (middle), and HCM (right). The color bar indicates the strain, blue: negative strain; red: positive strain. (Color figure online)

4   Conclusions

We have shown that 2D strain analysis of the 3Ch image series is potentially useful to distinguish between HCM patients and healthy volunteers. It also offers potentially important additional information about the different mechanical behavior of myocardium affected by hypertrophy, compared to otherwise normal-looking muscle tissue in HCM patients. Also, segments in HCM with normal wall thickness showed reduced strain and abnormal relaxation behavior compared to corresponding segments in healthy volunteers. Our data suggest that patients with septal knuckle may have qualitatively different motion patterns than patients with HCM. Specifically, although both normal thickness segment in patients with HCM and patients with septal knuckle develop normal amount of strain during systole, the normal thickness segments HCM patients show abnormally slow recovery of strain during diastole while the septal knuckle patients show more normal recovery (Fig. 3). This potentially provide a useful way to help distinguishing septal knuckle from HCM.

Tagged MRI is simple and fast to acquire in a single 3Ch plane and allows regional deformation quantification and local strain analysis. Acquisition of a single additional image series in this orientation offers a potentially useful trade-off between a minor increase in acquisition time and a large amount of additional potentially useful regional function data that can be extracted.

References

1. Gersh, B.J., et al.: 2011 ACCF/AHA Guideline for the Diagnosis and Treatment of Hypertrophic Cardiomyopathy: a report of the American College of Cardiology Foundation/American Heart Association Task Force on Practice Guidelines. Developed in collaboration with the American Association for Thoracic Surgery, American Society of Echocardiography, American Society of Nuclear Cardiology, Heart Failure Society of America, Heart Rhythm Society, Society for Cardiovascular Angiography and Interventions, and Society of Thoracic Surgeons. J. Am. Coll. Cardiol. **58**, e212–e260 (2011)
2. Maron, M.S., et al.: Hypertrophic cardiomyopathy is predominantly a disease of left ventricular outflow tract obstruction. Circulation **114**, 2232–2239 (2006)
3. Maron, M.S., et al.: Effect of left ventricular outflow tract obstruction on clinical outcome in hypertrophic cardiomyopathy. New Engl. J. Med. **348**, 295–303 (2003)
4. Sherrid, M.V., Pearle, G., Gunsburg, D.Z.: Mechanism of benefit of negative inotropes in obstructive hypertrophic cardiomyopathy. Circulation **97**, 41–47 (1998)
5. Ionita, C.N., et al.: Challenges and limitations of patient-specific vascular phantom fabrication using 3D Polyjet printing. In: Proceedings of SPIE–The International Society for Optical Engineering, vol. 9038, p. 90380 M (2014)
6. Sherrid, M.V., Balaram, S., Kim, B., Axel, L., Swistel, D.G.: The mitral valve in obstructive hypertrophic cardiomyopathy a test in context. J. Am. Coll. Cardiol. **67**, 1846–1858 (2016)
7. Axel, L., Dougherty, L.: MR imaging of motion with spatial modulation of magnetization. Radiology **171**, 841–845 (1989)
8. Tautz, L., Hennemuth, A., Peitgen, H.-O.: Motion analysis with quadrature filter based registration of tagged MRI sequences. In: Camara, O., Konukoglu, E., Pop, M., Rhode, K., Sermesant, M., Young, A. (eds.) STACOM 2011. LNCS, vol. 7085, pp. 78–87. Springer, Heidelberg (2012). https://doi.org/10.1007/978-3-642-28326-0_8
9. Ritter, F., et al.: Medical image analysis. IEEE Pulse **2**, 60–70 (2011)

Author Index

Abell, Emma 196
Abidi, Yassine 73
Achanta, Radhakrishna 187
Allain, Pascal 159
Ariga, Rina 361
Aslanidi, Oleg 11
Aung, Nay 304
Axel, Leon 352, 425
Ayed, Ibrahim 55

Bacoyannis, Tania 20
Bainbridge, Daniel 397
Banus, Jaume 285
Barbarotta, Luca 240
Barry, Jen 64
Bayraktar, Furkan 3
Bear, Laura 112, 196
Bendahmane, Mostafa 94
Benson, Alan P. 3
Bergquist, Jake A. 37
Bernardino, Gabriel 85
Bernus, Olivier 196
Bond, Raymond 112
Bonnin, Anne 187
Bourgault, Yves 46
Bouyssier, Julien 73
Bovendeerd, Peter 240, 249, 258
Bueno-Orovio, Alfonso 361
Bureau, Gautier 121, 405
Butakoff, Constantine 85

Camara, Oscar 85, 285
Carnahan, Patrick 397
Cedilnik, Nicolas 20, 55, 325
Chabiniok, Radomír 266, 379, 405
Chapelle, Dominique 266, 405
Chitiboi, Teodora 425
Chubb, Henry 11
Clarysse, Patrick 276
Cochet, Hubert 20, 325
Collin, Annabelle 121
Cork, Tyler E. 177
Corral Acero, Jorge 361
Coudière, Yves 121

Dangi, Shusil 415
De Craene, Mathieu 159
Degener, Franziska 230
Dejea, Hector 187
Dössel, Olaf 29, 147
Doste, Ruben 85
Dubes, Virginie 196
Duchateau, Josselin 325
Duchateau, Nicolas 159, 276

Eldredge, Jeff D. 294
Ennis, Daniel B. 177, 294
Eskandari, Mehdi 397

Ferguson, Sebastian 64
Finlay, Dewar 112

Gallinari, Patrick 55
Garfinkel, Alan 294
Genet, Martin 334
Gérard, Antoine 121
Gerardo-Giorda, Luca 139
Gharaviri, Ali 131
Gilbert, Kathleen 304
Gilbert, Stephen H. 196
Ginty, Olivia 397
Good, Wilson W. 37
Grau, Vicente 342, 361
Groth, Alexandra 405
Guerra, Jose M. 139
Guo, Fumin 64

Haissaguerre, Michel 196
Haliot, Kylian 196
Halpern, Dan 425
Hancox, Jules 11
Harloff, Andreas 370
Hennemuth, Anja 370, 425
Herz, Christian 397
Hitchcock, Robert W. 168
Holden, Arun V. 3
Huang, Chao 168
Huang, Qiaoying 352

Jaïs, Pierre 325
Jansson, Johan 139
Jarmatz, Lina 370
Jolley, Matthew A. 397

Kanski, Mikael 425
Karimkeshteh, Sahar 370
Karoui, Amel 94
Kaufhold, Lilli 370
Knighton, Nathan 168
Konukoglu, Ender 187
Krebs, Julian 20
Kühne, Titus 230

Lamata, Pablo 313, 361
Lasso, Andras 397
Le Gall, Arthur 266
Lee, Lik Chuan 334
Leoni, Massimiliano 139
Li, Mengyuan 64
Lin, Peter 64
Linte, Cristian A. 415
Liu, Wan-Yu 208
Loecher, Michael 177
Lorenzi, Marco 285

MacLeod, Robert S. 11, 29, 37, 147
Magat, Julie 196
Magnin, Isabelle E. 208
Mahjoub, Moncef 73
Malandain, Grégoire 388
Marrouche, Nassir F. 168
Martens, Johannes 219
Mclaughlin, James 112
Metaxas, Dimitris 352
Mirea, Iulia 208
Moireau, Philippe 121, 405
Moore, John 397

Nasopoulou, Anastasia 313
Neubauer, Stefan 304
Ng, Matthew 64
Niederer, Steven A. 313
Nordmeyer, Sarah 370
Nordsletten, David A. 313
Nuñez-Garcia, Marta 85

O'Neill, Mark 11
Oostendorp, Thom F. 29, 147
Ozenne, Valéry 196

Panzer, Sabine 219
Pearce-Lance, Jacob 46
Perez-Cruz, Fernando 187
Perotti, Luigi E. 177, 294
Pervolaraki, Eleftheria 3
Peters, Terry M. 397
Petersen, Steffen E. 304
Petras, Argyrios 139
Pezzuto, Simone 103, 131
Piechnik, Stefan 304
Piro, Paolo 159
Polejaeva, Irina A. 168
Ponnaluri, Aditya V. S. 294
Pop, Mihaela 46, 64
Potse, Mark 103, 131
Poulain, Emmanuelle 388
Pushparajah, Kuberan 379

Qi, Xiuling 64
Quaglino, Alessio 103
Quesson, Bruno 196

Rababah, Ali 112
Razafindrazaka, Faniry H. 230
Rjoob, Khaled 112
Rondanina, Emanuele 249, 258
Roy, Aditi 11
Ruijsink, Bram 379
Rumindo, Kenny 276

Sacher, Frédéric 325
Sachse, Frank B. 168
Saloux, Eric 159
Schaeffter, Tobias 11
Schneider, Jurgen E. 342
Schreiber, Laura M. 219
Schuler, Steffen 29, 147
Sermesant, Maxime 20, 55, 64, 285, 325
Sherrid, Mark V. 425
Siebes, Maria 219
Stampanoni, Marco 187
Suendermann, Simon 230
Suinesiaputra, Avan 304

Tanner, Christine   187
Tate, Jess D.   29, 37, 147
Tautz, Lennart   425
Tintera, Jaroslav   405
Trew, Mark L.   196

Upendra, Roshan Reddy   415

Vaillant, Régis   388
Vallée, Fabrice   266
van der Donk, Loes   258
van Osta, Nick   258
van den Wijngaard, Jeroen P. H. M.   219
Varela, Marta   11
Varray, François   208
Vellguth, Katharina   230
Verzhbinsky, Ilya A.   177, 294
Vinet, Alain   131

Wang, Shunli   208
Wasmund, Stephen L.   168
Weese, Jürgen   405
White, Kenneth L.   168
Wright, Graham   64

Xu, Hao   342, 361

Yamaguchi, Takanori   168
Yang, Dong   352
Yi, Jingru   352
Young, Alistair   304

Zacur, Ernesto   342, 361
Zemzemi, Nejib   73, 94
Zenger, Brian   37
Zhao, Jichao   85
Zugaj, Konrad   379

Printed in the United States
By Bookmasters